Kent
Clocks & Clockmakers

This edition limited to
1,000 copies

No . . . .

Colour Plate I An early eighteenth-century black and gold lacquer longcase clock by Thomas Elliott of Greenwich, about 1715. The 8-day five-pillar movement has internal countwheel striking (see also Fig 7/32)

Kent
Clocks & Clockmakers

Michael Pearson

Mayfield Books

To Denise and our son James,
with love.

Published by:
Mayfield Books
Matherfield House, Church Lane,
Mayfield, Ashbourne,
Derbyshire, DE6 2JR, England
Tel/Fax 01335 344472

Published in the USA by:
AutaMusique Ltd
Summit Opera House, Two Kent Place,
Summit, NJ 07901 USA
Fax (908) 273 9504

© Michael Pearson 1997

ISBN 0 9523270 7 4

The right of Michael Pearson as author of this work has been asserted by him
in accordance with the Copyright, Designs and Patents Act, 1993.

All rights reserved. No part of this publication may be reproduced, stored in a retrieval system, or transmitted in any form or by any means, electronic, mechanical, photocopying, recording or otherwise without the prior permission of Mayfield Books

British Library Cataloguing in Publication Data:
A catalogue record for this book is available from the British Library.

Printed in Spain by;
Keslan Servicios Graficos

Title page: Fine mahogany drop-dial wall clock with original 15in diameter convex wooden dial and 8-day timepiece movement with a-shaped plates. Both dial and backplate signed 'Godden, Malling'. Height 23in. About 1800

Contents

	Foreword	6
	Preface	7
	Acknowledgements	8
	Introduction	9
Chapter One	Turret Clocks in Kent	12
Chapter Two	Domestic Clockmaking in Kent	34
Chapter Tree	The Bakers of Maidstone & Town Malling	47
Chapter Four	The Cutbush Family of Maidstone	54
Chapter Five	Thomas Deale of Ashford	59
Chapter Six	The Greenhill Family	67
Chapter Seven	Kent Clockmakers & Watchmakers	85
Appendix I	Clockmakers & Watchmakers in Early Kent Newspapers	253
Appendix II	Kent Clockmakers & Watchmakers Listed by Town	303
	Bibliography	316
	Index	317
	Picture Credits	320

Foreword

Michael Pearson is already well known for his book *The Beauty of Clocks* written some twenty or more years ago, which introduced many people to horology and above all enabled them to appreciate not just the skill and ingenuity of the clockmaker but also that the best clocks are often works of art, highly individual in nature, and should be enjoyed as such.

Whereas *The Beauty of Clocks* was written in a comparatively short period, the research involved in producing this book on *Kent Clocks & Clockmakers* has taken no less than nineteen years, much of it devoted to searching county, town, ecclesiastical and indeed many other records. Although such work by its very nature can never be complete, Michael Pearson has done Kent proud in writing one of the finest and most thoroughly researched books on the makers of a particular county ever produced.

The early part of the book is particularly valuable in providing a backdrop for the rest of the contents. It explains the way in which the county was governed; the principal towns and their populations and the relative density of the population in Kent when compared with the other counties. It then goes on to discuss the guild system which governed clockmaking and it is interesting to note here that this prevented the influx of clockmakers from the Continent, particularly the Huguenots following the repeal of the Edict of Nantes, from pursuing their former craft in the county, whereas in London no such bar existed.

The book goes on to give a remarkably full account of all the turret clocks in Kent and concludes with a list containing all the known information on some 1,200 clock and watchmakers.

This book will be of great interest, not just to those who have Kent clocks but also to all those who study country clocks, dealing as it does for instance with the surprising fact that no seventeenth-century watches signed by a Kent maker (with the exception of one by Thomas Barratt) are known, although there were watchmakers living and working in Kent at that period. A somewhat similar situation arises with the bracket clock, although the explanation may well be different.

I would thoroughly recommend this book, not just as one of reference, although it is invaluable as such, but to all who have an interest in the history and practice of clockmaking.

Derek Roberts

Preface

In 1975 I purchased what is probably the definitive book on provincial clockmaking, *Suffolk Clocks and Clockmaking* by Arthur Haggar, FBHI and Leonard Millar, CMBHI. As someone who makes his living buying and selling antique furniture and clocks I was fired with enthusiasm to produce a similar volume on clockmaking in my own county, and in so doing make a small contribution to our knowledge of clockmaking in the provinces,. This has been vastly extended in recent years with the publication of various county books and the many volumes on the early history of the craft written by Brian Loomes and others.

Little did I realise the task I had set myself, and how many hours would be spent searching through the County Records Ofice and other archives attempting to unravel what became an ever-more fascinating subject. My particular interest is the seventeenth- and early eighteenth-century makers and I have tended to concentrate on the men of that period. I hope I have been able to provide an answer to some of the questions that have puzzled horologists, but I am sure that for every one solved a number of new ones have been raised.

Research into a subject such as this can never really be brought to a satisfactory conclusion and I felt that it was time to publish the material gathered so far, and trust that more information will come to light as a result of the publication of this book. It has not been my intention to go into great detail about the history of clockmaking in England, only insofar as it affects the development of the craft in Kent. Numerous volumes have been published on the subject and the reader wishing to learn more of the general history and technical details is recommended to consult some of the volumes mentioned in the comprehensive bibliography at the end of this book.

This book consists of an introduction and a general history of the development of the trade in Kent with a special emphasis on the many turret clocks to be found in the county. The second part is an alphabetical list of some 1,200 makers from the earliest period up to the late nineteenth century. It must be appreciated that many of the so-called makers from the beginning of the nineteenth century onwards, were merely suppliers of clocks and watches obtained from a factory source. In many cases the information on these later makers is confined to basic details gleaned from *Watchmakers & Clockmakers of the World* Vols 1 and 2 by Baillie and Loomes respectively and supplemented with information provided by the various trade directories published during the nineteenth century. As previously mentioned, my main research has been centred on the earlier clockmakers of the seventeenth and eighteenth centuries and as much biographical material as possible on each maker has been included, and in some cases individual chapters have been devoted to men of special interest.

In 1752 the beginning of the calendar year was changed from 25 March to 1 January. Prior to that year, dates between 1 January and 24 March were included in the same year as the preceding nine months. Therefore the system known as double-dating for dates in the first three months of the year is used, eg 24 February 1701/2.

Acknowledgements

No publication such as this would be possible without the assistance and help of a great number of people who have been generous with both their time and in some cases made available to me the results of their own research. While it would be impossible to thank everybody individually, I wish to acknowledge the help and information I have received over the years from various members of the Antiquarian Horological Society. If I have been dilatory in replying to their letters I ask for their forgiveness. They are too numerous to mention individually, but a special word of thanks is due to Michael Bundock of the Turret Clock Group for his generosity in sharing his knowledge and providing many of the photographs of turret clocks featured in the book.

The help I have received from many others is warmly acknowledged, in particular to John Marbrook for information on Sandwich makers and to Jeremy Ovenden for much useful assistance and photographs in connection with the Austen family of Challock and other Canterbury makers.

Many other of my fellow clock and antique dealers have helped by providing photographs, in particular John Mighell of Strike One, Brian Loomes, and Derek Roberts, who also read the manuscript and kindly wrote the Foreword. I must thank Sasha Luck of The Old Clock Shop, West Malling for her patience in putting up with me for so many hours over the years in her shop while taking photographs of the many local clocks which have passed through her hands. I must also express my gratitude to the Antique Collectors' Club for allowing me to reproduce photographs from two of their publications, *Early English Clocks* and *The English Lantern Clock*, which would otherwise have been unavailable. Likewise, to W. J. F. Hana for allowing me to reproduce the photographs of the Greenhill and Watts lantern clocks from his book *English Lantern Clocks*.

Much of the original research was carried out in the county's museums and archives and thanks are due to the following members of the Local Study units: Pat Salter at Rochester, Janet Adamson at Folkestone, Miss P. Stevens at Dartford, the staff of the Centre for Kentish Studies at Maidstone and the Canterbury Cathedral Archive all gave me invaluable assistance. The curator of Maidstone Museum, Henry Middleton, very kindly gave access to the museum files which provided much useful information, and special thanks are due to Claire Mason and Richard Stutley for arranging photography at Maidstone, and Kenneth Reedie at Canterbury Museum for similar assistance.

Invaluable help was provided by Jonathan Hills of Sotheby's, Phillips (various branches) and Tony Pratt of the Canterbury Auction Galleries, supplying photographs of clocks which have passed through their respective auction rooms, and their assistance is gratefully acknowledged, as is Miss Joan Jeffery who, in the course of her own researches into the eighteenth century, found numerous clock and watch references in the files of the *Kentish Post* and *Kentish Gazette*, and these form Appendix I.

Brian Loomes has provided extra information on a number of Kent clockmakers, while John Davis, who is a direct descendant of the clockmaker William Robbins and whose wife is descended from the clock case maker Charles Lepine, has provided information on Robbins, Lepine and William Goulden, the Canterbury maker of clock cases.

Finally, my thanks to Angela Gillmore for transferring the text to computer disc and handling the many revisions before the final draft and special thanks to John Robey for editing, designing and publishing this book.

Michael Pearson

INTRODUCTION

In order to fully understand the way that the craft of clockmaking developed in Kent it is necessary to know something of the historical background of the county and the manner in which its history and geographical features affected the development of the industry. In this book the old county boundaries, which existed until 1889 when the metropolitan areas of Deptford, Greenwich, Lewisham, Woolwich and Sydenham were incorporated into the County of London, have been used. Kent is the ninth largest county in England in size and has always been one of the most heavily populated, as a result it is rare to find a view in the county free of buildings. There are three reasons for this. Firstly, much of Kent consists of good farming land. Secondly, after the Dissolution of the Monasteries under Henry VIII, most of the land which had been previously owned by the two great abbeys in Canterbury, and which comprised of much of the county east of the Medway was sold or given away to private individuals and was not kept by the monarchy. The most important reason of all however was the unique system of inheritance, or *gavelkind*, whereby an estate did not go to the eldest son but was shared equally. By tradition the right to keep their ancient customs, including *gavelkind*, was granted by William the Conqueror when the county's inhabitants sued for a separate peace in return for recognizing his claim to the throne. Hence the county insignia and motto, 'Invicta'. Whether there is any real truth in this cannot be substantiated, but it is a fact that from early days the Kentish yeoman had been descended from a free peasantry and the villein class was virtually unknown in Kent, unlike the rest of England. In the same way the open strip system of agriculture was also unknown other than in isolated pockets in the east of the county, mainly in Thanet. Field enclosure had taken place at an early date and most of the land was either owned or rented in small parcels of a few acres by freemen.

Kent's location was also the basis for its importance, with its close proximity to London, and a large number of small ports along its long coastline giving easy access to the capital. These ports, together with its three navigable rivers, the Stour, Medway and Rother, meant that no town of any size was more than fifteen miles away from easy transport facilities. Its position between London and the Continent made it defensively England's most important county, which also accounted for the fact that there were more castles in Kent than elsewhere. The royal dockyards were sited there and Canterbury was England's religious capital.

During the seventeenth century, which saw the beginning of domestic clockmaking in the county, Kent was one of the most prosperous places in Britain. This prosperity was based on its varied agriculture, which provided the capital with large quantities of corn and livestock. The influx of Flemish refugees in the late sixteenth and early seventeenth centuries had seen the growth of market gardening in the Greenwich and Thanet areas, the majority of this produce finding its way to London and a ready market. This period also saw the introduction of hop growing and fruit farming for which the county became justly famous, earning the title 'The Garden of England'. Of equal importance was the county's industry. The main centres of shipbuilding were at Deptford, Woolwich and subsequently at Chatham. The ironworks of the Weald, which provided most of England's iron, extended into the western half of the county, resulting in the clearance of much woodland to make good grazing land, and supplying wool to the weavers of Cranbrook and Biddenden. This area was a most important supplier of cloth, second only to the West Country during the seventeenth century. The production of gunpowder at Faversham, lime and glass along the Thames estuary and paper-making at Dartford and Maidstone were also instrumental in adding to the wealth of the county.

A by-product of the system of *gavelkind* was that there were few large estates of more than 3,000 acres and most of the land was owned by a large gentry class numbering some 900 persons, underpinned by a rural middle class of yeoman and small tenant farmers. By the end of the seventeenth century the population of Kent numbered about 150,000 people and, while it was still a predominantly rural society, there were some twenty-five small towns or large villages with more than 400 inhabitants. Eight towns had more than 2,000 inhabitants, the largest of these being Canterbury which, with 7,000, was the tenth most populous town in the country after London. Although Rochester, with some 3,000 people, had always been the next largest after Canterbury since the middle ages, it was rapidly outstripped towards the end of the century by the dockyard towns of Deptford (6,500), Chatham (5,000), Greenwich (5,000) and the important market town of Maidstone (3,500).

Sandwich, with a population of some 2,000, of whom 500 were of Dutch descent, benefited greatly from the increased prosperity brought by the religious refugees who had arrived in ever-increasing numbers during the late sixteenth and seventeenth centuries, as also did Canterbury and Maidstone. These three towns operated a very strict guild system and the newcomers were not allowed to engage in any trade that already existed. These trading restrictions resulted in the rise of market gardening around Sandwich and the silk-weaving and thread-twisting trades in Canterbury and Maidstone respectively. An interesting side effect of the very strict trading conditions operating in these towns and the others with guilds, such as Dover, Hythe and Rochester, is that none of the clockmakers in Kent appear to be of Flemish or Huguenot descent, whereas in London they constituted a substantial proportion of the total number of makers, particularly in the early years of the seventeenth century.

By the end of the seventeenth century nearly a third of the county's population lived in the towns, and almost half that number, some 25,000, inhabited the north-west, the Medway towns and the parishes near the boundary with London. The numbers of clockmakers in this area are relatively few, probably because of the easy access to London, and the close proximity to the capital is the main reason for the noticeable lack of bracket clocks made in Kent before the 1750s. While the aristocracy would undoubtedly have patronized the London makers, the poorer members of the gentry and yeoman classes would have been unlikely to have been able to afford the more costly bracket clock, and longcase clocks housed in expensive cases made of walnut and marquetry are exceedingly rare. The traditional brass lantern clock and longcase made of oak were much more to Kentish taste.

During the seventeenth century the industry was centred mainly on Canterbury, with a lesser number of makers in Ashford, Rochester and Maidstone. The end of the century saw a rapid increase in the number of makers in the county and by 1740 virtually every town and large village supported one or more makers, although many of those from the smaller villages must also have been engaged in the allied trades of blacksmith, whitesmith and gunsmith. While it is true to say that Kent never produced a clockmaker of genius to rank with the likes of Thomas Tompion or George Graham, nor indeed gave birth to any technical idea or device which radically changed the course of clockmaking, it must be said that it is very rare to find a poor clock by a Kent maker. Almost all of those I have examined through the years have exhibited a fine degree of skill and workmanship and are equal in every respect to the majority of the output of London makers of the period.

Fig I/1 Kent in the seventeenth century by John Speed

Chapter One
TURRET CLOCKS IN KENT

The history of clockmaking in Kent begins, as in every other county in England, with the needs of the church and, in particular, the monasteries and cathedrals to find a better method of regulating their daily routine. Prior to the advent of the first truly mechanical clock towards the end of the thirteenth century, various methods of measuring time had been used, from the primitive sundials developed by the early civilizations of the Middle East, to the more sophisticated version used by the Greeks, which utilized the shadow made by a pointer on a marked plate. This became the standard type of sundial and remained in use for many hundreds of years. It can, of course, still be bought to this day, although it is used primarily as a garden ornament. While the sundial is a relatively simple device it is extremely accurate, and before the advent of the regulator in the later eighteenth century it was used by clockmakers to regulate the early mechanical clocks. Indeed, the inventories of many of the early makers include such a device among their effects.

Other methods used included clepsydra (water-clocks), which measured the dripping of water through a small hole, candles marked with the hours, and of course the hour glass, which measured the passage of sand between two glass bulbs joined by a narrow funnel. This method utilises the same principle as the water clock and was still used by the Navy until late in the eighteenth century.

It is now generally accepted that the first true mechanical clock utilising a weight to drive an escapement came into being at some time between 1277 and 1300. At one time it was believed that it originated either in Italy or Central Europe, but there is a body of opinion as outlined in *English Church Clocks* by the late Dr Beeson that subscribes to the premise that the first mechanical clock was made in England during the last twenty years of the thirteenth century. While this has yet to be proved, it is certainly true that church and cathedral records dating from this period document a number of references to '*horologium*', being the term used to describe such a device, in various ecclesiastical buildings throughout England from 1283 to 1306, ranging from Dunstable Priory, Exeter Cathedral, Merton College, Oxford, to Salisbury Cathedral, and others.

The first reference in Kent is to be found in the accounts of Christchurch Cathedral, Canterbury, which details the work carried out during the term of office of Prior Henry of Eastry from 1285 to 1331. This records the installation of a new clock in 1292 at a cost of £30: '*Anno 1292-Novum orologium magnum in Ecclesia … xxx li*'.

Much new building and reconstruction took place during these years with works being carried out in the choir, the chapter house and the great hall. At the same time a number of bells were installed, five in 1316 and three more in 1317. In his *History and Antiquities of the Cathedral Church* written in 1727, J. Dart gives his interpretation of the entries in the list of works as follows: '*una vocatur Thomas in Magno clocario … Tres alia in novo clocario longo … Thes Campanae in clocario sub angelo.*' In each case he translates '*clocario*' to mean clockhouse.

However according to Dr Beeson, the term actually means a belfry or bellroom. In other words we cannot be sure whether the '*orologium magnum*', which Dart obviously believed to be a real clock, was actually installed in any of these locations. It is interesting to note that in the same year, 1292, a tower was built at a cost of £10, and this may well have been the site of the clock.

Although the cathedral archives possess very detailed accounts, no trace of any record, either of the clock, or its maintenance, has been found other than a note in 1492 when mention is made of a clock with chimes housed '*in magno campanile*'. Whether the clock was subsequently moved to another position is not known, but it was definitely inside the cathedral as is made clear by Leland's description of it in his *Itinerary in England* of 1535-43 when he stated that it was in the south transept. This is confirmed by the only other reference found in the cathedral archives, namely, a copy of a letter written in 1973 from the archivist which refers to the 'great clock in St Andrews Chapel, formerly in the South West Transept'. It also states that this was the clock given by Prior Henry of Eastry in 1292 at a cost of £30 and that it was moved to the chapel in 1760. Regrettably there is no trace of this clock today and no other information has been elicited from the Dean and Chapter as to its whereabouts. However, it must be assumed that the clock would have been similar to the Salisbury Cathedral clock which was installed in 1386.

There has always been much dispute by horologists as to the exact meaning of the word '*orologium*' and many authorities have stated that it refers to a clepsydra or water-clock. However, Dr Beeson, who was considered to be the authoritative voice on the subject of turret clocks subscribes to the other view, that the numerous mentions in ecclesiastical records starting with Dunstable Priory in 1283 to '*horologium*' does not mean the acquisition of new clepsydra, but definitely refers to a mechanism controlled by an escapement. To add to the number of such accounts detailed in his book, an extremely interesting reference has been found in *Archaeologia Cantiana*, Volume X, 1876. An article on St Mary's Church, Elham, notes that this particular church was a responsibility of Merton College, Oxford. In 1290 the college spent £88 16s 0d on the church fabric etc, and among the items listed is an amount of £1 8s 10½d spent on the 'orologium'. According to Dr Beeson the bursarial accounts for Merton College suggest that an '*orologium*' or clock had already been installed at the college at some time between 1288 and 1290.

Is it possible that the college provided a clock of some sort for St Mary's Church during the same period? The article in *Archaeologia Cantiana* poses the question as to whether the device mentioned was a sundial, but as we have seen Dr Beeson's researches suggest that the term '*orologium*' does mean some form of mechanical device. What makes this proposition doubly interesting is that Richard of Wallingford was at Merton College during this period prior to his election as Abbott of St Albans. It is well known that he was interested in horology, being instrumental in devising an astronomical clock for St Albans Abbey, which was completed after his death. A miniature in the British Museum depicts the abbot standing alongside his clock which was placed at ground level. Is it too fanciful to suppose that he may have been responsible for installing some sort of mechanical device at Elham in 1290?

During 1361-77, Edward III had built a great new castle at Queenborough on the Isle of Sheppey as a defence against the French. At that time the main shipping lane was through the Swale estuary, between Sheppey and the mainland. The castle was unique in England and while it incorporated the latest defensive building design of circular towers around an inner circle, it had also been built for use as a palace. By 1366 Edward had already had a clock installed at Windsor by three clockmakers from Lombardy (1351) and in that year he ordered three large bells from the London bellfounder, John Belleyeterre. These were to be used on three clocks: one at Westminster Palace, another at Kings Langley and the third at Queenborough. Unfortunately there are no existing accounts for the clock, but in 1373-4 the clock belfry was roofed with lead, '*cooperiens postes berfridi del clokke infra dictum castrum*'. Repairs to the clock were also made at the same time by the clockmaker John Lyncolne (Nicole) who was responsible for looking after the Westminster clock. There are two other references to the clock in an inventory of 1393-6: '*i clokke infra quandam turrim*' and '*i parva clokke pendens in sancta capella*'.

By the end of the fourteenth century clocks were commonly installed in the great religious buildings in England and had begun to spread to secular buildings and smaller churches. In 1404 a clock was installed at Dover Castle although there is no trace of this mechanism today. The movement from the castle which is on display at the Science Museum in London dates from the first quarter of the seventeenth century, but regrettably there is no record of who made this clock (Colour Plate II, page 19). The Exchequer accounts of 1415-16 records the payment to a clock-keeper for a period of eleven years for the earlier mechanism as follows: '*Et solutum cuidam homini pro labore suo custodiente le clokke pro tempore hujus compoti capienti per Annum xlIIs. IIIId per idem tempris vll li vi. s. viii d.*'

The Hythe Corporation records note that St Leonards Church had a clock before 1412: '*Item in le vane pro le clokke solutus ... iiiid*', and while churchwardens' accounts from the late mediaeval period are particularly scarce in Kent it is clear that numbers of churches would have had clocks installed by the end of the fifteenth century.

At least one of the Canterbury churches, namely St Andrew the Apostle (now demolished), had a clock before 1485 for the accounts read as follows:

1485
It. paidd to Thomas Cook for kepyng of the Clok pertotum annum ... IIIjs.
It. paidd to Nicholas Bekelys for mendyng of the yerne werk of the Clok ... xxjd

A new clock was installed in 1494-95 as noted in the accounts:

1494-5
Item payd for payntyng of ye dyall ljs
　" 　pay'd to Bekylls for makyng of ye clok vjs.

A memorandum in the Canterbury City Accounts of 1428-9 records 'Johes Taillour [John Taylor], Clockmaker de Cantuar' renting a piece of land near the city wall between Burgate and Newingate (St George's Gate) for 6d per annum. He is the earliest known clockmaker traced so far. As this land was granted to him by the City Corporation it is reasonable to suppose that he was responsible for installing the clock in St Andrew's Church, which is known to have been used by the officials and city administration as their principal place of worship.

Another of the Canterbury churches, St Dunstan's, also has accounts dating from 1480-1580, but there is no reference to a clock so we can assume that this particular church did not possess one at that period.

One church with particularly close links with the cathedral and which undoubtedly did have a clock, maybe as early as 1432-48, is St Gregory and St Martin at Wye. John Kempe, Archbishop of York and later of Canterbury, was born in Wye and in 1432 founded a college in the town. An inventory of church goods taken at the Dissolution has luckily survived and it provides ample evidence of how richly endowed the church was, probably as a result of the generosity of the archbishop and others of his family who resided at the local great house of Olantigh. The 1529 accounts record that the church possessed a clock and an organ. The entry relating to the clock is as follows: 'paid to the clockmaker to amend the clock ... 6s. 8d.' It is reasonable to assume that the clock had been installed many years previously, as in 1542 a new clock was purchased financed by public subscription. According to the *Old Book of Wye* by G. E. Hubbard (1951) 34s 8d was raised, of which 30s was paid to the clockmaker and 4s 8d to the carpenter for his part in the installation. Other entries also record:

for setting up the dial ...	6d
for the clockmakers meat and drink at the same time ...	8d
for a lock for the clockhouse door ...	7d
for the clockmaker's earnest penny ...	4d

In 1572 the church steeple burnt down and whether the original clock was reinstalled is not certain, but the 1634 accounts record that the church clock was in need of repair and had to be taken to Ashford where it was mended by William Barrett at a cost of £2 14s 4d. Much more drastic measures were obviously needed than a straightforward repair as some four years later the churchwardens' accounts include the following item:

1638
to William Barrett of Ashford, for the top, being brass, of a sun-dial ...	5s.
to Fermer for laying it in oil and painting the sun-dial ...	6s.
to Richard Jarman for making of plates and bolts for the dial ...	3s. 4d
to John Coveney for a dial-post, and for planks and bolts for the dial ...	£1. 6s. 4d
to Thomas Austen's widow for her husbands help about the carting of the sun-dial ...	1s. 6d

In addition to the sundial, a new clock was provided at a cost of £9. Thomas Austen was the church sexton and had recently died. Very little is known of the clockmaker, William Barrett (see separate entry, pages 98-9) but his name also appears in the accounts for Bethersden Church.

By the end of the sixteenth century it is likely that a number of churches in the county would have had a turret clock. Many of these early mechanisms would not have had dials, but would have sounded the hours on a bell and would have been placed inside the church at ground level. Owing to the lack of surviving records it is not possible to give a complete list of churches that possessed a clock in the sixteenth century but it is known that in addition to those already mentioned, All Saints, Lydd, had a clock as early as 1520 and St Mary the Virgin in Dover has had a clock since 1539. This is believed to be the clock that was formerly in the church of St Martin le Grand in the Market Square before the Dissolution of the Monasteries. The present clock dates from 1866 and replaces a movement installed in 1733.

SOME EARLY TURRET CLOCKS IN KENT

By the end of the seventeenth century and during the first quarter of the eighteenth the vast majority of the village churches, together with a number of the great houses in the county had been equipped with a turret clock. During the seventeenth century the number of turret clocks installed both in churches and other public places, such as guildhalls, market places and sites such as the conduit in Maidstone High Street (Figs 1/1-2), increased dramatically, pro-

Fig 1/1 A drawing made in 1786 of the conduit and its clock in the High Street, Maidstone, prior to its rebuilding in the nineteenth century. It is recorded that a clock was on this site in the sixteenth century

Fig 1/2 A view of the High Street, Maidstone, in 1830, showing the conduit and its clock

viding a much-needed source of work and income for the growing numbers of clockmakers and clocksmiths. Members of the Barrett family and, in particular, the Greenhills of Ashford, Canterbury and Maidstone must have benefited greatly from this demand, as the wealth that they accumulated as detailed in their wills and inventories can amply testify. While church records in most cases are very scarce, other than those already detailed, at least two other churches in the county have comprehensive records. In the case of Herne we are fortunate that the references to the clock had already been extracted from the churchwardens' accounts, because they were destroyed when the vaults of Lloyds Bank in Herne Bay, where they were stored, were flooded some years ago.

To attempt to give a description of each clock is beyond the scope of this volume, but all the turret clocks known to the author or where relevant and interesting information exists are given below. It is to be hoped that the Turret Clock Section of the Antiquarian Horological Society will produce a book on the subject in the near future.

Ash, St Nicholas

The churchwardens' accounts for this parish only go back to 1633, but the church must have had a clock prior to this date as in 1636 Thomas Barrett, the early Canterbury maker, was paid £12 for repairing the movement. This was a considerable amount and must have represented a virtual rebuilding of the clock, when one considers that Richard Greenhill supplied what must have been a similar movement to Herne in 1677 for £14.

In 1677 the clock again needed repairing and the entry in the records reads as follows: 'Item payd Richard Grinhill for repairing ye clock and diall 5. 0. 0'.

A new clock was subsequently supplied to the church in the early years of the eighteenth century, but unfortunately the records do not state either the name of the maker or the date of its installation. However, it must have been before 1708 as the accounts note that John Barrett of Canterbury was to 'yearly and every yeare (about St. Michalmas) come over and view the clock and receive his five shillings and also make good what shall be needful'. The replacement clock may have been one of the last clocks made in the Greenhill workshop just before his death in 1705 and it is interesting to note that his inventory mentions a bell and dial and weight belonging to what is obviously a turret clock, and which may have been the original Ash church clock removed by him after the installation of the replacement movement.

In 1789, a third clock was purchased from John Mepstead, watchmaker of Ash for £61. 4. 9. Mepstead then sold the replaced movement to Lady Lynch of the nearby parish of Staple. She presented it to the village where it was installed in the church. It is still giving good service some 290 years after it was made, although its one-handed dial and mechanism have recently been converted to electric winding (Fig 1/16).

In 1931, the quoted cost of repairs to the third clock was compared unfavourably to the price of a new one from Gillett & Johnson of Croydon. Having sought the opinion of Sir John Prestige, the noted horological collector living at nearby Bishopsbourne Park and receiving his advice that it was of no interest and should be scrapped, it was sold and replaced by the present church clock.

Ashford, Godinton Park

This clock, dating from the mid-seventeenth century, is now in store at Canterbury Museum and is inaccessible to the public (Figs 1/3-4). It has end-to-end trains, original verge and

Fig 1/3 The turret clock from Godington Park, Ashford, now in store at Canterbury Museum. The small turret clock on the right was made by Thomas Shindler for the Poor Priest's Hospital in Canterbury

Fig 1/4 Another view of the Godinton Park clock. This mid-seventeenth-century clock probably came from the workshop of Richard Greenhill the elder.

foliot escapement and knob finials to the corner posts. It probably came from the Greenhill workshop in Ashford.

Ashford, St Mary

According to an account written in 1712 by William Warren, a clock was installed in the church in 1680. Because the churchwardens' accounts for this parish only go back as far as 1682 no record exists as to who may have been the maker of this movement, or indeed, if it replaced an earlier clock, which seems likely given the existence in the town of two leading families who made turret clocks, the Barretts and the Greenhills. While there is no evidence, the author suggests that the 1680 movement would have been supplied by the Greenhills, either Richard or his son John. The earliest note in the accounts is as follows: '1682 Paid to

Colour Plate II An early seventeenth-century turret clock, maker unknown, formerly in Dover Castle, now on display in the Science Museum, London

Fig 1/5 Turret clock supplied to St Margaret's Church, Bethersden in 1737. Because of the similarity to the Lydd movement it is thought to have been made by John Wimble of Ashford

John Greenhill for looking after the Clock and chimes £1. 0. 0'. The clock was replaced in 1885 by the present mechanism and there is no trace of the earlier movement. (See separate entries for Richard and John Greenhill, pages 67-84.)

Bethersden, St Margaret's

This church, according to the records, possessed a clock as early as 1625 and also has what is probably the most complete set of churchwarden's accounts in the county, covering the period from 1625 until 1796, with only a short break of 28 years in the middle of the eighteenth century. What is particularly fascinating is that the accounts detail the names of the various clockmakers who worked on the two clocks in question during that period. The list starts with William Barrett, then we find Richard Greenhill, followed by his son John of Ashford, who probably changed the clock's escapement from verge and foliot to anchor in 1682. After the Greenhills, the Punnetts of Cranbrook took over, succeeded by Kingsnorth, Flint, Thomas Wraight and finally, Wraight & Woolley.

Sadly, the two things that are not recorded are the names of the makers of either the original clock or the replacement (Fig 1/5), which was installed in 1737 and paid for by public subscription at a cost of £10. The installation date happens to coincide with a gap in the churchwardens' accounts, although we have a list of the subscribers with a breakdown of how much each of them paid. (The names of forty-three men are recorded, each donating varying amounts between 1s 0d and 15s 0d.) Owing to the similarity between this movement and the clock supplied to All Saints Church, Lydd in 1730, there is a strong possibility that they were made by the same maker, namely John Wimble of Ashford.

While it is unnecessary to reproduce the accounts over a period of 150 years, the most interesting entries are given below, especially where they relate to a particular clockmaker.

1643	Paid for a sun diall to Barret of Ashford		0. 5. 0
1651	Item to the clockmaker for new altering the clock and severall things about it, as by his bill appeareth		4. 0. 0
	Item for the clock diall		1. 2. 0
	for bringing ye clock home		0. 5. 0
	for helping about ye clock six days		0. 6. 0

Unfortunately, the accounts omit to mention the name of the clockmaker, who appears to have carried out major work to the clock, so we cannot be sure whether Barrett did the alterations, or whether Richard Greenhill of Ashford was responsible. The work involved suggests that at this date an outside dial may have been fitted, because there are earlier references to a dial being set up and varnished in 1635 and 1640. These entries lead one to suppose that the clock was already fitted with an inside dial prior to the year that Barrett supplied the sundial.

1652	Item Paid to Greenall for mending of the clock	0.	3.	0
1665	Pd to Mr Greenhill for mendinge ye clock	1.	10.	0
1671				
Apr. 7th	for carrying part of ye clock to Ashford	0.	2.	0
	To John Greenall for work about the clock	1.	00.	0
1675	It. paid to Richard Grinell for mending of ye church Clock as his bill appeares	1.	7.	0
	It. paid for beare when ye clock was had down & set up		3.	0

[The last item makes it clear that it was thirsty work manhandling a large amount of seventeenth-century ironwork!]

1682	To John Grinoll for makeing ye clock a pendulum	5.	0.	0

[This refers to the alteration of the clock from verge and foliot to a pendulum with an anchor escapement.]

1690	Mr. Greenhill repairing ye Church Clock as by bill for two yeares July 21 1690	1.	0.	0

[This must refer to John Greenhill, as does a similar entry in 1692.]

1696	To Mr. Poynett for cleaning & mending the clock	0.	10.	0
1696	To Thomas Poynett for cleaning & mending the clock	0.	10.	0

[This entry and another in 1701 must refer to Thomas Punnett of Cranbrook, although at this period he is believed to have been working in Rye, Sussex.]

1703	paid for cleaning ye clock	0.	5.	0
	paid more to Mr. Kingsnorth for work about ye clock as appeares by bill	1.	9.	0

[Possibly a reference to John Kingsnorth of Tenterden.]

In 1714 and 1715 more substantial repairs and alterations were carried out to the clock, probably including the addition of minute work, as there is mention of 'altering the clock and making a new hand'. What is interesting is that the maker mentioned is Tho.[mas] Kingsnorth who is previously unrecorded. Whether this is a mis-spelling for John seems unlikely so it appears that there was a son, Thomas, who also worked as a clockmaker.

1722	Pd to Mr. Woolley for mending ye clock	1.	1.	0

[This entry may refer to either the father or grandfather of Thomas Woolley of Tenterden, but no record of a maker of this name at such an early date has been found.]

1764/5	Paid Mr. Flint for looking after the Church Clock three years	0.	15.	0
1784/5	Paid to Thos Wraight for repairing the Church Clock		8.	6

There are various references to repairs carried out by Thomas Wraight between 1784 and 1791 and in 1792/3 the partnership of Wraight & Woolley was paid 14s. 0d for repairing the clock.

Biddenden, All Saints
This church has an early eighteenth-century movement.

Borden, St Peter and St Paul
An iron-framed movement with knob finials and dating from the early eighteenth century.

Canterbury, The Poor Priests Hospital
The early eighteenth-century movement made by Thomas Shindler in 1728 for the Workhouse is now in store at Canterbury Museum inaccessible to the general public (see Fig 1/3 and the entry for Shindler, pages 220-2).

Charing, St Peter and St Paul
A seventeenth-century movement was found by Mr Ken Stocker in an old barn and has since been restored and is now on display in the church (Fig 1/6). The clock is believed to date from the 1630-50 period.

Chilham, St Mary
A seventeenth-century movement which originally had end-to-end trains was altered in 1727. In 1790 minute work was added and the large blue octagonal dial was moved upwards by 8ft in the tower. There are two early chime barrels in the clockroom, but with no mechanism.

Dartford, Holy Trinity
This church must have had a clock installed in the early part of the seventeenth century as there is a note in the churchwardens' accounts referring to alterations carried out in 1693, presumably when the movement was changed to a long pendulum and anchor escapement:

Fig 1/6 A mid to late seventeenth-century turret clock at the church of St Peter and St Paul, Charing. This clock, with end-to-end trains, has been restored in recent years and re-installed in the church

Aprill ye 1 1693 pd. Mr. Stabes for altering ye church clock into a pendelom and wates ...
£6. 10. 6d.

Although Stabes is likely to have been a clockmaker, no other record has been found of him. However, this entry may refer to the London maker Thomas Stubbs who is referred to in the Clockmakers' Company records as a 'Great Clockmaker' (ie turret clockmaker).

According to the accounts the sum of £1 10s 0d was paid each year for 'looking after the clock' and this amount included payment for ringing the 6 and 8 o'clock bells. In 1715 Thomas Young was paid £2 for mending the clock and in 1728 Stephen Butterly, the Dartford maker, is mentioned for the first time. He was paid £2 10s 'for ye Clock as pr Bill' and his name appears in the accounts on a number of occasions until the middle of the century, receiving payments for minor repairs to the clock.

An outside dial may have been provided in 1728, for in that year Rowland Fry received eight guineas for gilding and painting the church dial and 3s 0d was paid to 'Giles Sedgwick for carrying up and bringing in down ye Church Diall from London'. During the latter part of the eighteenth century Gabriel Fowkes was responsible for the general maintenance of the clock and his name appears at regular intervals in the accounts between 1775 and 1785.

However, it is interesting to note that when the churchwardens decided to add quarter work to the clock in 1780, Fowkes was not entrusted with the task and a new movement was supplied by Thwaites of London at a cost of £124 10s 0d. Of this amount, £40 was raised by public subscription, the balance being paid by the churchwardens. Thwaites is mentioned a number of times in the accounts after the installation of the new clock in 1781, eg in 1792 he received £9 7s 6d for repairing the clock.

In 1910 it was felt necessary to replace the Thwaites movement and a third clock was installed in the church tower, paid for by public subscription. In recent times an additional dial has been provided for the clock to make it visible from all sides of the church. (See separate entries for the clockmakers Butterly [page 117] and Fowkes [pages 139-41].)

East Malling, St James
A late seventeenth-century movement with knob finials from this church is in store at Maidstone Museum.

Fig 1/7 A mid eighteenth-century movement with side-by-side trains and countwheel striking, at St James Church, Egerton. The iron frame has furled finials. The setting dial, dated 1854, is signed by Bartlett, Maidstone

Egerton, St James

A mid eighteenth-century birdcage movement with side by side trains, anchor escapement and countwheel strike housed in an iron posted wrought-iron frame with furled finials (Fig 1/7). A silvered setting dial attached to the movement is signed, Bartlett & Sons, Maidstone, dated 1854. The clock was probably overhauled by them at this time.

Goudhurst, Glassenbury Park

An unusually small early seventeenth-century movement only 15in by 12in, with end-to-end trains, now converted to operate a dial.

Goudhurst, St Mary

A late seventeenth-century/early eighteenth-century movement.

Headcorn, St Peter and St Paul

A seventeenth-century birdcage movement with knob finials. Now in store at Maidstone Museum.

Herne, St Martins

This clock was supplied to the church by Richard Greenhill of Canterbury in 1677-8 at a cost of £14 soon after his move to the city. It is typical of seventeenth-century design, although it has suffered alteration through the years (Figs 1/8-9). The clock originally had end-to-end

Figs 1/8 and 1/9 The turret clock in St Martin's Church, Herne, made in 1677 by Richard Greenhill of Canterbury. Originally with verge and foliot escapement and end-to-end trains, it was converted to anchor escapement in the eighteenth century. The corner posts are nineteenth-century replacements for the originals

trains, which were probably converted to the present side-by-side arrangement at the same time as the original verge and foliot escapement was altered to anchor, some time in the eighteenth century. Other modifications include the addition of vertical bars of twisted section held by nuts and bolts; the original bars would have been held in position by iron wedges, as was the practice before the introduction of nuts and bolts. Although the clock possesses its original iron striking train the going train has been replaced with one with brass wheels and steel pinions. The original finials have disappeared, so that it is difficult to make comparisons with other clocks in the area which in all probability came from the Greenhill workshop.

The following extracts are taken from the churchwardens' accounts between 1663 and 1728.

1677-8	It. Pd to Mr. Grinall for a new clock to the Church	14.	0. 0
1678-9	To John Overy, Clerk, for looking after the church clock		10s. 0d
	It. to Mr. Greenhill for one yeares repaires of the church clock due at Michalls 1678		5s. 0d
1679-80	It. to Mr. Greenhill concerning ye clock		5s. 0d
1680-1	" " " " "		5s. 0d
1681-2	It. to Greenhill for the clock 5s.0d		
1687	It. to Mr. Greenhill concerning ye clock 1.15.0		
1693	Paid to Mr. Greenhill for 6 yeares repaires for the clock til Michalmas last	1.	10. 0
1698-9	Mr. Greenhill for 5 yeares repaires of the church clock due at Michalmas last		
	It.; 5s for new lines as per account	1.	10. 0
	pd. to John Buckhurst for Communion Bread & Oyl for ye clock		7. 0
1707-8			
June 11	Paid for mending & cleaning ye clock		14. 0
	Paid for painting the diall of ye clock Lampblack it and Lowances		7. 9
June 13	for to lines for the clock and duson of trushes of the pues		13. 0
1709-10			
Nov. 22	Paid Mr. Greenhill Eight years arrears for the Church Clock	2.	0. 0

[NB Richard Greenhill died in 1705. The money must have gone either to his widow or his son, Samuel.]

1710-11			
July ye 10	for cleaning ye clock		5. 0
	paid for Ile for ye Clock		7.
1714-15			
Apl.20	Pd to Wm.Badcock for Wages, washing the surplice & keeping the clock	3.	10. 0

[Note: Badcock was the Parish Clerk.]

Apl.20	More for bread for the Communion and Oyle for the clock		2. 6
1715-16			
Oct.25	for a pynt and a Quarter of Oyle for ye clock		1. 3
Feb.4	Paid Goodlad for cleaning the Church Clock		7. 0

[This extremely interesting reference must refer to Richard Goodlad, who was the first of Greenhill's apprentices. See entry under his name, pages 146-7.]

1717	Paid the painter for the Clock	2.	10.	0
	For a clock line for ye Church Clock		3.	6
	Pd John Baret for Church Clock	2.	0.	0
1719-20				
Mar.22	Pd John Barret for cleaning of the Church Clock		10.	0
1720-1				
Sept 26	Pd Moyse for a clock line		3.	6
1722-3				
June 7	pd Jos Reeve for cleaning the clock		10.	0
1724-5				
July 9	pd for cleaning the clock		7.	0
Apl.17	Oyle for ye clock		1.	0
1727-8				
Oct 16	pd for cleaning the clock		10.	0
Apl 20	pd for wiar for the clock			4½

John Barrett, another of the Canterbury makers, probably took over the task of looking after various clocks that Greenhill had supplied after the latter's death in 1705. Barrett was also mentioned in the accounts for Ash Church and there is a possibility that he may in fact have supplied their new clock in 1708.

Hever, St Peter
A seventeenth-century birdcage-framed clock, originally with a verge and foliot escapement, now converted to pendulum with anchor escapement and side-by-side trains (Fig 1/10).

Knowlton, St Clement
A late seventeenth-century movement with knob finials to the iron frame.

Leeds Castle
A typical seventeenth-century turret clock with end-to-end trains, the corner posts with knob finials, no external dial. Probably dating from 1660-70.

Lydd, All Saints
The clock in All Saints, Lydd may have been installed towards the end of the fifteenth century

Fig 1/10 St Peter, Hever. A seventeenth-century movement, originally with end-to-end trains and a verge and foliot escapement. Subsequently converted to anchor escapement with side-by-side trains and now fitted with electric winding.

Fig 1/12 Close up of the setting dial of the clock at Lydd

Fig 1/11 The turret clock made in 1730 by John Wimble of Ashford for All Saints, Lydd

as the entries in the churchwardens accounts start with a payment to Thomas Papworde for keeping the clock in 1520. This particular clock was replaced in about 1599/1600, and the existing clock was ordered from John Wimble of Ashford in 1730 (Figs 1/11-12). The extracts from the accounts relating to the clock make interesting reading:

1520	It[em] pd to Thomas Papworde for kepyng of the clokke the last yere …	20s.
1521	Item paid to Thomas Papworde for kepyng of the clokke this yere endyng nowe at the accompte …	20s
1522	Ffurst paid to Thomas Papworde in full payment of his wage for kepyng of the clokke for the last year	2s. 4d
1524	Paid to Roger Clokemaker for makyng of yecloke	10s. 0d
1526	Paide to Robert Fletcher the Sunday after Seynt Andrewes Day in parte of payment of the cloke kepyng	12d
1526	Payde to Thomas Hewett for kepyng of the clock in Lent	5s.
1527	Paide to Thomas Hewett for kepyng of the clokke	5s. 10d
1528	Payde to Thomas Hewett for a qrt wage kepyng of the clock and tollyng of the bell	5s. 10d
1529	" " "	5s. 10d
1530	Payde for wyer for the clock	2d
1530-8	[Records of payments to Thomas Hewett at 5s a quarter for clock-keeping.]	
1539	Pd to John Loker for a claspe for the clock [He was a blacksmith and also made clappers for the bells]	2d
1544	Pd to John Locker mendyng the clock	6d
1546	Pd for mendyng of the clock	6d
1546	Pd to the clockmaker in part payment 5s for his yerely ffee for kepyng the seid clock and the chymes	12d
	Pd for the new makyng of the clock and the chymes	26s. 8d

This account presumably refers to an overhaul of the quarter-striking mechanism which in church accounts throughout the sixteenth and seventeenth century is invariably kept separate from the clock costings. Unfortunately there is no trace of the name of the clockmaker involved.

There are other references in the accounts to sums paid out for minor repairs, but as stated earlier this clock was replaced about 1599/1601 as there are notes in the records concerning a new clock platform and dial supplied by Goodman Ellis of Rochester. In 1730, John Wimble of Ashford supplied a new one-handed clock and dial for an unspecified amount, but the following references appear in the accounts up to 1741, the year of Wimble's death.

1734/5			
1st Feb	for carrying dial to Romney		3s. 0d
7th March	Fetching dial from Romney		3s. 0d
1st April	Paid Mr. Rooks for painting dial	5.	15s. 6d
	Paid Mr. Wimble as his receipt		15s. 0d
1735/6	Paid for clockline		5s. 0d
1738/9	Paid John Wimble for cleaning the town clock		10s. 0d
1739/40	Paid John Wimble for cleaning, mending the church clock as per bill	1.	3. 0d
	[NB The town clock was the same as the church clock.]		
1740/1	Paid John Wimble as per bill		11s. 6d

In 1839 a new copper dial was fitted to replace the original wooden one and minute work was added in the same year. At the same time the clock was raised to a greater height in the tower. It is still in use today, but has been converted to electric winding in recent years.

Lyminge, St Mary and St Ethelburga
Remains of an early eighteenth-century clock are in the church.

Fig 1/13 Early eighteenth-century movement from the Town Hall, Maidstone, now on display in Maidstone Museum (the missing barrel is in store)

Fig 1/14 Seventeenth-century clock from St Nicholas, Pluckley. The movement, with end-to-end trains, in need of restoration, is in store in Maidstone Museum

Maidstone, Town Hall
An early eighteenth-century movement with side-to-side trains is on show at Maidstone Museum (Fig 1/13). There is some confusion as to who made this particular clock, but it bears a striking similarity to other Kent turret clocks of the period.

Minster-in-Thanet, St Mary
The church has a late seventeenth-century movement.

Pluckley, St Nicholas
Another early to mid seventeenth-century clock with end-to-end trains in store at Maidstone Museum (Fig 1/14). Also fitted with knob finials.

Rochester, The Guildhall and Corn Exchange
A clock had been installed at the Guildhall some time before 1663, but unfortunately there is no record of the name of the maker. A minute in the city records dated 1663 states:

> It was agreed that Wm Chatbourne keeper of the Jail at the Dolphyn shall not for the future receive the yearly salary of 40s for looking to the clock in the City but he shall look after the same in good and careful manner without any allowance.

Fig 1/15 The Corn Exchange, Rochester. The present movement was supplied by Edward Muddle in 1771 to replace an earlier clock (possibly supplied by Joseph Windmills, the noted London maker, and presented to the city by Sir Cloudesly Shovel). It utilised the original bracket, which may have come from Grinling Gibbon's workshop, or possibly may have been carved by one of the many skilled craftsmen working in the Royal Dockyard at nearby Chatham

The following entry also appears in the minutes for 1677.

> 1677 Paid Simon Lambe as appeares by his acquittance for new wheels and other work by him done to the Town Clock   £1. 0. 0

In 1706 Rear Admiral Sir Cloudesly Shovel, who represented the city in parliament as its MP, provided the finance for a new Corn Exchange and ceilings for the Guildhall and presented the city with a replacement clock. Once again it is not clear which maker was employed. There is a strong possibility that it was made by a leading London clockmaker, as soon after its installation a minute dated 16 October 1708 records that 'Mr. Windmill was ordered to be sent for to repair the town clock'. Further repairs were carried out by the same man in 1718. These references presumably relate to either the well-known maker, Joseph Windmills or his son Thomas. Joseph is another of those makers admitted to the Clockmakers' Company in London as Free Brothers and referred to as 'Great Clockmakers', a term now thought to mean turret clockmakers (see Richard Goodlad, pages 146-7).

Soon after the installation of this clock Shovel was drowned when his ship the *Association* was wrecked on the Scilly Isles. His body was subsequently recovered by order of Queen Anne and he was buried in Westminster Abbey. The clock he provided was in use until 1771 when the city council decided to replace it with a new mechanism made by Edward Muddle of Chatham (see entry for Muddle, pages 192-4). The dial of Muddle's clock utilises the same ornate bracket overhanging the High Street as the earlier clock (Fig 1/15). Unfortunately, the city records make no mention of the name of the carver of this fine example of craftsmanship, typical of the Queen Anne period, which has overhung the High Street since 1706. It is interesting to speculate whether it may have been carved in Grinling Gibbons's workshop, or at least under his supervision. The memorial tablet to Shovel in Westminster Abbey was carved by Gibbons and the tablet commemorating Sir John Narborough and his brother (both tragically died in the same shipwreck) which can be seen in St Clements Church, Knowlton, Kent, has also been attributed to him. The brothers were Sir Cloudesley Shovel's stepsons.

Fig 1/16 An early eighteenth-century movement of about 1705, formerly at St Nicholas Church, Ash, and moved to St James, Staple in 1789. This clock is likely to have been made by Greenhill of Canterbury, and has recently been converted to electric winding

Fig 1/17 St Nicholas, Sturry. A late seventeenth-century movement with end-to-end trains, that is thought to be from Greenhill's workshop

Sevenoaks, Knole Park
An early turret clock exists here, but the details are not known.

Smarden, St Michael
In the clockroom are the remains of a seventeenth-century birdcage-framed turret clock with a great wheel and strike lever surviving from the original movement. The only records so far discovered in the churchwardens' accounts are as follows:

1704	Item	paid Henry Philpott for clock line	6 d
1705	Item	paid Thomas Perrin his bill for looking after the clock	£2 5 0d

Staple, St James
See St Nicholas, Ash (Fig 1/16).

Sturry, St Nicholas
A very interesting late seventeenth-century movement with end-to-end trains, not at present in working order.

Fig 1/18 St Mary the Virgin, Thurnham. A late seventeenth-century movement with end-to-end trains and finials of a type that are rarely seen in Kent

Fig 1/19 St Mary the Virgin, Wingham. A movement of about 1670-80, originally with a verge and foliot escapement and end-to-end trains, converted to long pendulum and anchor escapement in 1726. Another clock that may have come from the Greenhill workshop

Thurnham, St Mary the Virgin
A movement with end-to-end trains dating from the middle of the seventeenth-century.

Wingham, St Mary the Virgin
This church has an early birdcage movement dating from circa 1670 with a large octagonal dial (Fig 1/19). It was originally fitted with verge and foliot and end-to-end trains but has since been converted to long pendulum and anchor escapement. Unfortunately, the churchwardens accounts only date from 1719 onwards so that it is not possible to state who made this clock. However we do know that John Brice was paid varying sums between 1719 and 1726 for repairs and general maintenance to the clock. An entry in 1726 also records John Gardner, 'altering, repairing and cleaning clock in ye church ... £2. 15. 0'. This entry probably refers to the change made to an anchor escapement.

In addition to the churches already mentioned, the following parishes also have (or had) turret clocks fitted at an early date:

Birling
Cobham
Cranbrook (with Denison's prototype three-legged gravity escapement, developed for the Houses of Parliament clock)
Edenbridge
Eynesford
Farnborough
Folkestone (on display in Folkestone Museum)
Groombridge
Leeds (made by William Gill in the early eighteenth century, Figs 1/20 -21)
Lenham
Lynsted
Maidstone
New Romney
Rainham (supplied by William Gill in 1713)
Tonbridge
Yalding

Fig 1/20 St Nicholas Church, Leeds. An early eighteenth-century clock made by William Gill of Maidstone

Fig 1/21 Close-up of the setting dial of the clock in Leeds church, with similsr engraving to contemporary lantern clock dials

Chapter Two
DOMESTIC CLOCKMAKING IN KENT

As we have seen, the number of turret clocks installed in both churches and other public places, such as the four-sided clock on the tower of the old market building in Tonbridge, now sadly demolished, and the Guildhall in Rochester, increased dramatically during the seventeenth and early eighteenth centuries. Concurrently there arose a demand for a relatively cheap domestic clock and the late sixteenth and early seventeenth centuries saw the introduction of the lantern clock into England.

Prior to the beginning of the sixteenth century small versions of the large turret clock, commonly called chamber clocks, had been in use in the great palaces and houses of the aristocracy and members of the wealthy merchant class. The one disadvantage possessed by these clocks was that they were not portable, and at some time in the early sixteenth century spring-driven table clocks and clock-watches, which could be hung from a person's belt, came into use. Most of the early table clocks were in the shape of round drums with the dials placed on top, and unlike the earlier chamber clocks, the mechanism was mounted between two metal plates, rather than within an open frame. These clocks were not only very expensive, but almost without exception they were made on the Continent in the Low Countries, France and Germany. These early table clocks were extremely intricate pieces of work, with cases of brass and silver-gilt often superbly carved and engraved, and because of their great cost were only available to the most wealthy section of the population. Needless to say, the skills necessary to produce such clocks were not to be found among the provincial craftsmen of Kent. The clockmaking industry in England at that period was mainly in the hands of the Blacksmiths' Company in London, while the makers of turret clocks in Kent and the rest of the provinces were engaged in the ancillary trades, as blacksmiths, locksmiths and whitesmiths, for it would not have been possible to exist solely by making and supplying turret clocks in the early sixteenth century.

The Tudor and Elizabethan period saw England enjoying a greater period of prosperity than she had ever known and, inspired by the Renaissance, the English Court had become a major centre of learning and scientific discovery. The inevitable result of this new prosperity was an increase in the standard of life for the upper classes in particular, and a consequent growth in an ever more affluent middle class with a growing demand for the luxury goods that were finding their way into the country from the mainland of Europe. This period also saw the beginning of much religious persecution on the Continent driving many Protestants to leave the Low Countries and France to find refuge in England, which by this time had broken with the Roman Catholic Church and had seen the establishment of the English Church following the Dissolution of the monasteries. These refugees included large numbers of craftsmen of all trades who thus brought their superior skills to the benefit of this country. Among them were a number of clockmakers who settled in London and the last quarter of the sixteenth century saw the introduction of the lantern clock, which became the principal method of timekeeping for the next 100 years and indeed, its use and manufacture lasted into the middle years of the eighteenth century in the provincial counties, including Kent.

During the early years of clockmaking in England the majority of the craftsmen were foreigners and the industry was largely in the hands of and under the control of immigrants. Their dominance caused growing resentment among the increasing numbers of English-born makers in London and a group of them banded together to petition the king to form a Clockmakers' Company separate from the Blacksmiths, who had controlled the industry hitherto. Their aims were to regulate the methods of working and conditions of employment and to restrict the growing numbers of foreigners who, they claimed, were taking away their livelihood. At the second attempt a Charter of Incorporation was granted by Charles I on 22 August 1631. This charter laid down the basic rules by which the industry was governed for the next hundred and fifty years, with the main emphasis on condition of work and the standards required in order to maintain a high quality of workmanship. However, it must be pointed out that the company's authority did not extend beyond a 10 mile radius of the capital.

Lantern Clocks
While it is beyond the remit of this book to go into great detail about the early history of the trade in London, two books by Brian Loomes: *Country Clocks And Their London Origins* and *The Early Clockmakers Of Great Britain* are required reading for anyone requiring a full account of the early years of the craft in England, and in London in particular. Sir George White's book *English Lantern Clocks* is also necessary reading for those wishing to learn in detail about the lantern clock, which by 1620 had become the staple output of the London makers and had begun to be made in the provinces by this date.

It is quite probable that clocks from this period were made by Kentish makers, particularly in Canterbury, but sadly, none from this early date have survived, so far as is known. However, references to 'old ballance clocks' are to be found in a number of wills and inventories from the 1680s and it can be assumed that they were a common possession of the yeoman farmer and the more prosperous tradesmen from the towns. Thomas Deale's inventory taken in 1687 mentions 'an old ballance clock' value £1 5s 0d, and this entry obviously refers to an early lantern clock. Numerous Kent makers are known to have made such clocks, and examples dating from the last quarter of the seventeenth and early eighteenth century have survived by the following: Richard Greenhill of Ashford, John Greenhill of Ashford, John Greenhill of Maidstone, Edmund Grigsby, John Baldwin, Thomas Shindler, Edward Cutbush, Robert Silke, John Mercer, John Kingsnorth, John Bishop, George Thatcher, John Dodd, Samuel Greenhill and William Gill.

The early lantern clocks used the balance wheel escapement (hence the name 'ballance clock') giving very unreliable timekeeping, but the invention in 1657 of the pendulum used in conjunction with the verge escapement, by the Dutch scientist Christiaan Huygens Van Zulichem, revolutionised clockwork, and made accurate timekeeping possible for the first time. He gave the right to use his invention to a clockmaker from the Hague, Salomon Coster, in whose employ at the time was John Fromanteel, the son of Ahasuerus Fromanteel, one of the leading London makers. In 1658 Ahasuerus introduced the pendulum to England and with it the longcase clock. After this date lantern clocks were also fitted with a verge escapement, but many of the surviving examples have since been altered to a long pendulum using the anchor escapement which was introduced in 1675, giving even more accurate timekeeping.

The earliest surviving Kent lantern clocks are two superb large examples (Figs 6/9-13) fitted with quarter striking by Richard Greenhill of Ashford dating from 1680 or possibly a

Figs 2/1 and 2/2 (above, right) The dial and movement of a good quality lantern clock dating from about 1680/90 by John Greenhill III of Maidstone

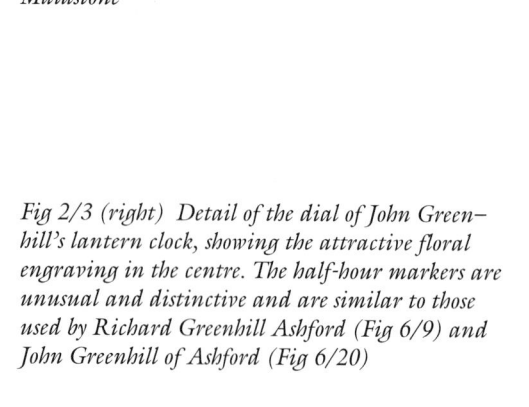

Fig 2/3 (right) Detail of the dial of John Greenhill's lantern clock, showing the attractive floral engraving in the centre. The half-hour markers are unusual and distinctive and are similar to those used by Richard Greenhill Ashford (Fig 6/9) and John Greenhill of Ashford (Fig 6/20)

Colour Plate III Early-eighteenth-century lantern clock by George Thatcher of Cranbrook, about 1725. This clock is a rare example of a lantern clock with a rack-striking movement (Fig 7/109)

Colour Plate IV A 30-hour wall clock by Stephen Harris of Tonbridge, about 1715, in a lacquered hooded case

little earlier, and a fine small clock with a balance wheel escapement made by his eldest son, John, in Ashford a few years later (Fig 6/20). Although the lantern clock had been largely superseded in London by the longcase and bracket clock it still survived in the provinces, including Kent, until the 1740s and, in fact, a derivative from it, the small brass posted-frame wall clock with alarm work, was particularly popular in Kent until the very last years of the eighteenth century.

The early clocks by the Greenhills, with their fine central engraving of tulips, bear a striking resemblance to the longcase movement by Thomas Deale, and may possibly be decorated by the same engraver, and, as discussed later, the engraving may be by Deale's own hand.

Watches

One of the most striking results of the researches made in the course of preparation of this book is the distinct lack of watches from the seventeenth century, in spite of the known existence of a number of makers during the period in question. As far as can be ascertained an early example by Thomas Barrett of Canterbury is the only survivor that has been traced to date (Figs 7/10-11). The records show that there were a number of watchmakers, particularly in Canterbury, working from as early as 1647 when Ambrose Bliss received the Freedom of the City by redemption. He subsequently moved to London in 1654 and is recorded as taking three apprentices, and yet no work signed by him appears to have survived. The same apparent lack of surviving workmanship applies to James Ellis, also of London, and his apprentices, who are all believed to have worked in Canterbury during the last quarter of the seventeenth century, without leaving a trace; as also in the case of Thomas Dennis, the Maidstone maker who had been trained in London, no watches by him have been found.

While one expects the survival rate of clocks and watches from the seventeenth century to be relatively small, it seems strange that absolutely nothing from this period by these makers can be traced. Is it possible that they were engaged in making movements for makers in London who subsequently signed the work with their own name? This practice was common in other parts of the country, such as Liverpool and Birmingham, but to date there has been

Fig 2/4 and 2/5 An early verge watch movement by John Baldwin of Faversham, No 1026 of about 1710

no evidence to suggest that it happened in Kent, but there seems to be no other explanation for the distinct lack of watches surviving from this period.

With the onset of the eighteenth century there is an increase in the number of watches made in the county with some relatively early examples by such makers as John Baldwin of Faversham (Figs 2/4-5) and Mercer of Hythe (Fig 7/83), both of whom were working in the first quarter of the century. By the middle of the century there appears to have been a flourishing trade in watches in the county. The evidence for this comes not from surving examples, which are few, but from the numerous advertisements relating to lost watches which appear in the pages of the two Kent newspapers, the *Kentish Post* and the *Kentish Gazette*.

Many of the watch and clockmakers appear to have also been engaged as goldsmiths and silversmiths, judging from the advertisements in the *Kentish Gazette* in 1776 placed by such men as John Vidion and Richard Bayley relating to the exchange of light gold coin following the Proclamation of George III in that year (see Appendix I).

While a large number of watches dating from the 1770s onwards are known, many of these may have been supplied from London workshops and merely been finished by the Kent makers whose names appear on them. The same applies to the many hundreds of watches bearing the names of numerous nineteenth-century makers who would in all probability have had nothing to do with their actual manufacture but would have bought in the finished or roughly finished movements from the great centres of watchmaking in Liverpool and London.

Bracket Clocks
The earliest bracket clock traced by the author is the example in the Canterbury Museum collection by John Watts dating from the 1740 period, and although other clocks do exist signed by such makers as James Warren from the latter part of the century, they were never made in large quantities. There were two reasons for this. Firstly, bracket clocks were much more expensive than other types of clock and would probably only be owned by the gentry and the more prosperous merchant class. Secondly, the ease of access to London and its geographical closeness made it likely that someone wealthy enough to afford a bracket clock would purchase it in the capital. It is no coincidence that the number of early bracket clocks in the provinces increases the further away from London, and it probably has nothing thing to do with a lack of skills, as some writers have suggested in the past. John Greenhill's will of 1706 makes mention of a spring (bracket) clock and it is interesting to speculate as to who may have made this. That the skills did exist is only too apparent when one looks at other surviving clocks from the Greenhill and Deale workshops and other makers, such as James Jordan of Chatham.

Longcase Clocks
While the lantern clock was the most common form of timekeeper in the seventeenth century in Kent, the advent of the verge escapement and pendulum saw the introduction of the longcase clock, and after the invention of the anchor escapement in 1675 by, it is thought, William Clement or Joseph Knibb, the longcase became pre-eminent in London and by the end of the century had spread throughout the rest of the country. While it is likely that examples of longcase clocks by the top London makers would have found their way into the houses of the local aristocracy at an early date, the number of clocks dating from the pre-1700 period known to have been made by Kentish craftsmen is very small. The earliest of these are undoubtedly two fine 8-day clocks by Richard Greenhill (Figs 6/14-16) and Thomas Deale

Fig 2/6 The only known longcase clock by Samuel Greenhill of Canterbury, with a five-pillar movement, in its original laquered case, about 1705-10

Fig 2/7 The 12in square dial of the Samuel Greenhill 8-day clock

(Figs 5/1-2), both of Ashford, dating from the 1680 period, together with a 30-hour example by Greenhill of Canterbury dating from a few years later, about 1685 (Colour Plates V and VI).

A number of clocks by John Greenhill of Maidstone, made between 1685 and 1700, have also survived (Figs 6/3-8) together with a fine 8-day movement in a lacquer case by John Bishop of Maidstone (Fig 7/17). Other examples by James Jordan of Chatham (Figs 7/71-73) and John Wimble of Ashford (Colour Plates XII and XIII) from the 1700-10 period have also been noted. An interesting feature of these later clocks is the continued use of the convex moulding to the trunk, which had gone out of fashion with London makers by the turn of the century. Another notable feature of longcase clocks by Kent makers from this period is the use of lacquer which appears to have been particularly fashionable, and was in common use by other makers such as Henry Baker of Malling (Fig 3/6), Thomas Jenkinson of Sandwich (Colour Plate X) and Gabriel Fowkes of Dartford until late in the eighteenth century.

A vogue for lacquer furniture had swept Britain with the founding of the East India Company, and the Dutch East India Company had initiated a similar craze in Holland from whence it gave added impetus to the fashion following the restoration of Charles II and the subsequent arrival of William of Orange in 1688. That year also saw the publication of a book by Stalker and Parker entitled *An Illustrated Treatise of Japanning and Varnishing* and as a result the art of japanning soon became a popular pastime in the wealthier households of London and the provinces.

The popularity of lacquer decoration in Kent and East Anglia may also have had much to do with the fact that large numbers of immigrants from the Low Countries had settled in those counties nearest to the Continent following the religious persecutions of the late sixteenth and seventeenth centuries. Sandwich and Canterbury benefited greatly from the influx of these refugees.

The number of immigrants in Canterbury amounted to about 1,000 out of a total population of approximately 7,000 by the end of the century, and in Sandwich the proportion of immigrants was some 500 out of a total population of 2,000 people. That the Dutch influence was very strong in the region can be seen by the number of farmhouses with Dutch-type gables that are still a common feature of the landscape of East Kent to this day. The Dutch influence is also to be found in some of the furniture which regularly turns up at local auctions, with quite distinctive features traceable to that country but obviously made by English craftsmen of the period. Lacquer was much more commonly used for clock cases than previously realised, as many of the plain oak and pine cases dating from the early years of the century now to be found, either ebonized or in a stripped state, were more than likely to have originally been lacquered. However, owing to its fragile nature a much lower proportion of laquer cases in their original condition have survived than of other types.

In addition to the Flemish immigration during the early years of the seventeenth century a number of Huguenot refugees arrived in Canterbury following the revocation of the Edict of Nantes in 1685 and the resultant religious persecution by the Catholic majority in France. They formed the backbone of the silk-weaving community in Canterbury, which by the end of the century was among the most important centres of the trade in Britain. However, they do not appear to have exerted any other influence in the decorative field and it is almost impossible to find any trace of such an influence in either furniture design or indeed in the style of clock cases which in Kent, as in other southern counties, tended to follow the current London fashion.

Whether the Flemish community included among their numbers skilled craftsmen in the art of carving is not known, although, of course, it is a fact that Grinling Gibbons was discovered by John Evelyn working in a cottage near his estate in Deptford in the far west of the county, and Gibbons was of Flemish extraction. However it is interesting to note that the area in the vicinity of Ashford centred on Bethersden produces a type of marble which has been used to good effect on various tomb-chests in the local churches, eg Chilham (Fig 2/8), and also appears on the fireplace surrounds of a number of the great houses of the district, particularly at Chilham and Godinton. The marble in each case is invariably carved with an all-over pattern of flowers with trails of roses, vines and honeysuckle, combined with strapwork de-

Fig 2/8 The early seventeenth-century tomb-chest of Margaret Palmer in St Mary's Church, Chilham. It is a typical example of the superb craftsmanship to be found in many of the churches and great houses of the Ashford and Canterbury area. The quality of such work is reflected in the finely engraved dials of the early Kentish clockmakers

signs. Unfortunately, there appears to be no record of the craftsmen responsible for these fine carvings which, with the contrast obtained by polishing the areas in relief against the cut carving, gives the impression of damask cloth. The Dutch influence is even more pronounced in the two fine monuments at Wingham and in the tower of St Mary Magdalene in Canterbury, dating from 1682 and 1691 respectively. The first of these is believed to have been carved by another leading Dutch emigré, Arnold Quellin, a contemporary and associate of Gibbons. He may also have been responsible for the superbly carved wooden frieze at Godinton depicting the pike and musket drill of the Honorable Artillery Company also dating from this period. Once again we have no knowledge as to the name of this fine carver, but the owner of the house, the late Mr Alan Wyndham Green, says that there is a strong tradition that it was carried out by Flemish craftsmen, for the owner of the estate, Captain Nicholas Toke, in the mid-seventeenth century. Another possibility is that the carver was Jacobus Deale, the grandfather of Thomas Deale of Ashford, and that the tradition of carving and artistry continued in the family with the superbly engraved dials on the Greenhill and Deale clocks, which the author believes may very possibly have been engraved by Thomas. There is also a striking resemblance between the figures on the frieze and the jacks on Greenhill and Dodd's lantern clocks (Figs 6/1-2 and 6/12-13), and it may be significant that three jacks (which may be clock jacks, rather than domestic jacks) are itemised in Deale's workshop inventory. It was hoped that the Toke family papers in the archives at Maidstone would provide the answer to this fascinating question. Unfortunately, only the agricultural accounts appear to have survived, and there is no trace of any material relating to the refurbishment of Godinton or the relevant accounts.

Further evidence of the artistic skills that existed in the region are to be seen in the interior furnishings of many of the churches and other great houses. Notable examples are the superbly carved coats of arms of Charles II to be found hanging in the churches at Knowlton and Ashford. The late seventeenth century saw a vast amount of rebuilding and refurbishment, together with new building, particularly in brick, almost certainly the result of the increasing wealth and prosperity of the gentry, tradesman and yeoman farmer.

This period saw the refitting of the Cathedral Choir at Canterbury in a baroque style by John Davis, another of the seventeenth century's master woodcarvers, partly in an attempt to repair the ravages caused by Cromwell's soldiers during the Civil War, but also as a result of the new decorative ideas spreading throughout England following the vast rebuilding programme in London under the direction of Sir Christopher Wren after the Great Fire in 1666.

Much of Davis's work, together with part of the superbly carved wooden archbishop's throne by Grinling Gibbons, was swept away during the 'gothicising' by the Victorian restorers. While the majority of this rebuilding would have been carried out under the supervision of master-builders, such as Davis, it seems reasonable to suppose that there was a skilled pool of local craftsmen to assist them.

While we may never discover the names of the craftsmen responsible for these fine carvings and engravings dating from this short period between 1630 and the 1680s, enough evidence has been provided to dispel the notion held by some notable horological writers that the skills necessary to produce the superbly engraved dials of the Greenhill and Deale clocks could not have existed in the Ashford area, and by inference must have been supplied from London. It is also true to state that the tradition of fine engraving in the region was carried on well into the eighteenth century, as many of the clocks made by the Bakers and other makers from the Maidstone area can amply testify. This tradition lasted through the period of the silver dial with its associated rococo style engraving and its influence extended to the dial painters of the 1790-1820 period. Much of this painting is in a similar style and would appear to be the work of, at most, two or three craftsmen. Compared to the dials of clocks from other counties it is very restrained, usually consisting of sprays of flowers and birds, scrolling tracery, and very occasionally with a landscape or hunting scene in the arch in the case of arch-dial movements. The only exceptions to this style are the simple 30-hour dials with either shell or geometric designs to the spandrel areas. While painted dials were supplied from Birmingham, as a rule, those used in Kent appear to be different from those of other counties. Is it possible that the dial painters were influenced by the clockmaker and encouraged to produce these simpler styles?

It is also surely no coincidence that the superb engraved silver dials from the Maidstone area appeared at the same time as one of the finest engravers of the eighteenth century, namely William Woollett, who himself was of Flemish descent. He is known to have worked for Joseph Harris of Maidstone (Fig 7/49) and quite possibly worked for other makers in the area before moving to London.

The style of clock case used by the Kent makers to house their longcase movements remained fairly standard throughout the eighteenth century, and in other counties in the southeast they were strongly influenced by the designs that were fashionable in London. This particularly applies to the case design used in the first Kent longcase clocks during the 1680-

Fig 2/9 The hood and dial of a small 8-day longcase clock by John Buckley of Canterbury, about 1790. The plain silvered dial has date and seconds indicators, a strike/silent facility in the arch and fine cast-brass spendrels. The arched-top mahogany case, 7ft 1in tall, has the earlier style of Kent cresting on the top of the hood

1730 period, where they are virtually indistinguishable from their London counterparts. The only exception to this is the slightly later use of the convex moulding under the hood, which was used by both Canterbury and Maidstone makers until about 1710. The majority of cases from the 1700-50 period were made of oak, and some with lacquer decoration. Mahogany was rarely used until later in the century, with the exception of one or two examples from Maidstone and Chatham. Walnut cases are relatively unknown, and only a couple of cases have been recorded with seaweed or arabesque marquetry decoration.

The introduction of the arch dial brought with it the pagoda-topped case, but whereas London cases from this period are invariably made in mahogany, in Kent they are again usually in oak or a lacquer finish. With the advent of the silver and painted dial, 30-hour movements continued to be housed in simple, elegant oak cases of standard design, with little variation. However, it is possible to recognize a particular maker's clocks by variations in hood design. Clocks by the Thatchers and especially the Baker family can be readily identified, as can be seen by the example pictured in Fig 7/112.

While it has proved impossible to trace the names of the clockcase makers from the early period, one of the later workshops known to have supplied cases at the end of the eighteenth and during the early part of the nineteenth century was owned by the Goulden family in The Borough at Canterbury. Labels advertising this firm have been found in a number of cases from the Canterbury/Faversham area and the style of case is very distinctive, especially in the

Fig 2/10 Timepiece/alarm wall clock with painted dial, by Thomas Sutton of Maidstone, about 1795. These wall clocks, with both earlier brass dials and later painted dials, were common in Kent in the second half of the eighteenth century

Fig 2/11 The 30-hour movement, showing the similarity to those used in lantern clocks. The wall bracket is original and does not appear to have been fitted with a hood

design of the hood with its characteristic wavy fret and reeded mouldings. As these features appear on the majority of cases from that area of East Kent from Faversham to the Thanet coast, it appears likely that they supplied cases to almost every maker in the locality. This style of case is usually in mahogany of extremely high quality and mostly found with painted dials, as is to be expected at this late date. The excessively wide and inlaid decorated case so common in northern clocks is never seen in these late Kent cases. which still retain the elegance and restrained workmanship of an earlier period. Moreover it is rare to see religious subjects appearing on the later painted dials in contrast to those from Wales and the north of England where this style of decoration was particularly popular.

Another type of clock that was common in Kent during the last half of the eighteenth century was a timepiece/alarm movement in a brass posted-frame bearing a striking resemblance to the earlier lantern clocks. Some makers of these clocks obviously preferred to use a framed pillar construction similar to a longcase movement, and these are invariably found housed in a wooden case designed to be hung on the wall and are commonly called hooded wall clocks. Among the many makers of this type of clock were Thomas Sutton of Maidstone (Figs 2/10-11) and William Flint of Charing (Fig 7/34).

Numbers of tavern clocks have survived from the 1760-90 period. These are often inaccurately described as Act of Parliament clocks, following the imposition by William Pitt of a tax on clocks and watches in 1797 in an effort to raise money to help to pay for the Napoleonic War. This type of clock had come into common usage in the middle of the century with the expansion of the coaching routes and were to be found in the staging posts centred on inns on the main roads between London and the coast. Fine examples by such makers as Owen Jackson of Cranbrook (Fig 7/63), Wraight and Woolley of Tenterden (Fig 7/130) and William Chalklin of Canterbury (Colour Plate VIII) have survived.

While the tax was soon repealed it resulted in a rapid and marked decrease in the number of makers of clocks and watches and this decline was exacerbated during the early nineteenth century by a huge increase in the importation of mass-produced movements from America and the Continent, and the industry never really recovered. The majority of those makers noted as working in the early-to-mid nineteenth century were, in most cases, merely purveyors of bought-in foreign clocks and watches. The trend towards buying in clock parts and merely assembling and then selling a manufactured product to which the so-called maker added his name had started in the last quarter of the eighteenth century with the advent of the painted dial. Most of these dials were made in Birmingham and were fitted to the plated movements by means of a false plate. However, it is interesting to note that very few of the early 8-day painted dials dating from the 1780s in Kent are found with false plates, particularly on clocks made by the more skilled craftsmen, such as James Warren of Canterbury. None of these dials appear to have the name of the dial maker stamped on them as is the usual practice with those from the Birmingham area The majority of the later painted dials examined by the author have false-plates signed by Wilson. So far there is no trace of any record of dials being made and enamelled in Kent so it must be assumed that these unmarked early dials also came from the Midlands. Other clock parts were purchased from the workshops of Clerkenwell in London and other centres throughout the country.

The rapid increase in the Industrial Revolution and the subsequent mechanisation and introduction of the factory system brought with it the demise of the small watch and clock makers workshop and the end of an era during which English clockmakers had reigned supreme for over 150 years.

Chapter Three
THE BAKERS OF MAIDSTONE & TOWN MALLING

Although the Baker family were prolific clockmakers who were actively working from the late seventeenth century until a hundred years later, very little is known of their background. There is no record of them in the parish registers of St Marys Church in West Malling, or Town Malling as it was more commonly called in the eighteenth century. Nor is there any trace of a will or inventory by any member of the family held at the extensive archive collections of either the Cathedral in Canterbury or the County Records Office. One theory put forward by a local historian is that they were Quakers, but no evidence has come to light to substantiate this, nor do any of the many surviving clocks bearing their name exhibit any of the characteristics usually associated with Quaker workmanship, ie in the manner of their dial decoration.

It is known that Henry I was a clockmaker, or more probably a clocksmith, and was elected

Fig 3/1 Dial of an 30-hour birdcage movement, by Henry Baker of Maidstone, about 1715

Fig 3/2 An unusual single-handed 30-hour timepiece/alarm hanging wall clock by Henry Baker with very distinctive engraving, about 1730

Fig 3/3 An 8in wide arched dial single-handed 30-hour clock by Henry Baker, about 1720

Fig 3/4 The original slim oak case housing Baker's 8in arched dial clock

a Freeman of Maidstone in 1622. A search of the parish registers has revealed virtually nothing, so far, other than an entry in All Saints, Maidstone of the marriage of Henry Baker to Joan Burgese on 14 November 1624. Although we cannot be certain that this entry relates to the clockmaker, Henry, it is possible, given that Freemen's marriages are often recorded as taking place either just before or after the date the craftsman obtained his Freedom. However, it is unlikely that any of the surviving lantern clocks bearing the name 'Henry Baker, Maidstone' are by him. They are more likely to be the product of the second Henry. Much more research needs to be done into this family, but they have proved to be extremely elusive, and we can only hope that the registers of All Saints, Maidstone will eventually reveal more about the lives of this family of clockmakers.

Fig 3/5 Superb dial of a five-pillar movement by Henry Baker, with very fine engraving and well-finished spandrels, about 1730. Note the strike/silent lever at 9 o'clock

Figs 3/6 and 3/7 A fine 8-day longcase clock of about 1730 by Henry Baker in a tortoishell lacquer case

Some of Henry II's earliest work is to be found in the excellent local collection of clocks and watches in Maidstone Museum. All his early work is signed as from Maidstone and it is assumed that after being apprenticed to his father he carried on working there before moving to West Malling in the early years of the eighteenth century. Numerous examples of his clocks survive, from lantern to hooded wall clocks, and both 30-hour and 8-day longcase clocks. All of them exhibit a high standard of workmanship and finish to dials, spandrels and movements, and it must be said that this high quality of production is common to all the makers originating from the Maidstone area.

The cases of almost every longcase examined examined by the author has been constructed

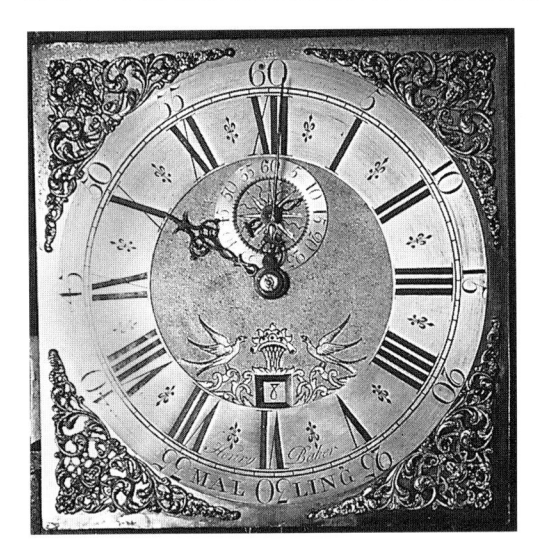

Fig 3/8 A 30-hour two-handed 11in square dial by Henry Baker of Malling, about 1735

Figs 3/9 and 3/10 Hood and dial of an 8-day longcase clock in an oak case by Henry Baker, Malling. The centre pillar of the movement is latched and the date 1745 is engraved on the great wheel

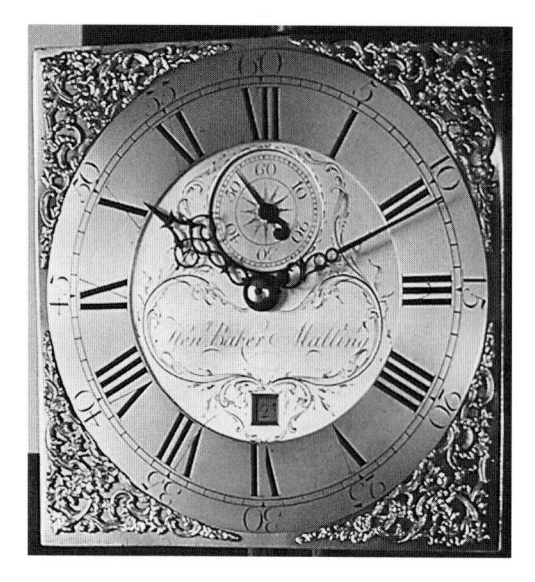

Fig 3/11 The fine dial of a plated 30-hour movement with date and seconds dials, by Henry Baker of Malling, about 1760

Fig 3/12 A dial with fine rococo-style engraving from an 8-day five-pillar movement by Henry Baker of Malling, about 1760

of oak with little ornamentation, other than frets to the square or arched hoods and either brass or giltwood capitals fitted to the pillars (eg Fig 3/4 and 7?112. All the cases are very similar and would appear to emanate from the same workshop. The only exceptions to this are a couple of lacquer-cased examples (eg Fig 3/6) and two other oak-cased clocks (eg Fig 3/9). The latter have the peculiar feature of the glass in the hood door not being fixed in the usual way, ie with putty into a rebate in the rear of the hood door, but merely held by a rebate at the base of the door, fixed only with pins and with the top of the arched piece of glass totally free and held against the hood door by pins with no rebate. Not a very satisfactory method of securing the glass, which surprisingly, was the original crown glass in both cases, having survived some 250 years.

The only other clock seen with the hood glass secured in this manner is by Thomas Ranger of Chipstead, with an exactly similar case to the two by Baker, the only difference being the addition of an inlaid marquetry star to the trunk door. Chipstead is a village near Sevenoaks, some 12 miles from West Malling, and for some reason Baker used another cabinet maker for these two cases, which also have different ornamental mouldings to the hoods than the usual Baker fashion. The case of the Ranger clock is almost certainly by the same carpenter as the hood mouldings are exactly similar.

One of the finest 8-day clocks by Henry Baker II is shown in Fig 3/5. This dial is beautifully engraved, particularly in the arch surrounding the moonphase work and the ubiquitous feature of the birds and basket of fruit motif above the calendar aperture on a well-matted centre, common to many Kent clocks of the early eighteenth century. This particular movement probably dates from the 1730-40 period and is of the usual five pillar construction. The case is of oak with a flat top above the arch dial, but unfortunately a photograph of the case is unavailable. This fashion was relatively common in the early part of the eighteenth century before the introduction of the pagoda-topped case. It had the added advantage of being admirably suited to the low-ceilinged rooms of the time, being that much smaller than the later fashion which became so popular with the advent of the taller rooms of the mid to late Georgian period.

While a number of 8-day clocks are known, it appears that the members of the Baker family particularly specialized in 30-hour clocks of all types, and some of the numerous examples in existence today are shown in Figs 3/1-4 and 3/8. Typical of his work is the clock pictured in Fig 3/11. The dial is extremely attractive with its date and seconds indicators, finely finished spandrels and superb engraving in the rococo fashion. A very desirable clock indeed, and exhibiting Baker's high standards of workmanship. In common with the majority of other southern makers, all of the 30-hour movements examined have been of birdcage construction, apart from the movement of the clock shown in Fig 3/11. Another very interesting 8-day clock by Henry Baker is housed in one of the three unusual cases previously referred to (Figs 3/9-10). This also has a superb dial, fitted with five pillars, with the centre pillar fastened by latches, with the additional unusual feature of the date 1745 inscribed on the great wheel.

At least one tavern clock signed 'Baker, Malling' exists and this may well be the work of Thomas Baker, who is presumed to be the son of Henry II. He is known to have been working in 1768 until the late eighteenth century and was made a Freeman of Maidstone in 1775. It may be that the earlier date, which is taken from Loomes, refers to the beginning of his apprenticeship. Is it possible that because of the length of time separating the known working dates of the three Bakers, ie 1622-1784, there was a third Henry Baker? Clockmakers, as a general rule, are long-lived, but 160 years covering three generations seems unlikely in an era when to reach the age of sixty was something of an achievement. However, this can only be supposition until further evidence is discovered.

Towards the end of the eighteenth century all of the Baker clocks that the author has seen, whether of 30-hour or 8-day duration, have very plain all-over silver dials with little or no engraving, other than a finely-executed name, simply signed 'Baker, Malling'. All of these are thought to be the work of Thomas Baker. Both Henry II and Thomas are known to have made watches, and examples are to be seen in the Maidstone Museum.

In conclusion, it is to be hoped that after publication of this volume more information will come to light concerning this very interesting family of clocksmiths and clockmakers who, while not the best of the Kentish craftsmen, were certainly the most prolific, judging by the number of their surviving clocks.

Chapter Four
THE CUTBUSH FAMILY OF MAIDSTONE

Any attempt to unravel the history of this most prolific and interesting family, of whom at least twelve members were actively engaged in the craft of clockmaking and its associated trades during a period spanning some 150 years, is no easy task, and the author is not at all sure that the mystery has been completely solved.

As in the case of the Greenhill family, one of the main problems is the recurring use of John, Richard and Robert as Christian names and the entries to be found on the Corporation Apprentice Lists do not make it easy to differentiate between them.

The early records are extremely confusing in that the Clement Taylor Smythe manuscripts in Maidstone Museum note that an Edward Cutbush, locksmith, took his son Robert as an apprentice in 1652. He is almost certainly the Edward, son of Edward Cutbush, who was christened in July 1614. However, according to the corporation lists Robert is mentioned as beginning his apprenticeship to his father Edward, a locksmith, in 1702. Is it possible that there was an Edward Jnr and a Robert Jnr, or is the reference in the Clement Taylor Smythe manuscripts the date that Edward either entered his apprenticeship or gained his Freedom rather than his son Robert? This theory is lent credence by a note in the County Library referring to a 30-hour clock made by Edward Cutbush Jnr. Unfortunately this has not been substantiated, nor is there any indication from where they obtained the information.

To confuse the matter further there is a reference to a Thomas Cutbush, locksmith, in the 1702 apprentice list. The ratebook for the year 1699 records that Edward Cutbush was living next door to The Bird In Hand in Bank Street, and there is a fine lantern clock by Edward in Maidstone Museum (Fig 4/1). A thorough search of the county archives reveals no trace of either a will or an inventory by any member of the family, and the only record found in the All Saints parish registers is a note of the burial of Thomas Cutbush on 21 January 1678 and the marriage of John Cutbush to Priscilla Champ on 23 November 1680. No other reference exists for Thomas, who is noted on the burial records as a locksmith. He does not appear on any apprentice or freemen list, nor is his relationship to Edward known. Could he have been the father? A more extensive search of the parish records may reveal the answer.

The next record we have is John II placing his son, John, as an apprentice to William Jeffery, carpenter, in 1722. It has not proved possible to confirm whether he finished his apprenticeship or if he subsequently trained as a clockmaker. He is possibly the same John who appears on the 1734 apprentice lists as a clockmaker. It would appear unlikely that any of the offspring mentioned between 1722 and 1734 are the sons of the first John Cutbush, who, as we know, was married in 1680, presumably in his twenties, and would therefore have been in his sixties before any of his sons were apprenticed, which just does not seem feasible. The only definite record we have is that George, William and Charles were indentured to their father, John, in 1725, 1727 and 1734 respectively, and the author is convinced that John II is the son of the John who was married in 1680.

Fig 4/1 A fine lantern clock by Edward Cutbush of Maidstone, dating from about 1705

Fig 4/2 12in square dial of a five-pillar 8-day movement by Edward Cutbush of Maidstone, about 1705. Note the unusual spandrel design, used mainly by Thomas Tompion and his apprentices

   The next problem encountered is with Robert Cutbush. After the note of the entry of his apprenticeship in 1702 there is no record of a Robert Cutbush until the apprenticeship of George Jeffery to him in 1748, followed by William Southgate in 1751. George Jeffery is almost certainly the son of William Jeffery, the carpenter, mentioned above. After gaining his freedom at an unknown date he moved to Chatham and founded a clockmaking business there.

   In 1761, Robert's son Richard was apprenticed to him and his second son, John III, followed suit in 1772. We have, therefore, exactly the same problem with Robert as with John II, ie that it is extremely unlikely that someone apprenticed in 1702 at, say, 15 years of age, would be taking sons as apprentices in 1761 and 1772 when he would be in his seventies at the very least. To confuse the matter further is an entry on the apprentice lists for another John being indentured to his father Robert in 1765. The only other explanation is that the

Colour Plate V The dial of a single-handed 30-hour clock by Richard Greenhill of Canterbury, made about 1680. This clock, his only known domestic clock, is one of the earliest surviving Kent longcase clocks

Colour Plate VI The birdcage movement of the Richard Greenhill clock with brass corner posts and a vertical bell. Note how the horizontal arbor for the anchor escapement is cranked to clear the fly of the striking train

Fig 4/3 The dial of a five-pillar 8-day movement by Robert Cutbush of Maidstone, about 1760. The clock is in a lacquer pogoda-topped case

entry for John in 1765 is correct and that the year 1772 refers to the end of his apprenticeship rather than the beginning. When one considers that an apprenticeship normally lasted seven years (the time between 1765 and 1772) this explanation does go some way to clarify the mystery, although it does not explain the question relating to Robert's age at the date of his sons' indenture. Until such time as new evidence is uncovered in the parish the author remains convinced that there is another Robert Cutbush.

One other possibility is, of course, that there were two branches of the family descended from Edward and Thomas (if they were brothers) using the same Christian names, but this possibility only seems to compound the confusion!

Towards the end of the eighteenth-century the position becomes much clearer. After Richard gained his freedom and began to work on his own account he took Thomas Sutton as an apprentice in 1775 and William Whaley in 1779. The final record is that William Cutbush appears in the *Universal British Directory* of 1791 as a watchmaker. If this is the same William who was apprenticed in 1727 it means that, if he was aged 15 at the beginning of his indentures, he would have been in his seventy-ninth year in 1791. A great age for the time, and if we are to believe the directory entry that he was still actively working, quite remarkable!

What it also means, of course, is that if the earlier members of the family had also survived to similar ages all the previous theories have come to nought, and maybe the extra Edward, John and Robert did not exist after all. Hopefully the All Saints registers will eventually provide the answer to a fascinating mystery.

The other tantalizing problem we are faced with is the lack of surviving clocks by them. So far, only nine have been traced by the author. Maidstone Museum has a good 8-day oak-cased example by Richard Cutbush, in addition to a lantern clock by Edward (Fig 4/1). A fine 8-

day movement with five ringed pillars by Edward also exists (Fig 4/2). This particular dial has a rare type of spandrel normally only found on clocks made by Thomas Tompion and his apprentice, George Allett, and is certainly not common on provincial dials. An 8-day movement by Edward Cutbush, housed in its original black and gold lacquer case dating from about 1700 was stolen in Canterbury in February 1995. An interesting 30-hour posted-frame clock in an ebonized case with a 10in dial with two hands by Charles Cutbush has been recorded. The chapter ring of this clock is signed 'Charles Cutbush in Stone Street, Maidstone fecit'. Richard Cutbush also worked from these premises as the same address appears on a 30-hour brass dialled longcase with date and seconds dials inscribed 'Richard Cutbush, Stone St. Maidstone' on a cartouche, A longcase clock by John Cutbush, reputed to be about 1700 and then in the possession of Welbeck Abbey (Nottinghamshire) is mentioned in Britten's *Old Clocks and Watches*. The following inscription is carved in relief on the trunk door beneath a downward pointing index hand:

> Master, Behold me here I stand,
> To tell ye hour at thy command,
> What is thy will 'tis my delight
> To serve thee well by daye and night
> Master, be wise and learn from me
> To serve thy God as I serve thee.

Whether this inscription is contemporary with the clock or a later fanciful nineteenth-century addition is not known. It is not known if this clock is still in Welbeck Abbey.

Two longcase clocks by Robert Cutbush are known: an arched-dialled clock (Fig 4/3) in a lacquer case and another in a mahogany pagoda-topped case (Fig 4/4).

While many clocks will have been lost or destroyed, particularly from the late seventeenth-century, one would have expected far more examples to have survived from such a large number of makers, who, judging by the quality of the work that does still exist, were highly skilled craftsmen, more than equal to their average London counterparts of the early eighteenth-century. This tradition was very competently continued by their apprentices, in particular Thomas Sutton and George Jeffery, who in their turn produced many fine clocks.

Fig 4/4 A five-pillar brass-dial clock by Robert Cutbush of Maidstone. The dial, engraved with birds and scrolls, with a sunburst on a boss in the arch, mask-and-foliage spandrels, and ringed winding holes, appears to be be about 1720-30. The mahogany case, with an unusual carved urn on the pagoda-topped hood, is later

Chapter Five
THOMAS DEALE OF ASHFORD

One day in the early 1980s a superb month-going movement by a totally unknown (at the time) Kentish maker shown to the author by a clock-dealer friend, initiated a fascination with the enigmatic Thomas Deale. The movement had been bought from an antiques 'runner' without a case and it had no hands, but it did have its original brass-cased weights (Fig 5/1). About a year later the movement was sold to another dealer and was subsequently exhibited at a fair in London. By this time it had acquired a fine pair of hands and was housed in a good walnut inlaid case typical of the 1680-90 period. It is open to question as to whether or not the case that now houses this movement is the original one, although if it is a marriage between case and movement then it is a very good one. However, there is no doubt that it does bear a strong resemblance to a case housing a movement by Threkeld of Newcastle which was on sale at about the same time. Regrettably, at the time it was common practice among some members of the clock trade to switch movements from case to case in order to increase their value. Happily, the rise in value of clocks by lesser-known makers has led to the cessation of this reprehensible practice, which is abhorred by most dealers and collectors.

After the fair the clock was bought back by the dealer, since retired, who had first bought the movement and was offered to the author at a time when funds were low and very reluctantly the the purchase had to be foregone — a decision regretted ever since! During this period attempts had been made to trace the maker, without success, and it was only after its sale to a retired detective, who, putting his former skills to good use, managed to trace Thomas Deale's will and the inventory taken for probate after his death, in the archives at Maidstone.

There was no mention of the maker in any of the recognized listings and this clock appears to be his sole surviving movement. He must have been one of the finest of the Kent makers of the seventeenth century. The clock was subsequently sold at Sotheby's in 1986 and its present whereabouts are unknown.

During the course of research for this book a little more about Thomas Deale has been discovered. We know from his probate that he died in September 1687 and the parish registers for St Marys in Ashford note that he was buried on 11 September 1687, leaving a widow, Jane, and three children, Samuel, who was christened in 1680, and Elizabeth and Mary. It has not been possible to find his exact date of birth as he was born during the Interregnum at which time few church records were kept. However he was almost certainly born in 1655 as his age is given as 25 when he applied for a licence to marry Jane Philpot of Willesborough, aged 17, on 17 May 1680, some three months after the birth of their son.

One of the witnesses to his will is named as Thomas Deale of Boughton Aluph, and a search of the parish registers of this church reveals other members of the family who almost certainly originated in the village and were descended from Jacobus Deale, a carpenter, possibly of Flemish origin. Various other references note a Thomas and Mary Deale having a daughter, also Mary, christened in 1685. Because of the gaps in the registers between 1641 and 1660 it is not possible to state with certainty, but Thomas and Mary may have been his parents. In his will he mentions his mother, Mary, and implies that she has remarried, so it is unlikely that the witness named Thomas Deale was his father. However, there was another branch of the family in Boughton Aluph with the head of the household, also called Thomas Deale, married to Ann, and to add to the confusion with a son called Samuel who was christened in 1707. So he is likely to have been the witness to the will. So far, there is no trace of a record of the father's death in the register.

In his will Thomas Deale left his house, together with the outbuildings, yard and gardens, to his son Samuel. He notes in his will that the house was divided into two dwellings, with one half occupied by a Thomas Burbage. Burbage was one of the assessors when the inventory was taken. He also left the income from twenty-three acres of marshland in the hamlet of Hope All Saints near New Romney to be divided equally between his two daughters and an annuity of £10 to be paid to his mother, Mary Tyler, if she was widowed again. As can be seen from the inventory, the Deales lived in some comfort; the house was more than adequately furnished, and they also possessed a quantity of silver. The total value of his estate in Ashford was £138 14s 11d.

Although the mystery of his origins has now been solved there is no definite knowledge of where he gained his undoubted skills as a clockmaker, as he does not appear on any apprentice lists, either in Canterbury or Maidstone. He may have been trained by one of the Greenhills and may well have worked for the Ashford branch of the family, which would account for the lack of surviving clocks signed by him. We know that John Greenhill, the eldest son of Richard, was married to Mercy Colt at Boughton Aluph in 1668. So the two families would almost certainly have been well known to one another. Is it possible that the Deales made clockcases for the Greenhills or carved the jacks which appear on two large lantern clocks signed by Richard Greenhill and another similar clock made by John Dodd? There is an intriguing reference to three 'Jacks' in the inventory of goods in his workshop, 'two finished and one almost', but it is not clear whether they are of metal or of wood, or indeed if they are clock jacks or domestic jacks.

Very few clocks exist by the Greenhill family, but the two lantern clocks signed by Richard, the lantern by Dodd, the lantern made by John Greenhill, a 30-hour movement by Richard of Canterbury, and the fine longcase made by Richard Greenhill of Ashford all have one thing in common with the Deale movement: superbly engraved dials. A close examination of these dials shows a marked similarity and one is led to the conclusion that the engraving is the work of the same craftsman.

The very full inventory taken after Thomas Deale's death gives a comprehensive list of working tools and other items which could only be the stock-in-trade of a busy clockmaker and moreover someone who was engaged in every stage of clock manufacture. Engraving tools and compasses to mark out dials are mentioned among these and the author is quite convinced that Deale was the person responsible for the engraving of these clocks, if not for their complete manufacture. All of these clocks date from the period when we know that Deale was working, ie the 1680s, and it is interesting to note that the early 8-day longcase by John Greenhill of Maidstone in the museum collection also has a similarly engraved dial.

In *Early English Clocks* Dawson, Drover and Parkes discuss the 8-day movement by Rich-

Fig 5/1 and 5/2 The dial and walnut case of a month-going longcase clock by Thomas Deale of Ashford, of about 1675-80, possibly the finest example of a movement and dial by a Kent maker. The seven-pillar movement is fully latched, with five wheels in each train and an outside countwheel. The 12in square dial is superbly engraved and the drapery cartouche enclosing the signature is very similar to that on the dial of the 30-hour clock by Richard Greenhill of Canterbury. The faded walnut case is inlaid with lines of boxwood and ebony, and stands 7ft 7in tall, including the cresting

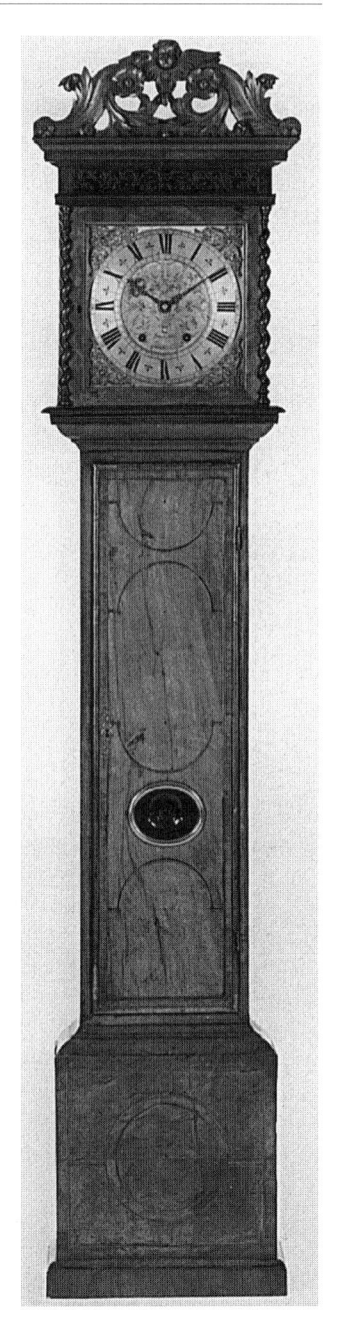

ard Greenhill of Ashford and assert that it is a provincial attempt to copy the London style of 1680 at least a decade later, an argument that the author does not agree with. We know that Greenhill died in 1687, the same year as Deale. They also assert that as Ashford was a small market town, some fifty miles from the centre of London and its specialist craftsmen, the skills necessary, particularly with regard to the dial engraving, would have no chance of developing in a rural community. They have fallen into the same trap as many other writers before them of dismissing the skills which abounded in the provinces. As shown in a previous chapter the economy of Kent was closely tied to London and the county was far from being the remote place that they would have us believe; with a strong tradition of carving in this area of Kent, amply manifested by the numerous church monuments and fireplaces in the local great houses, made of Bethersden marble which is superbly carved and engraved in patterns reminiscent of damask cloth. This carving also dates from the seventeenth-century and was in many cases contemporary with the dial engraving, so to suggest that the skills were not available is misguided at the very least.

At the present time only seven surviving clocks by the Greenhills are known in which Thomas Deale may have played some part in the making. It is to be hoped that other clocks signed by Deale will be discovered to enable more light to be throw on this, hitherto, unknown but superb craftsman, who died at the tragically early age of 32.

WILL of Thomas Deale of Ashford made on June 2nd 1687

IN THE NAME OF GOD AMEN I Thomas Deale, Clockmaker of Ashford in the County of Kent being well in body, mind and memory praised be Almighty God for the same but considering the uncertainty of this life and that it is appointed for all men also to die doo therefore in my health make and ordaine this my last Will and Testament in manner and form following and I doo willingly and with a free heart render my soul into the hands of Almighty God hoping through the Death and passion of Jesus Christ my Saviour to receive pardon and remission of all my sins and my body I remitt to the earth from whence it came to be decently buried according to the descretion of my Executrix hereafter named. And touching my estate both real and personal wherewith it hath pleased Almighty God to bless me with all in this world I give devise and bequeath as followeth that is I give and devise unto my son Samuel Deale all that my Messuage and Tennament now divided into two dwellings with the Outhouses, Buildings, Yards, Closes, Gardens, Banksides Water and appertanences thereunto belonging situate lying and being in Ashford aforesaid and now in the common occupation of myself and Thomas Busbage to have and to hold the same unto him and his heirs for ever. Item I give and devise unto my two Daughters Mary and Elisabeth Deal all those my two pieces or parcells of Fresh Marshland containing by estimation three and twenty Acres and odd portions more or less with their appertanances lying and being in Hope All Saints in Romney Marsh in the County aforesaid and now in the Tenure or occupation of Robert Hamms or of his assignes which I lately purchased to — of Thomas and Jonathan Jones Gentld. To have and to hold the said Marsh Land with their appurtennce unto my said two Daughters Mary and Elizabeth Deal and their heires forever equally to be divided between them But if it shall happen that Mary Tyler my Mother shall be a widdow again then my will and mind is that shee shall have and I doo give and devise unto her one Annuity or yearly Rent Charge of Tenn pounds of lawfull money of England to be issueing and goeing out of all my said Marsh Lands lying and being in Hope All Saints aforesaid to have and to hold to her and her assignes during her widdowhood and noe longer at four feasts or Terms in the year that is to say att the feast of the annunciation of the Blessed Virgin Mary, the nativity of St John the Baptist, St Michael the Archangels and the Nativity of our Lord and Saviour Jesus Christ by equall portions quarterly to be paid the first payment to begin and be made in and upon that feast of the feasts aforesaid that shall next happen after the said Mary my mother shall be a widdow again and if it shall happen that the said Annuity or yearly rent of Ten pounds or any portion thereof to be behind and unpaid by the space of one and twenty days — or after any of the said feasts on whitness aforesaid the same ought to be paid that then and from thenceforth and att all times after it shall and may be lawful to and for the said Mary my Mother into all singular my said Marsh Land to enter and destroy and the distresse and distresses then and there found and taken from thence to Lead, Drive, carry, and bear away and the same to withold ipoind, strain and keep untill shee shall be fully satisfied and paid with reasonable costs and charges in that — sustained And my mind and will further is that Jane my wife shall receive the Rents Issued and Proffits of my said Marshlands deducting the said Annuity if that shall happen for the mainteynance of my said two Daughters untill they shall ... attaine unto their ... ages of one and twenty years this is the last Will and Testament of me the said Thomas Deal touching the disposal of my Messuages Lands and Tennements and as touching the disposall of my goods and chattells of what nature or kind soever they be I give and bequeath the same unto the said Jane my Wife for the use of her selfe for her life and afterwards I give and bequest the same to all my children to be equally divided between them and I make and ordain my said Wife Jane Sole Executrix of this my last Will and Testament desiring her to educate my children carefully and to perform this my Last Will and Testament in every particular in witness whereoff I the said Thomas Deal to this my last Will and Testament have sett my hand and seal containing two sheets of paper dated the Day and Year in the first Sheet first written Thomas Deale signed Sealed published and declared to be the last Will and Testament of me? the said Thomas Deale in the presence of William Botting, Thomas Deale of Boughton Aluph, John Marsh.

PROBATUM made by the above attested Thomas Deale of Ashford Sept 1687. Jane Deale, executrix.

The following inventory (Kent Record Office PRC 11/51/30) taken after Thomas Deale's death, together with the Greenhill inventory of 1706 (pages 81-2), provides a fascinating glimpse into the living conditions of two very prosperous craftsmen of the late seventeenth century. Of particular interest is the very detailed list of the contents of his workshop, and is possibly the most comprehensive such record to have survived. It shows that he must have been involved in every stage of clockmaking, from the casting of brass components, to the designing of dials and their subsequent engraving.

It is worth noting that although he did not possess a *wheel-cutting engine* he did have a dividing plate. The latter would have been used in conjunction with a pair of dividers (of which he had two pairs) to mark out the position of the wheel teeth, which were then cut by hand and filed to shape. Even when the teeth were cut with an engine, they were invariably just gashed out and then filed to shape by hand. It is significant that Thomas Deale had 134 files of various sorts — they must have been one of the main tools of his trade.

Even at this early stage in the history of clockmaking it was common practice, both in London and the provinces, to buy in both finished and partly finished clock parts from other craftsmen. Many of these specialized in making particular components, such as wheels, barrels pinions and dial plates. Engraving was also often carried out by specialists, many of whom are known to have travelled throughout the country practising their craft.

It seems unlikely that Deale availed himself of these services and probably made every component himself. The sheer number and value of his tools — over £30 in total — leads one to suppose that he may also have been supplying other clockmakers in East Kent with parts, and gives added credence to the author's supposition that he was involved in the manufacture of many of the Greenhill family's output.

It is interesting to compare the quantities of raw materials, ie iron and brass, with those listed in the Richard Greenhill inventory. Deale owned a large quantity of brass, suggesting a man specializing in lantern and longcase movements, whereas Greenhill had a huge stock of iron, some 7½ tons, inferring that the latter was primarily a maker of turret clocks.

An Inventory of the goods and chattels and creditts of Thomas Deale late of Ashford in the County of Kent clocksmith deceased taken and apprised by us whose names are given under written, the 16th day of September Anno Dom 1687.

Imprim[is] his purse and ready mony	5.	0.	0.
Itt[em]: his wearing Apparell	5.	5.	0.
Itt: ten pairs of sheets; five pairs of pillowcoats	5.	10.	0.
Itt: 6 tablecloths 3 duzon of napkins; 1 duzon of hand Towels	2.	2.	6.
Itt: one Silver Tumbler 6 silver spoons; 2 silver tasters and one small silver cupp	3.	13.	6.

In the best Chamber
One feather bed and the furniture therunto belonging	5.	15.	0.
one case of drawers and 2 tables	2.	0.	0.
one pair of Andirons and one pair of creepers fire shovell and tongs	0.	8.	2.
One looking glass	0.	7.	6.
4 chairs two stools 6 cushions	0.	6.	9.

In ye middle chamber
One feather bed and thefurniture therunto belonging	4.	2.	6.
One old trunk 1 chest	0.	4.	0.

In the outer Chamber
one bed and furniture	2.	10.	0.
one trundle bed stedde and the bed only; two chests, 3 pewter chamber potts	0.	12.	6.
one stool	0.	0.	6.

In the Clossett
two duzon of glass bottells	0.	5.	0.
books	0.	5.	0.
one pillion and cloath and footstools	0.	13.	6.

In ye Garrett
one bed and its furniture	0.	16.	6.
one cradle one wicker chair and one flaske	0.	5.	0.
two old chests and one old cupboard	0.	4.	6.

In the fire room
two tables	0.	10.	0.
6 rush leather chairs	0.	6.	0.
6 rush chairs small and great, 2 hanging presses		3.	0.
three box-irons 8 heaters		5.	0.
one folding board one grate		5.	0.
one jack and ye chains and three spitt	1.	0.	0.
one pair of Andirons one pair of creepers, fire shovel, one pair of potthangers; 2 Trivett one Spitt iron & the bellows		16.	9.
one spice box	0.	2.	6.
all ye earthen ware	0.	2.	6.
two Clocks in Cases	7.	0.	0.

In the great Buttery
14 pewter dishes 7 plate 6 spoons, one drippin pan one Tinn cover & bread grate	1.	14.	2.
one brass ladle & other small things		2.	6.
one table one spiningwheel, one bowl and one brine tubb		9.	9.
4 barrell one scalder one tilter one pair of skales		10.	3.
one meale bagg one wooden cupp one ratt trap		2.	4.

In ye little Buttery
3 duzon of Trenchers , 1 duzon of bowls & dishes 6 spoons		4.	1.
3 Barrels & one skalder		7.	1.
two iron pott one frying pan and Gridiron & 1 warming pan		15.	9.
one lanthorne and other lumber		2.	8.

In ye Brewhouse
The Copper and cover	2.	2.	0.
2 pair of pothangers		1.	6.
3 small brass kettels 3 skilletts one brass morter and one brass pott	1.	2.	0.
Tubbs and peelers and pails	2.	0.	0.
2 pair of books one jack one saddle and Malepillion		14.	0.
Cordwood and brushwood in the backside	4.	10.	0.

In the Fore Shopp
Two vises and one vice board	1.	6.	6.
4 hand vices		6.	0.
8 hammers		2.	6.
one pair of compases & two pairs of Dividers		1.	6.
Two paire of pliers			10.
one flatt gunlock		3.	6.
14 large smooth & bastard [files] of divers sorts		6.	0.
15 splitting fyles without shafts		2.	6.
10 of ye largest fyles of divers sorts		3.	4.
26 of smaller files of divers sorts and 3 pollishers		7.	0.
30 or more old fyles of all sorts		4.	0.
17 new fyles rough and smooth great and small		7.	0.
10 large rough fyles newcutt		7.	6.
12 more small fyles of all sorts		1.	6.
3 screwplate and 7 tapps		3.	0.
the drill tools and bonding and Jack squares and the stake and stake tools		10.	0.
for the ingraving tools and one bagg of wax and the Sun Dyall tools and a plane		2.	6.
84 old screws		7.	0.
one old Ballance Clock	1.	5.	0.
3 jaks two finished and one almost	2.	0.	0.
a borax box and borax, putty a small bickhorne		2.	6.
two oyle lamps a brass oyle bottle		5.	0.
stallgrate		2.	0.
garnish pins & kats gutt		12.	0.
one dividing plate			10.
A small dyer		4.	0.
6 jack lines		17.	0.
34 pairs of clocklines and one small board vise and one pair of scalls		1.	6.

4 yards of double jack chaine at 5d ye yard and 9 yards of small		4.	0.
two Ballance Clocks	3.	8.	0.
30 fyle shafts with firrols		3.	0.
one large form plate tapps	1.	5.	0.
two pair of stockes and one rift		1.	0.
one pair calleppers and one large gymblott		1.	2.
one rasp and 1 scraper and one other pair of storks		11.	4.
19½ of Shruff Brass		11.	0.
33 pounds of filings of brass, 71 of newest brass	3.	11.	0.
4 bells at 4s apeece		16.	0.
2 bells at 4s.6d apeece and one at 2s and 3 at 1 appeece		14.	0.
one pair of brass pullies for a clock			6.
for old things in the two drawers		2.	6.
10 old bells		14.	0.
for new and old nutts		3.	0.
for one shelf two frames and three drawers	0.	4.	6.
for iron and two shopp candlesticks		4.	0.

Old trade in ye Garrett

Old armour plating old jacks wheelsdrawers	0.	15.	0.
two pull up jacks; two windup jacks	0.	18.	0.
old musketts and gun barrells	1.	0.	0.

In the Forge

The bellows posts and cross shaft	1.	12.	6.
the turning lathe and things belonging to itt, the grindstone trough, spindle and post		3.	0.
The forge trough and the board to lye over ye Grindstone		10.	0.
A pot and ladle to melt brass in		3.	0.
The Anvell weighing one hundred, one quarter & 13 pounds at 5d the pound	3.	5.	2.
2 upright hammers and three hand hammers		7.	0.
one vise b [?] iron and vise boards	1.	1.	0.
one square, four boulsters, three pairs of tongs, chissels and punches and things belonging to ye fire		1.	0.
5 pounds of new steel at 6 pence ye pound		3.	0.
64 pounds of new iron at 2p ye pound		10.	8.
Old armour plate and old Iron and other things		5.	0.
and old kettle weighing 8 pounds at 8d ye pound		5.	4.
Book debts	8.	0.	0.
mony upon Bond	30.	0.	0.
Sum is	138.	14.	11.

Executors   Thomas Burbidg   his mark
             Georg Knight

[Is it possible that the George Knight named as one of the executors, is the same man as the clockmaker from Faversham?]

Chapter Six
THE GREENHILL FAMILY

The members of this prolific family of clockmakers, gunsmiths and locksmiths number among them the finest craftsmen working in Kent during the seventeenth century. It is also fortunate that owing to the survival of various wills and inventories they are probably the most well-documented, although the number of their surviving clocks is very small, fifteen in total, together with an indeterminate number of turret clocks.

The earliest references to them in the county archives are in the marriage licence records which suggest that they came from the Romney Marsh area, as numerous Greenhills from the region appear in these lists during the period 1490-1520. In the late sixteenth and very early seventeenth centuries they moved to the village of Stockbury near Maidstone where three members of the family are known to have prospered as typical yeoman farmers owning land in and around the village. The earliest reference that has been found is a Richard Greenhill in 1615, mentioned in the record of his son Robert's baptism in that year. Inventories taken after the deaths of these three, which took place within the relatively short period of time of 1695-7, are further evidence of just how prosperous and wealthy the Kentish yeoman farming class were in comparison with the rest of the population.

The three who died were brothers: John, Robert and Richard, who left in addition to land and buildings, estates worth £95, £258 and £309 respectively. Of the £258 left by Robert no less than £215 was in cash. This was a large sum when one considers that the average tradesman and shopkeeper's income was about £50 per annum, and to obtain some idea of what these figures mean today one has to multiply by at least 400 times.

Because of the tendency of this family to use the names John, Richard and Robert, it has not proved possible to solve the question of the exact relationship of the first Greenhill from Maidstone, namely John, who is known to have worked as a smith, to the early members of the Stockbury branch of the family. It is interesting to note however that Elizabeth, the daughter of Richard Greenhill of Stockbury, married William Dodd of Stockbury in 1635. William is the grandfather of John Dodd, the clockmaker from Faversham who may have learnt his craft from Richard Greenhill of Ashford.

The earliest mention of John Greenhill in the Maidstone records appears in the Freemen lists for 1607 in which he is noted as a smith. It is not clear whether he was a blacksmith, locksmith, gunsmith or clockmaker, but he is known to have lived in Stone House, Lower Stone Street near to the Judges Lodgings. According to notes in the Clement Taylor Smythe manuscripts held at Maidstone Museum, John I appears to have been of a very volatile nature and was often in conflict with his fellow freemen. For example, the 1624 records state that 'John Grennel, smith was in trouble for declaring that it was no matter if the town charter was overthrown'.

As we shall see, his rebellious sentiments were handed down to his descendants. By his

Figs 6/1 and 6/2 A superb late seventeenth-century quarter-chiming lantern clock by John Dodd of Faversham., who was related to the Greenhill family. The similarity of this clock to the two clocks shown in Figs 6/9-11 and 6/12-13, together with the late style of the chapter ring suggests that this clock may have been finished by Dodd after Greenhill's death

marriage to Elizabeth (surname unknown), which probably took place in 1607 although the exact date is uncertain, he had five children. The eldest son John II was born in 1608 and he appears in the Freemen lists for 1636. He married Elizabeth Platt on 15 October 1635 and is described as a locksmith. The second son, Richard I, was born about 1616 and his marriage licence records him as a clockmaker. After Richard's marriage to Mary Sammon of Ashford on 9 February 1642/3 he moved to Ashford founding that branch of the family. No trace in the records has been found with whom Richard served his apprenticeship, so it must be assumed he was trained by his father and later by his brother, for it is known that John I had died prior to 1632/3. This is made clear on his daughter, Elizabeth's, marriage licence application to marry George Dennis, locksmith, in February 1632/3, her mother consenting to the marriage as the father was deceased. Among the children of this marriage were George and Thomas Dennis who subsequently became clockmaker and watchmaker, respectively.

Whether John I and II were involved in clockmaking is not certain, but to date no domestic clocks bearing the Greenhill name can be attributed to them. There is a possibility that some of the turret clocks made for the churches in Maidstone and the surrounding district

Figs 6/3 Hood and 12in square dial of a large ebonised longcase clock by John Greenhill III of Maidstone, about 1695. This clock, which is over 8ft tall, may have been made for the Council Chamber, as Greenhill was a member of Maidstone Council and also served as mayor

Fig 6/4 (top right) Close-up of the dial centre showing the high quality of the matting and engraving

Fig 6/5 The striking side of the John Greenhill movement, showing the decoratively finned pillars characteristic of the period

Fig 6/6 An 8-day longcase clock by John Greenhill III of Maidstone, with four-pillar movement and inside countwheel striking. The Dutch-influence lacquered case may not be original

Fig 6/7 The 11in square dial is superbly engraved, but lacks its original hands

were by their hand, but research in this area is hampered by the lack of surviving records, many of the Churchwardens accounts being either missing or incomplete for the seventeenth century.

During the Civil War and the Interregnum very few church records were kept, so it is difficult to trace the children of John II and his wife, Elizabeth. However, from the marriage licence of John III and Alice Harris it is possible to deduce that John III was born in 1655. On the licence he is described as a gunsmith, presumably trained by his father, and it is recorded that he obtained his Freedom in 1674. Alice was 22 years old when they married in 1680 and the author believes that there is a connection with Walter Harris, another Maidstone clockmaker who was apprenticed in Canterbury to John's cousin Richard Greenhill in 1693. She is thought to have been his sister, or because of the difference in ages, maybe his aunt.

Fig 6/8 An 8-day longcase clock by John Greenhill III of Maidstone. The 11in square dial and five-pillar movement are in the original ebonised case of about 1690

John Greenhill III was a man of some substance in Maidstone for, in addition to being in business as a gunsmith and clockmaker he was a member of the Corporation and had substantial land holdings, owning property and land in Stone Street, farms and woodland at Boxley and farmland at Linton, in addition to properties situated near 'The Great Bridge'.

In 1687 he was expelled from the Corporation together with other council members by order of James II. Following the granting of the new Town Charter he again served on the council and was elected Mayor of Maidstone in 1702. He died in 1712. Unfortunately, there is no trace of an inventory, but his very detailed and long will leads one to suppose that he must have been worth a quite considerable sum, probably in excess of £1,500. His properties and land in Stone Street and Upper Stone Street were left to his eldest son, John, while his second son, Arthur, inherited the remainder of his lands at Linton and Boxley together with other properties near the bridge in Maidstone, with other bequests being made to his two daughters, Elizabeth Harris and Mary Morris, and to his mother-in-law, Elizabeth Harris. It is believed that his daughter Elizabeth married either Walter Harris or Walter Harris Jnr, also clockmakers, sometime before 1711, although to date the marriage in the records of All Saints, Maidstone has not been traced.

Of the clocks made by John III, at least one lantern clock from the 1680 period has survived (Fig 2/1-2), together with five longcase clocks: one lacquer example, one oak, two in ebonized cases and a fine marquetry eight day clock, which is now believed to be in a Swiss collection. The lacquer-cased clock is in Maidstone Museum, and though having only a four-pillar movement has a very finely engraved dial, but it is missing its original hands (Figs 6/6-7). One of the two ebonized cased clocks is housed in a much taller and larger case than is normal for domestic clocks and it may possibly have been made for the council chamber (Figs 6/3-5). Baillie also records a longcase by this maker in the Virginia Museum, USA but enquiries to the curator have revealed no trace of the clock, which presumably had been on loan many years ago and its present whereabouts are unknown. It is to be hoped that more clocks by this early Maidstone maker will come to light in the future. The death of John III in 1712 marks the end of the Greenhill involvement in clockmaking in Maidstone as neither of his sons carried on the family tradition.

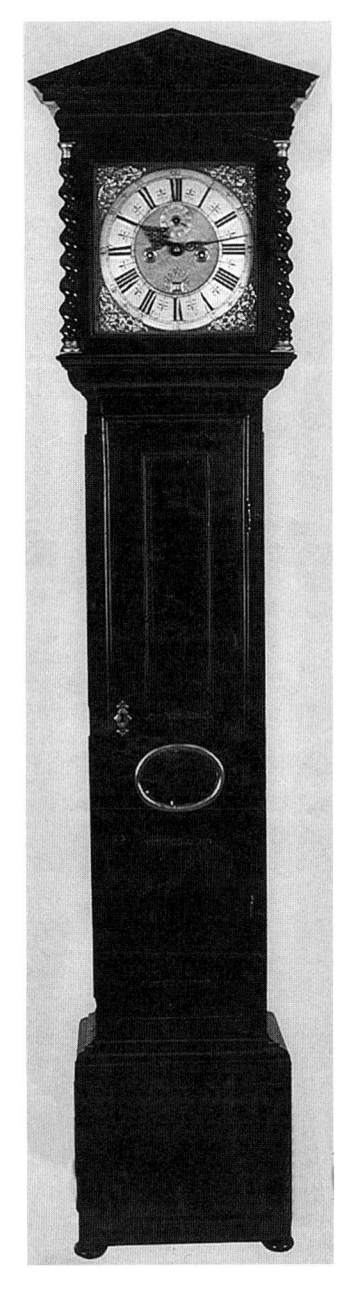

The Greenhill connection with Ashford began following Richard I's marriage to Mary

Sammon in 1642 and their subsequent move to that town soon after. There are numerous references to the children of this marriage, both in the registers of St Mary's Church, Ashford and in the records of marriage licences in the Cathedral archives. As a result it has been possible to compile a detailed family tree (Fig 6/23).

Richard and Mary had four children between the years 1644 and 1653, two of the three sons, namely John and Richard, carrying on the family clockmaking tradition and the second son, Robert, becoming a tanner, a trade which his son, also Richard, followed very successfully after Robert's death in 1694.

The eldest son, John, was presumably trained by his father as a clockmaker and must have made his living in the trade as he describes himself as such in his will written in March 1705. However, to date only one domestic clock by him has been traced, a fine early balance wheel lantern clock dating from the 1680 period (Fig 6/20). It must be assumed that like his younger brother Richard, and maybe his father, he was mostly engaged in making turret clocks. He is known to have carried out repairs to the clock which was newly installed in 1680 in St Mary's Church, Ashford. Unfortunately we do not know who made this as the churchwarden accounts only go back as far as 1682. However it seems reasonable to suppose that it was manufactured by one or other of the family. This particular mechanism was replaced in

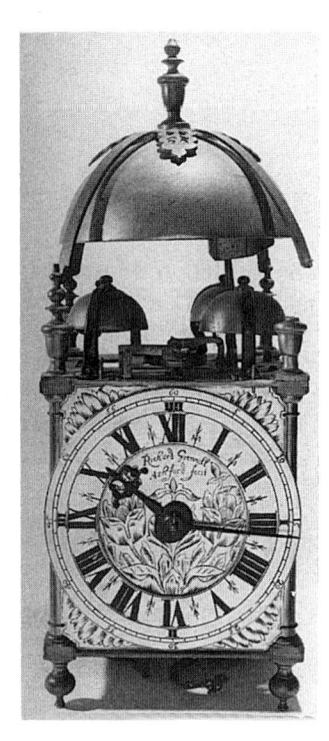

Figs 6/9-6/11 A massive quarter-striking lantern clock signed 'Richard Grennell [Greenhill] Ashford fecit', dating from about 1680. The pendulum of the verge and crownwheel escapement swings between the going and chiming trains. Note the very large hammer spring. Though the frets are missing and a finial is damaged the dial is superbly engraved, with similar half-hour markers used on the lantern clocks by John Greenhill of Maidstone (Fig 2/3) and John Greenhill of Ashford (Fig 6/20)

Figs 6/12 and 6/13 A very rare quarter-striking lantern clock by Richard Grennell [Greenhill] of Ashford, dating from about 1670 and standing 16½in tall. The quarter jacks are in early Jacobean dress and a subsidiary dial below the main one shows the date. Note the very unusual 'eared' chapter ring

the nineteenth century. There are also numerous references to him in the very detailed accounts of Bethersden Church (see Chapter 1, pages 20-1).

On April 15 1668 he married Mercy Colt at Boughton Aluph. The Colts were a family of woollen drapers and came from the same small village as the Deale family of carpenters. Thomas Deale, the clockmaker, was a grandson of Jacobus Deale, the earliest known member of the family, and the theory as to the possible close working relationship between the Deales and the Greenhills has already been discussed. The connection between the Greenhills and the Colts was further strengthened when Richard I's daughter Mary was married to George Colt in 1675, her brother John acting as bondsman.

That Richard Greenhill of Ashford was a fine clockmaker cannot be denied when one examines the three surviving examples of his work. Two of these are superb early large lantern clocks with quarter-striking movements (Figs 6/9-13), the other is an 8-day striking movement with a superbly engraved 10in dial very similar to the clock by Thomas Deale (Figs 6/14-19). These three clocks must date from before 1687/8, the year of Richard's death, and the lantern clocks are probably no later than about 1680, and possibly, much earlier. The engraving on the second of these bears a striking resemblance to the dial of the only known domestic clock by Richard II of Canterbury (Colour Plate V) and I am of the opinion that the engraving may be the work of Thomas Deale.

Although no inventory of Richard I's effects has been traced, his will gives a great deal of information. There must have been trouble between him and his eldest son for his will makes

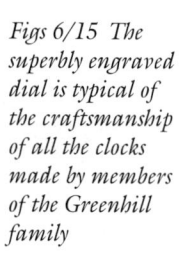

Figs 6/15 The superbly engraved dial is typical of the craftsmanship of all the clocks made by members of the Greenhill family

Fig 6/16 The case is much more provincial in character than any other seventeenth-century Kent longcase, and is probably the earliest surviving example

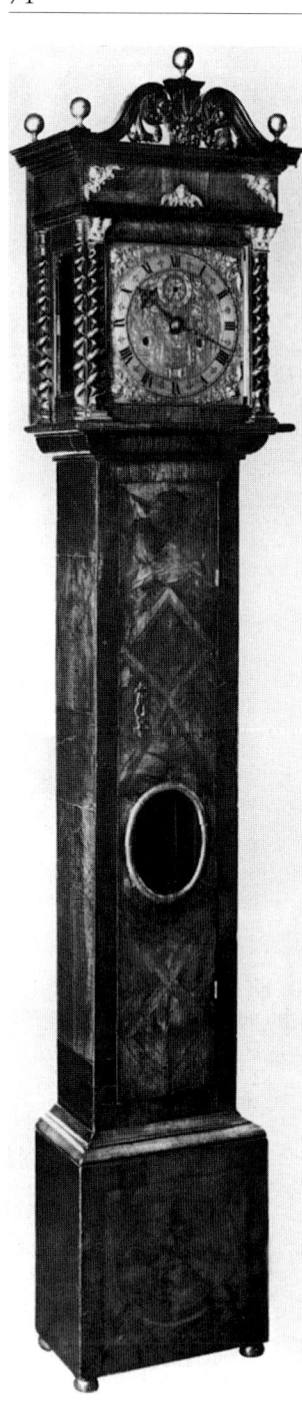

Figs 6/14 An 8-day walnut-cased longcase clock by Richard Greenhill, Ashford, about 1680

Fig 6/17 The under-dial work of the Richard Greenhill longcase clock. The movement is obviously the work of a lantern clock maker. The plates are, very unusually, of a horizontal shape. The bolt-and shutter maintaining power is a replacement

Fig 6/18 A back view of the movement showing the large bell stand, similar to the one on the 30-hour clock (Colour Plate VI) by Richard Greenhill, his son

Fig 6/19 The partly dismantled movement of the Greenhill clock. Note that although the plate pillars are pinned, the dial feet are latched. With the exception of the escapewheel, which is fitted to its arbor with a collet, all the other wheels are shouldered onto their pinions

it plain that nothing is to be left to John, but various legacies are left to his grandchildren, the sums to be paid out of the £320 owed to him by his second son Robert for land in Ashford purchased from the father. Robert is also left copyhold property including a house, barns, stables and farmyard. The children of his son-in-law, George Colt, were left properties in the churchyard at Ashford, and Samuel Greenhill, his other grandson, was left buildings and land at Charing which had only recently been purchased. All of these properties were occupied and must have produced a sizeable rental income. The rents from the Charing property were to be paid to his son Richard II until Samuel was twenty-three years of age.

His final instructions are that his remaining house and premises in Ashford, which at the time of his death were in the possession of his eldest son, should be left to Richard II. Further evidence of the bad feeling existing between the father and John, and also between the brothers, is the rider in the will stating that if John fails to pay the rents accruing from these premises or hinders his brother Richard in any way, the previous legacies made to John's children shall be deemed void and the total sum of £260 shall be paid to Richard instead.

Richard I obviously owned the greater part of the property around the churchyard in Ashford and it is interesting to note that many of the Greenhill tombs can still be seen immediately in front of the east window of the church. The hearth tax returns for Ashford in the year 1662 record that Richard I was liable for tax on four hearths although we cannot be sure which particular house this refers to.

John Greenhill of Ashford also left a very detailed will before his death in 1706 and the bequests made are further evidence of just how prosperous and successful the family had become. All the furniture, beds, household effects and a silver tankard were to be left to his wife together with the 'new house', and the list also includes a spring clock (bracket clock) and a weather glass (barometer) plus a quantity of pewter and brass, all subsequently to pass to his son John after her death. After making financial bequests to his daughters amounting to £220 he also left the sum of £200 to purchase land and tenements, the rents from which were to provide an income for his daughters. Other freehold property in Ashford was left to his son John with the remainder of his estate, comprising a millhouse and brewhouse, being left to the executor of the will, his nephew Richard, tanner of Ashford.

His son John does not appear to have followed his father's calling as a clockmaker, but Lydia, his youngest daughter, had married Arthur Hurt, a clocksmith, in December 1704. It is believed that a lantern clock by Hurt has survived and it is assumed that he had worked for his father-in-law. Whether he carried on the business after John's death is not certain.

While John was busy as a clockmaker, his brother Robert had been building up a successful tanning business, marrying one of the Chittenden sisters from Smeeth in 1672. The other sister married into the yeoman farming branch of the family from Stockbury. After Robert's death in 1694 his son Richard continued in business as a tanner and, according to the inventory taken after his death in 1708, his household effects were worth £1,073 4s 0d. If one adds to this the probable value of his properties and business premises, tannery, etc it is obvious that he was an extremely wealthy man.

Meanwhile, Richard II, the youngest son, who had presumably also been trained by his father as a clockmaker, at the age of 28 had moved to Canterbury where he gained the Freedom of the city by redemption (purchase) on 2 June 1676. Whether his move from Ashford had been precipitated by the ill-feeling which seems to have existed between himself and his eldest brother John, which is only too apparent when one reads the contents of his father's will, or whether it was merely a desire to found his own business, can only be conjectured at. However later that same year in December he married Lydia Caffinch, and settled in the area of the city covered by the parish of St George The Martyr, the same church at which the playwright Christopher Marlowe had been baptized more than a hundred years earlier.

Their two surviving children were born some years after the marriage: a son, Samuel in August 1684 and a daughter, Sarah, who is recorded as being christened on 17 July 1687. Samuel was apprenticed to his father at the age of 15 on 30 September 1699. He is referred to again later.

At some time after the beginning of July 1687 Lydia, the wife of Richard II, died, possibly in childbirth, and in January 1689/90 he married his second wife, Margaret Nethersole of Barham. She was a member of a well-known family of lawyers, one of whose predecessors, John Nethersole, had risen to prominence in the late fifteenth and early sixteenth centuries, serving on the Priors Council at Christ Church. By his marriage to Margaret, his status in the city must have risen considerably.

Fig 6/20 Another superb lantern clock from the Greenhill workshop at Ashford, signed by John Greenhill. This small balance-wheel clock, based on a 'third period' London frame, dates from about 1680. The finely engraved dial, similar to those of clocks by his father and his brother (and also Thomas Deale's longcase clock, Fig 5/1), were possibly the work of the same engraver

Some eighteen months after setting up in business in Canterbury Richard took his first apprentice, one Richard Goodlad who was indentured on 17 February 1677/8. He was the first of six, possibly seven, apprentices who served with Richard during the years he worked as a clockmaker in the city. The others were:

John Gee	apprenticed 20 March 1680/1
Isaac Milles	apprenticed 20 March 1680/1
Edward Chen	apprenticed 29 April 1687
Walter Harris	apprenticed 29 April 1693
Samuel Greenhill	apprenticed 30 September 1699.

The seventh man noted in the archives is Walter Harris Jnr on 14 May 1700. This entry is confusing as on the same day as the apprenticeship is recorded, a Walter Harris, Gentleman of Maidstone, is also noted as receiving the Freedom of the city by gift. If the records have been read correctly then this man must be the Walter Harris apprenticed on 1693 and who subsequently worked in Maidstone in 1704. As previously mentioned there is also thought to have been a relationship by marriage with the Greenhill family.

No other maker in Canterbury is on record as having such a large number of apprentices and one is led to presume that Richard must have had the busiest workshops in the city, if not the whole of East Kent. He must certainly have been the most prosperous, judging by the inventory taken after he died intestate in 1705, leaving effects worth a total of £1,543, approximately three-quarters of a million pounds at today's value! To this figure, of course, must be added the value of any property he may have owned, which is impossible to estimate. That he and his wife lived in some style is evident from the list of goods mentioned in the inventory which is reproduced here.

What is inexplicable is that as far as is known only one domestic clock has survived bearing his name, a fine 30-hour birdcage movement with a superbly engraved dial, probably dating from the 1680s (Colour Plates V and VI). He may have been predominantly a maker of turret clocks but at present only one church clock can definitely be attributed to him, that is the new clock supplied to St Martins, Herne in 1677/8 at a cost of £14. He may also have been involved in supplying the new clock for Staple Church which was installed after his death, and it is highly likely that the original clock in the tower of St George The Martyr, Canterbury, may have been from his workshop. Unfortunately the churchwardens accounts were destroyed during the blitz on Canterbury in World War II, but it is interesting to note that he is buried with his second wife and their two children immediately beneath the clock tower, and their tombstone can still be seen here today. Another interesting piece of evidence to support this supposition is the large quantity of iron mentioned in the inventory, seven and a half tons, surely such an amount would only be in the possession of a maker who habitually used large amounts of the metal, ie a maker of turret clocks. The list of items in the yard also include the weights, bell and dial of an old turret clock, one which he had presumably replaced.

This theory is given added credence by the fact that his first apprentice, Richard Goodlad, is almost certainly the same man as the maker who was admitted as a Free Brother to the Clockmakers' Company in London in September 1689. In the company records he is described as a 'Great Clockmaker', a term which many now believe was used to describe makers who specialized in turret clocks. As has already been mentioned, he was not heard of again in Canterbury until five weeks after Greenhill's death when he applied for the Freedom of the city on 13 April 1705. Subsequently his name also appears in the Herne accounts in connection with a payment for cleaning the clock in 1715/16. Many of the local church clocks must have been supplied by the Greenhill workshop, but this is very difficult to prove in view of the lack of surviving documentation and the fact that many of the early mechanisms have been replaced so that comparisons with the Herne movement are not possible.

Neither of the two sons Richard Greenhill II had by his second wife lived long enough to carry on the family business. Recent research has discovered that before his death in 1710, the eldest of these, John, had intended to follow in his mother's family traditional occupation. After attending the King's School in Canterbury he went up to Trinity College, Cambridge, aged 16, in 1707. He was admitted to the Middle Temple in the same year, a place having been reserved for him in advance, as was the common practice at the time. In 1699 Richard took Samuel, a son from his first marriage, as an apprentice.

Unfortunately for Samuel his father died suddenly without leaving a will. Owing to the intestacy it seems likely that he received nothing from his father's estate. Probate was finally granted in 1710 to his stepmother with Letters of Administration being given to Stephen Nethersole, Gent, of Womenswold, and William Nethersole, a lawyer. These two are believed to have been Margaret's brothers and it would seem that they looked after her interests at the expense of her stepson, for in spite of setting up in business on his own in Wincheap, just outside the city walls, he is recorded as bankrupt by 1723, and no record of his death in Canterbury has been found.

Two clocks signed by Samuel Greenhill have survived: a lantern clock dating from the early eighteenth-century which is now in Canterbury Museum, and a good 8-day longcase clock with a 12in dial dating from about 1705 (Figs 2/6-7). When first inspected this clock was housed in its original lacquer case which interestingly showed a late use of the convex moulding. However, at the time of writing this movement has been for sale in a country walnut case of the 1710 period and the whereabouts of its original case are unknown. The lacquer case was of a similar design to those housing movements by Wimble of Ashford and Bishop of Maidstone, both dating from the same period.

Richard and his wife Margaret were closely involved with St George's Church, where he served as churchwarden, and his signature appears on the Bishop's Transcripts for the year 1688. His children were all christened there and, as stated earlier, he was buried beneath the clock tower, to be joined later by his sons from his second marriage and finally by his wife Margaret in 1753 (Fig 6/21). She survived to reach the great age of 94.

Until further proof is found some of these theories must remain purely conjectural, but having studied the Greenhill family in some detail the author is convinced that they must primarily be considered as makers of turret clocks and that they were responsible for many of the clocks in the churches and large houses in central and East Kent. It is to be hoped that in the future someone will investigate the possible relationship and physical links, if any, between the remaining examples to be found in the region. In view of the sad lack of surviving

Fig 6/21 The gravestone of Margaret Greenhill, widow of Richard Greenhill of Canterbury, beneath the clocktower of St George's Church, Canterbury

examples of domestic clocks by them there is no other explanation for the way in which they were able to amass the considerable wealth they undoubtedly possessed.

With the deaths of John Greenhill in Maidstone and Richard in Canterbury, added to Samuel's bankruptcy, we come to the end of the story of this extraordinary clockmaking family. Hopefully, more clocks bearing the Greenhill name will come to light and help to add to our knowledge of a family who by their marriages with other clockmakers must have exerted a major influence on the industry throughout Kent during the seventeenth and early eighteenth centuries.

A Copy of the Inventory Taken After Richard Greenhill's Death in 1706 (Kent Record Office PRC/11/70/29)

A true and perfect Inventory and Apprizement of all and singler the Goods Chattells and Creditts of Richard Greenhill late of the City of Canterbury Clockmaker deceased made taken and apprized this twenty-seventh day of March Anno dom.1706 by us whose names are hereunto subscribed: vizt.

	L	S	D
Impris in ready money and for the wearing Apparell of the said deceased of all sorts	56	00	00

In the Chamber called the Green Chamber of the said deceaseds late dwelling house.

	L	S	D
Item One bedstead with Sacking Bottom, one Suit of Sarge Curtains and Vallance one feather bed and bolster, Rugg & Blanketts	004	10	00
Item one case of drawers seven chairs & stools one table one small looking glass, one pair of Andirons firepan & tongs, a Suit of window curtains and Rodds, the old hangings of the room & some earthenware	002	06	00

In the best Chamber

	L	S	D
Item one bedsted with sacking bottom one Suit of Camblett Curtains and Vallence one feather bed & bolster with the blanketts One silk Counterpane and pillowes, Seaven Cane Chairs, six silk pads, one olive table and Stand, one dressing box, one looking-glass, one pair of brass andirons and ffender, firepan & tongs, a pair of with brass heads, one spring clock and the old hangings of the room	015	17	06

In the Passage

	L	S	D
Item one case of drawers, two old trunks one old press, one little cloth, fifteen pair of sheetes, & nine dozen of napkins, one dozen of table Cloathes, two dozen and half of towells and ten pairs of Pillow	013	11	06

In the Garrett

	L	S	D
Item One bedsted with sacking bottom and curtains and Vallence, one feather bed & bolster, three pillowes with the blanketts & Coverlid, two old chests & two old Stools	005	01	06

In the Maids Chamber

	L	S	D
Item one bedsted with the flockbed & bolster Curtains & Vallence as it stands	000	16	00

In the Hall

	L	S	D
Item ten old turkey work Chairs, one hall Table, one pair of Iron Andirons one pair of ffirepan, tongs and ffender	001	06	00

In the Kitchen

Item one pair of Coleirons, one ffender, Pott Racks, firepan & tongs, one pot iron, one Worm Jack with the wights Line & Pulleys, one dish ring & Plate, Chaffin dish, two old tables, eight old chairs a small round table, one clock and case, one old watch, two irons and heaters and one pair of old bellowes   007 = 11 = 06

In the Buttery

Item one pair of brass candlesticks & three Dozen & five pewter plates, one pair of Bellowes, one warming pan, five Skilletts, one Saucepan, one brass frying pan, one small brass mortar, one pair of brass and one pair of iron Candlesticks, one & twenty Pewter dishes and pye Plates, one bason, one Chopping Knife, six pieces of tin ware, one old Gridiron, one Dripper, one Porringer, two skilletts, one brass pott & two iron potts.   004 = 19 = 06

In the Washhouse

Item one brewing copper, one washing iron Kittle, of all sorts   005 = 16 = 00

In the Cellars

Item Eight Beer Casks, one old brine tubb, and three Stillings   000 = 19 = 03

In a Closet in the best Chamber

Item Two silver Salvers, one silver Plate, two silver tankards, one silver Cup, two silver porringers, three silver Salts. one silver Caster, one pair of Silver Candlesticks, one set of silver Casters and ten silver spoons   027 = 10 = 00

Item a parcell of Earthenware & China ware and a small Case of Drawers.   000 = 12 = 00

Item a Parcell of old books   000 = 15 = 00

In the Yard

Item a large beam and scales, five hundred weights, & a quarter, one Bell & Dyall of the old clock   004 = 10 = 00

Item Seaven tuns and an-half of iron.   112 = 10 = 00

In the Workshop

Item Nails, ffiles, working tooles, old clock cases & some lines   009 = 00 = 00

Item Due to the said deceased upon Bills Bonds and other Securities and for Arrears of Rent in all   910 = 00 = 00

Item due to the said deceased in Book Debts good and bad   362 = 00 = 00

      1543 = 15 = 3

Matthias Gray
Sam [—]
John Elvey
  his mark
Will Harris
} Apprisers

The Greenhill Family

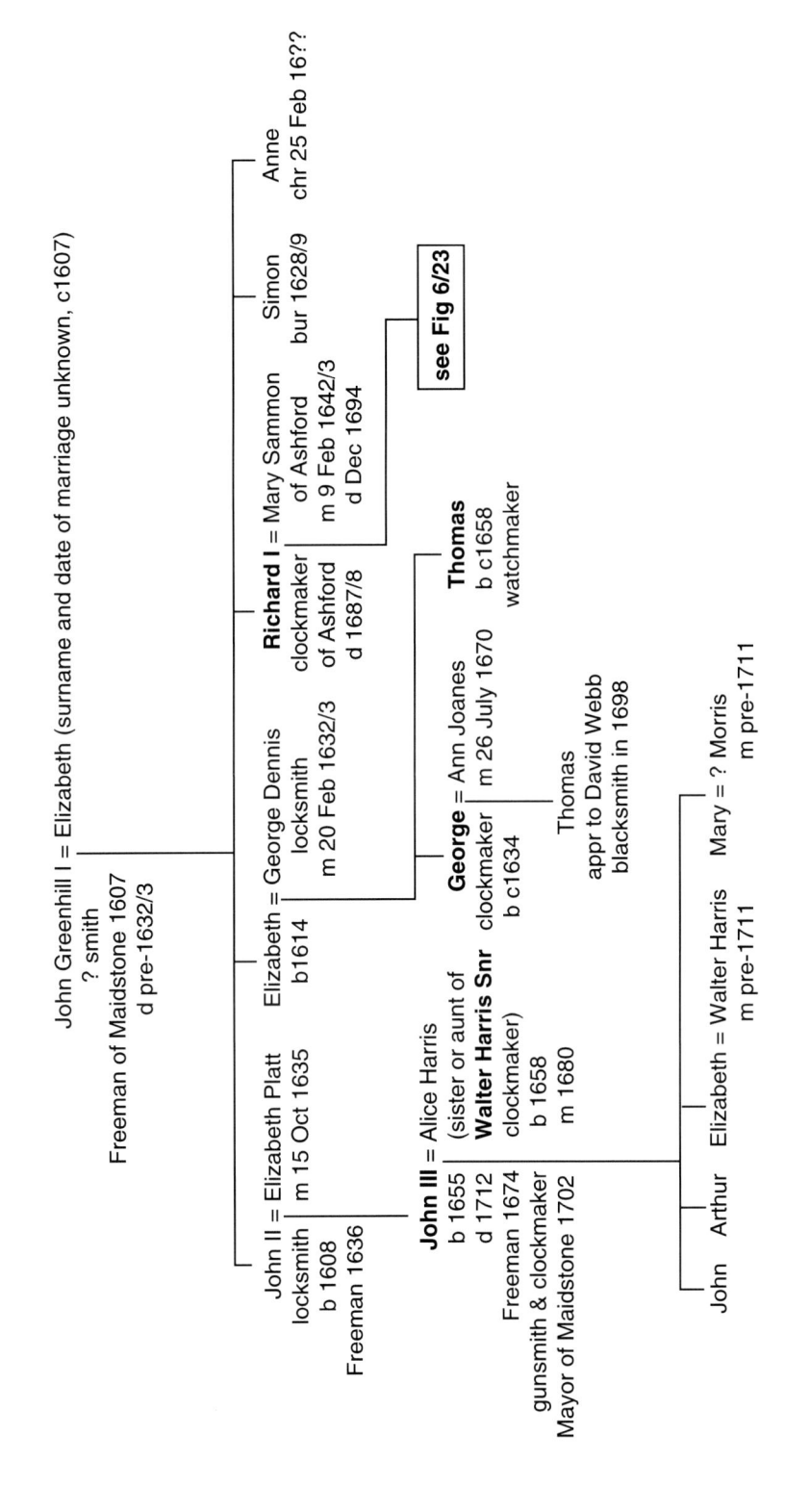

John Greenhill I = Elizabeth (surname and date of marriage unknown, c1607)
? smith
Freeman of Maidstone 1607
d pre-1632/3

- John II = Elizabeth Platt
  locksmith m 15 Oct 1635
  b 1608
  Freeman 1636

  - **John III** = Alice Harris
    b 1655 (sister or aunt of
    d 1712 **Walter Harris Snr**
    Freeman 1674 clockmaker)
    gunsmith & clockmaker b 1658
    Mayor of Maidstone 1702 m 1680

    - John
    - Arthur
    - Elizabeth = Walter Harris
      m pre-1711
    - Mary = ? Morris
      m pre-1711

- Elizabeth = George Dennis
  b1614 locksmith
  m 20 Feb 1632/3

- **Richard I** = Mary Sammon
  clockmaker of Ashford
  of Ashford m 9 Feb 1642/3
  d 1687/8 d Dec 1694

  - **George** = Ann Joanes
    clockmaker m 26 July 1670
    b c1634

    - Thomas
      appr to David Webb
      blacksmith in 1698

  - **Thomas**
    b c1658
    watchmaker

  see Fig 6/23

- Simon
  bur 1628/9

- Anne
  chr 25 Feb 16??

Fig 6/22 The Maidstone branch of the Greenhill family. The clockmakers are in bold

see Fig 6/22

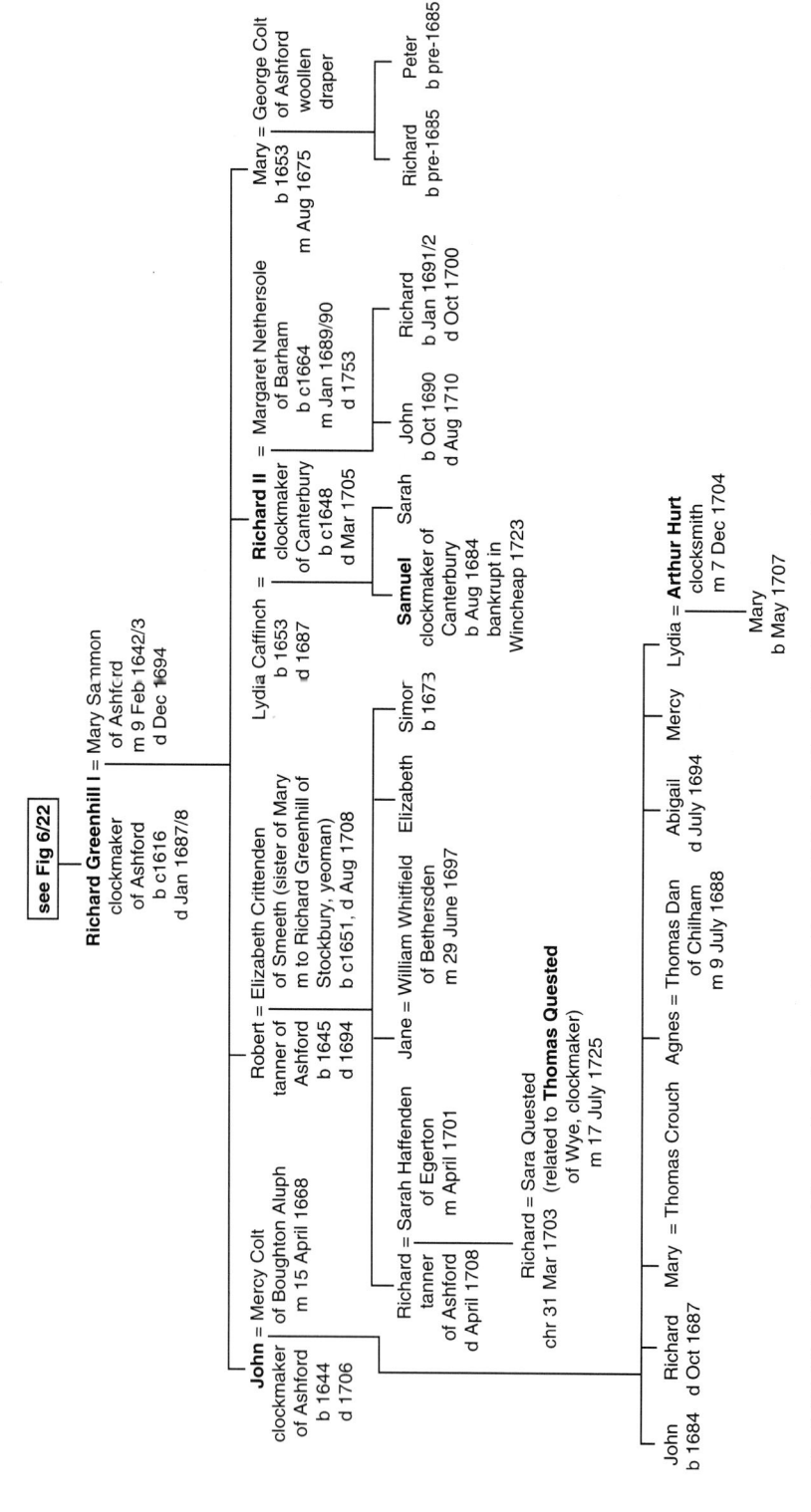

Richard Greenhill I = Mary Sammon
clockmaker   of Ashford
of Ashford   m 9 Feb 1642/3
b c1616   d Dec 1694
d Jan 1687/8

├── **John** = Mercy Colt
│   clockmaker   of Boughton Aluph
│   of Ashford   m 15 April 1668
│   b 1644
│   d 1706
│   │
│   ├── Richard = Sarah Haffenden
│   │   tanner   of Egerton
│   │   of Ashford m April 1701
│   │   d April 1708
│   │   │
│   │   └── Richard = Sara Quested
│   │   chr 31 Mar 1703 (related to **Thomas Quested**
│   │   of Wye, clockmaker)
│   │   m 17 July 1725
│   │
│   ├── Mary = Thomas Crouch   Agnes = Thomas Dan
│   │   of Chilham
│   │   m 9 July 1688
│   │
│   └── John   Richard
│   b 1684   d Oct 1687
│
├── Robert = Elizabeth Crittenden
│   tanner of of Smeeth (sister of Mary
│   Ashford   m to Richard Greenhill of
│   b 1645   Stockbury, yeoman)
│   d 1694   b c1651, d Aug 1708
│   │
│   ├── Jane = William Whitfield   Elizabeth   Simon
│   │   of Bethersden   b 1673
│   │   m 29 June 1697
│
├── Lydia Caffinch = **Richard II** = Margaret Nethersole
│   b 1653   clockmaker of   of Barham
│   d 1687   Canterbury   b c1664
│   b c1648   m Jan 1689/90
│   d Mar 1705   d 1753
│   │
│   ├── **Samuel**   Sarah   John   Richard
│   │   clockmaker of   b Oct 1690   b Jan 1691/2
│   │   Canterbury   d Aug 1710   d Oct 1700
│   │   b Aug 1684
│   │   bankrupt in
│   │   Wincheap 1723
│   │   │
│   │   └── Mercy   Lydia = **Arthur Hurt**   Abigail
│   │   clocksmith   d July 1694
│   │   m 7 Dec 1704
│   │   │
│   │   └── Mary
│   │   b May 1707
│
└── Mary = George Colt
    b 1653 of Ashford
    m Aug 1675 woollen
                draper
    │
    ├── Richard   Peter
        b pre-1685   b pre-1685

Fig 6/23 The Ashford and Canterbury branch of the Greenhill family. The clockmakers are in bold

Chapter Seven
KENT CLOCKMAKERS & WATCHMAKERS

References to Baillie and Loomes are to volumes I and II respectively of *Watchmakers & Clockmakers of the World*, and CC indicates that an apprentice gained his Freedom of the London Clockmakers' Company.

It should be appreciated that many of the watchmakers listed would be mainly retailers of watches, jewellery and similar items, and not makers — or even repairers — of watches. Even well known 'makers' of watches in London usually only finished rough movements supplied by others, and the components of these were actually made by dozens of very specialist workers, each of whom made just one small part. In many cases the name on the dial just indicates the retailer, particularly after the middle of the eighteenth century.

Where a maker signed his clocks with more than one place, they may have merely been sold at different towns — a number of Kent clockmakers advertised that they would be attending neighbouring towns at specifed times — and so signed their work to encourage local business. It does not necessarily mean that they had two places of work, or moved premises.

ABRAHAM, ISAAC   Sheerness   1802
Watchmaker.

ABRAHAM, JOSEPH   Greenwich   1812-24
Recorded in various directories as a watchmaker between these dates.

ABRAHAMS, ABRAHAM   Canterbury   1838-74
Bagshaw's Directory (1847) records him at 52 Northgate. A fine 8-day loncase clock with a painted dial by Wilson in a superb Goulden/Lepine-type mahogany case was sold in 1997. The dial was signed by Abraham, but the clock would appear to date from no later than 1790/1800. Possibly from the workshop of James Warren Jnr and resigned in the 1840s.

ABRAHAMS, HENRY   Hythe   1806

ABRAHAMS, SAMUEL   Sandwich   1783-1803
Silversmith, jeweller and watchmaker, near the Flower-de Luce. In August 1786 he advertised for sale jewellery, watches, hardware as well as clothes, hosiery hats and shoes (see Appendix I).

ABRAHAMS & SON   Ramsgate   1790
Watchmaker. A pair-case silver watch with the 1790 assay marks is known by the Ramsgate Abrahams and given the close proximity of Sandwich to Ramsgate it is likely to refer to Samuel. It is also probable that all of these makers were related and formed part of the thriving Jewish community of the late eighteenth/early nineteenth centuries that was centred on

Canterbury. The synagogue there, built in the guise of an Egyptian temple, can still be seen in King Stteer, and now forms part of the Kings School. A number of other Jewish watchmakers and clockmakers were also active in Canterbury during this period, notably members of the Heitzman and Solomon families.

ACRES, EDWARD　　　　　　　　Sittingbourne　　　　　　　　1767
Recorded as having died at this date. A watch by 'Edward Acurs [*sic*]' was reported stolen in January 1773 (see Appendix I).

ADAMS, ROBERT　　　　　　　　Deal　　　　　　　　nineteenth century
Watchmaker.

ADAMS, SAMUEL　　　　　　　　Bromley　　　　　　　　pre-1775-92
Recorded in Baillie as working between these dates. A watch signed by him exists.

ADKINS, G. B.　　　　　　　　Deal　　　　　　　　nineteenth century
Watchmaker.

AIANO, C.　　　　　　　　Canterbury　　　　　　　　c1840
Also known to have made barometers.

ALDERSLEY, JOHN　　　　　　　　Chatham　　　　　　　　1832
Presumably related to the following.

ALDERSLEY, GEORGE　　　　　　　　Whitstable　　　　　　　　1838

ALDRIDGE, EDWARD　　　　　　　　Deal　　　　　　　　apprenticed 1704/5
The only known clock by this maker is an early eighteenth century 8-day longcase clock with an 11¾in brass dial, silvered chapter ring, wheatear engraving to the edge of the dial, maskhead spandrels, a fully engraved dial centre with birds and foliage, and concentric rings surrounding the winding holes. The movement is of good quality with five ringed pillars and inside locking-plate striking. The chapter ring has fleur-de-lis half-hour markings and a narrow minute ring. The dial also has what is probably the most typical decoration to be found on Kentish clocks of the 1720s, namely a basket of fruit with a bird on either side surrounding the calendar aperture. Although this particular form of engraving is found on some clocks from other southern counties, nearly every Kent clock from this period that has been examined is decorated in this manner. The case of this particular clock is constructed of pine with a lenticle in the trunk door and was probably either ebonized or lacquered originally. (See *Antiquarian Horology*, Vol XII No 2, by M. D. J. Ward.) A silver watch signed 'Edward Aldridge, Deal' was reported lost in August 1783 (Appendix I).

The Apprentice Book of the Clockmakers' Company lists an Edward Aldridge bound for seven years on 5 March 1704/5 to Zachary Mountfort. He was the son of Edward Aldridge, an oatmeal maker from Sevenoaks who had died prior to 1704/5. The Clockmakers' Company records also show a Thomas Aldridge, the son of Thomas, a victualler of Canterbury, deceased, entering an apprenticeship on 9 February 1753 to Christopher Potter and gaining his Freedom on 9 October 1769. These two Aldridges are almost certainly the same as the makers from Deal and probably related, either grandfather, grandson or uncle and nephew. In all probability Thomas took over Edward's business.

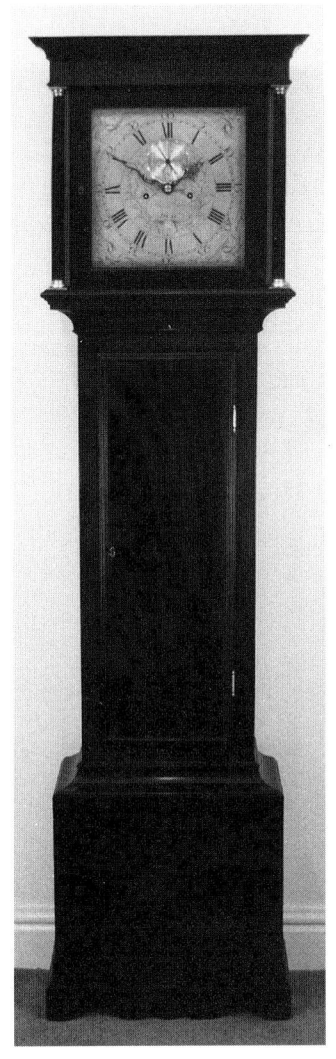

Fig 7/1 A small mahogany longcase clock by Thomas Aldridge of Deal, about 1785, with an 8-day five-pillar movement. The single-sheet silvered brass dial has a sunken seconds hand and a square date aperture. The 6ft 4½in tall case has reeded pillars, a blind fret above the hood door and stands on a double plinth

Fig 7/2 A five-pillar 8-day clock of high quality by Andrews of Dover, in a superb mahogany case 7ft 7in tall, about 1775. The engraved and silvered dial has a moving ship in the arch

Fig 7/3 Another high quality 8-day clock by Andrews of Dover, about 1790. The silvered arched dial is in an inlaid Sharaton-style pagoda-topped mahogany longcase

ALDRIDGE, THOMAS   Deal   apprenticed 1753-91

Although he is noted in Baillie as a watchmaker working 1784-91, an 8-day brass and silver dial movement with four pillars, engraved dial, silvered chapter ring, typical of the 1770s, housed in a slim oak case was sold at Canterbury in 1991, and a mahogany longcase of about 1785 is known (Fig 7/1). A watch signed 'Aldridge, Deal' was reported stolen in August 1777 (see Appendix I). He was working as early as 1769 following the end of his apprenticeship to Christopher Potter. (See Edward Aldridge.)

ALLANSON, JOSEPH   Canterbury   late eighteenth century

Recorded in the Canterbury archives as a watchmaker, but nothing else is known.

ALLEN, THOMAS   Deptford   pre-1789

Watchmaker.

ALLEN, THOMAS   Tenterden   1823

Noted at this date in Loomes, but nothing else known.

ALMAN, W.   Woolwich   1817-24

Watchmaker.

ANDERSON, —   Gravesend   c1820

Clockmaker.

ANDERSON, FREDERICK B.   Gravesend   1823-51

ANDERSON, HUGH   Gravesend   1839-66

ANDREWS, JOHN EDWARD   Whitstable   1866-74

ANDREWS, RICHARD   Dover   Free 1797

Little is known of this maker, who is believed to be the son of and apprenticed to Thomas Andrews.

ANDREWS, THOMAS   Dover   1773-1802

He is listed in both Baillie and Loomes as a watchmaker and a number of his watches survive. However, a verge dial clock and at least three good longcase clocks in mahogany cases are known by him and a fine example is shown in Fig 7/2. This clock is of the highest quality with a number of very individual features. The five-pillar hourly-striking movement has a well engraved and silvered centre with a large seconds dial and semicircular date aperture with the signature engraved below. The rococo pattern spandrels are well cut and engraved as is the arch which has a wavy cut out to the bottom and leaf spray decoration to the edge. Within the arch is a finely painted and detailed man o'war rocking to and fro.

    The case, of highest quality with good figuring to the mahogany veneers, is typical of so many of the late eighteenth/early nineteenth century emanating from the area of Kent east of Faversham. The author has seen several such cases with the trade label of '[William] GOULDEN of Canterbury, The Boro, cabinet maker and clockcasemaker', pasted inside (Fig 7/40). These cases have the same characteristic fretting to the base and hood and while this example is a little more ornate than usual it is almost certainly from the same workshop.

    Another longcase signed Andrews (Fig 7/3) is probably slightly later, dating from the

1790s, and could possibly have been made by his son, Richard. The case is again of finely figured mahogany in a pagoda top arched-dial case. The inlays, while in typical Sheraton style, may have been done at a later date. The dial itself is a good example of the later type of silver dial with its engraved floral drapes and swags and simple lettering that had become so popular by the end of the eighteenth century. Fig 7/4 shows a further longcase clock with a single-sheet silvered brass dial in a Goulden-style case.

In May 1773 he advertised the opening of a shop near the Market Place, but in July 1775 he moved to the 'QUAY at the PIER' (see Appendix I). In common with many of the makers from this period he also worked as a silversmith and had a flourishing business, judging by the following advertisement he placed in the *Kentish Gazette* on 26 Feb 1780:

> WANTED IMMEDIATELY,
> A JOURNEYMAN, In the CLOCK-MAKING BUSINESS
> A good Workman may have constant Employ by applying to THOMAS ANDREWS, Clock and Watch Maker, and Silversmith, at Dover.
> Also wanted An APPRENTICE, A Premium will be expected.

ANDREWS, WILLIAM   Westerham & Edenbridge   1847

ANNOTT, CHARLES   Canterbury   born c1659-82

Watchmaker. Born in about 1659 he was apprenticed in September 1673 to James Ellis, watchmaker in London. After completing his apprenticeship he moved to Canterbury and was made a Freeman on 5 October 1680. On 9 June 1682 he took Stephen Gray as his apprentice. Nothing else is known of this maker other than these entries in the Canterbury records.

ANSELL, HENRY   Chatham 1839-55

ANSELL, JOHN   Ramsgate   1845

Nothing known other than an entries in Loomes; they were presumably related to each other.

APPS, WILLIAM   Maidstone   1839-74

In *Bagshaw's Directory* of 1848 he is recorded as being in partnership with Susanna French working from an address in King Street. Other records show the partnership lasting from 1839 to 1874.

ARDGRAVE, GEORGE P.   Canterbury   1846-74

ARNOLD, CHARLES   Eltham   CC1824

The son of the famous London maker John Roger Arnold and grandson of John Arnold, who was a noted chronometer maker and founder of the family firm that subsequently became Arnold & Dent in 1830 and was finally taken over by Charles Frodsham in 1843. Although John Arnold and his son John Roger lived at Well Hall in Eltham in the late eighteenth century and are believed to have had a manufacturing department in additional wings built on to the old house, they have not been included as Kentish makers as they were primarily London clockmakers.

ASHDOWN, FRANK   Tonbridge   1874

ASHDOWN, G.   Sevenoaks   1866

ASHER, MORDECAI   Rochester   late eighteenth centuty

A 12in silvered dial wall clock with a concave beaded brass bezel of about 1790 is known.

AUNALE, —   Gravesend   pre-1743

Watchmaker. The entry in Baillie under this name may very well be a mis-spelling of Avenell (*qv*), and they are likely to be one and the same.

AUSTEN, ROBERT   Challock   1730- died 1793

Baillie notes a longcase clock by this maker in the Virginia Museum, Richmond, USA and at least two other longcase clocks by him have been sold in the 1990s. Both of them were 8-day striking arched brass and silvered dials with calendar and seconds rings. In each case the maker's name is inscribed on the chapter ring with a sun engraved on a boss in the arch together with a number, 35 on one and 95 on the later example. Both have five-pillar movements in pagoda-topped arched oak cases, athough the case of the clock No 95 has been heavily carved at a later date. The later clock also has an unusual chapter ring, typically Dutch with very florid half-hour markings, a rare feature on an English clock.

However, as already explained, there was a great deal of Dutch influence in East Kent in the late seventeenth and early eighteenth centuries due to the influx of refugees from the Low Countries and Canterbury had a thriving population of a 1,000 'strangers' engaged in the silk-weaving industry, so this particular clock may well have been a special order for one of the resident Huguenots.

The other unusual feature of these clocks is the numbering, which is very rarely found on southern movements, the only other Kentish makers to do so being Thomas Hall of Maidstone and John Baldwin of Faversham. Two other longcase clocks by Austen with numbered dials have been recorded, an oak cased example No 140 (Fig 7/5) and a mahogany cased clock number 202 which was sold at the Dale Hall, Suffolk sale in 1971. A numbered lantern clock in the later eighteenth-century style is also thought to exist.

Unfortunately, not much more information has been discovered about this maker, who was obviously reasonably prolific from the evidence of his numbering. Challock is a small village midway between Ashford and Canterbury, but he does not appear on either the Canterbury apprentice lists or the list of marriage licences in the Cathedral archives. It must be assumed therefore that he settled in Challock some time before 1730 as the parish registers record the baptisms of seven children between 1730 and 1737, three of them dying either soon after birth or in their childhood. One of the most striking features of this period is the very high infant mortality rate, which is only too evident in the parish registers. In many cases the number of deaths in any given year tend to cancel out the number of births and this is the main reason for the very slow rate of increase in population in the seventeenth and early eighteenth century.

Robert's second son, Thomas, was baptized in 1733 and was presumably trained as a clockmaker by his father. He was buried in Challock in 1805 having survived his father by twelve years. Robert Austen may have been related to the Austen family of nearby Godmersham Park. The owner of the estate, Edward Austen, was Jane Austen's brother and some of her novels were written at Godmersham. The family connection has yet to be proved, but it is interesting to note that the same spelling is used for the surname in both cases.

AUSTEN, THOMAS   Challock   born 1733-died 1805

Thomas, the fourth child of Robert and Elizabeth Austen, was baptized at Challock on 14

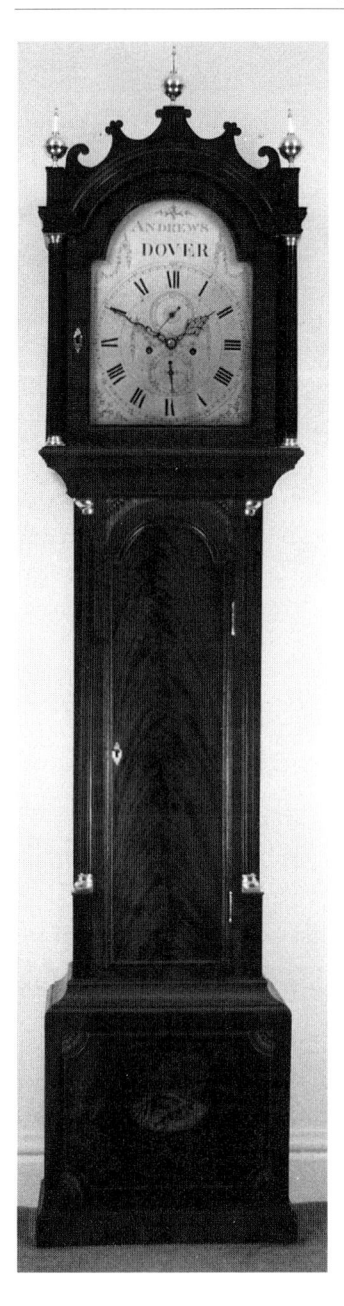

Fig 7/4 A fine quality loncase clock, 7ft 1in tall, about 1790, with 'Andrews Dover' displayed prominently in the arch and typical Kent cresting. The flame mahogany case, inlaid with shells and oak leaves, is typical of the Goulden workshop

Fig 7/5 A five-pillar 8-day clock, No 140, by Robert Austen of Challock, in its original pagoda-topped oak case, about 1750

Fig 7/6 A 12in x 16½in painted dial signed by Thomas Austin, Challock, about 1780. There is no dial-maker's name on the falseplate

October 1733. He and his wife, Mary (née Clark), subsequently had six children. The marriage registers record that he did not marry until 14 June 1785 at the surprisingly late age of 52.

An archdial oak longcase with an early style 8-day painted dial and a five-pillar movement is known (Fig 7/6). This clock probably dates from about 1780 and is very similar to those made by Thomas Sutton of Maidstone at the same time. It is interesting to note that Thomas had adopted the more usual spelling of Austin, or was this a mistake by the dial painter? He

Fig 7/7 The enamel dials of verge watches by Robbins of Canterbury (left) and Avery of Smarden (right), showing rural scenes

was buried on 19 December 1805 and neither of his two sons appear to have followed their father's trade as clockmakers.

AVENELL, WILLIAM   **Gravesend**   1720-70

He is listed in Loomes as a clockmaker working about 1770, but he was probably actively working at a much earlier date judging by the style of two clocks signed by him which have appeared on the market in recent years. Both examples had typical features of the 1720-30 period: five-pillar movements with 11in brass and silvered dials and birds and basket of fruit engraving surrounding the calendar aperture as commonly found on clocks of this period. A watch by him was reported stolen in November 1745 (see Appendix I).

AVERY, W.   **Smarden**   nineteenth century

An attractive painted dial watch decorated with a farming scene is in Maidstone Museum (Fig 7/7), otherwise nothing else is known of this watchmaker.

BAILEY, WILLIAM   **Maidstone**   pre-1785

Watchmaker.

BAILEY, W.   **Bromley**   c1830

Clockmaker.

BAKER, ALFRED   **Boughton**   1855-66

BAKER, A.   **Sittingbourne**   1874

Possibly the same man as in the previous entry.

BAKER, EDMOND or EDWARD   **Gravesend**   1858-74

BAKER, GEORGE   **Farnborough**   1874

BAKER, HENRY   **Boughton**   1826-57

A number of clocks, both longcase and bracket, have survived. All of those seen have had painted dials and have been housed in mahogany cases (Fig 7/8). The longcases have all been made in the style typical of East Kent at this period and were very probably made by Goulden of Canterbury. Henry is assumed to be related to Robert Baker of Boughton and is possibly the father of Alfred.

BAKER, HENRY I   Maidstone   Free 1622
Clockmaker.

BAKER, HENRY II   Maidstone & Town Malling
                           late seventeenth & early eighteenth centuries

See Chapter 3.

BAKER, JOHN   Sevenoaks   mid-eighteenth century

The only known reference to this maker is in Loomes. It is highly likely that he was related to the Bakers of West Malling given the proximity of the two towns. His only known clock is a shield-dial tavern clock in Maidstone Museum. Though it has been subject to some repairs, it quite clearly resembles other noted examples by Kentish makers such as Owen Jackson of Cranbrook, albeit of a type common to other southern makers of the eighteenth century.

While clocks of this type are commonly called Act of Parliament clocks, following the

Fig 7/8 An 8-day painted dial with a four-pillar movement by Henry Baker of Boughton. The finely painted dia has a shooting scene in the arch and hunting dogs in the spandrels, about 1825

Fig 7/9 A lantern clock by John Baldwin of Faversham, No 57, about 1700

introduction of the tax on clocks, instituted by William Pitt in 1797, they had been in existence for many years before then, the majority being found in the coaching inns along the main stage-coach routes. This particular example dates from the 1770s and was probably housed in either the Rose and Crown or The Royal Crown in Sevenoaks on the London to Hastings road. Both of these have long since disappeared to be replaced by other commercial premises, the latter by a cinema in the 1920s. Also in Maidstone Museum is the brass and silvered dial only of a 30-hour movement. The silvered dial centre is engraved in typical rococo style together with the name and place of origin.

BAKER & SON, M.   **Boughton**   1836

A silver pair-case watch with the assay mark for 1836 is known. In all probability this particular Baker was related to Robert Baker and presumably succeeded to the business.

BAKER, ROBERT   **Boughton**   1827

Clockmaker and watchmaker.

BAKER, THOMAS   **West Malling**   1768-late eighteenth century

See Chapter 3.

BALL-BAKER   **Gravesend**   1851

BALDWIN, JOHN   **Faversham**   pre-1739

The date given above is taken from Baillie, but there is a strong possibility that he is the same maker as the John Baldwin, born about 1677 who was apprenticed in London to Stephen Rayner in May 1691 until 1698, and who is not recorded as obtaining his Freedom. As far as can be ascertained only one longcase, a lantern clock and a watch by this maker have come to light. Both the lantern clock, numbered 57 (Fig 7/9), and the watch (Figs 2/4-5) are in Maidstone Museum. The oak longcase is an 8-day arch-dial example housed in a good case with cross-banding to the trunk door and is typical of the 1730s period. What is quite untypical is that the clock is numbered 294 on the dial, which leads one to suppose that there must be others in existence. The numbering of clocks by southern makers is very unusual and this feature is normally only to be found on Midlands, Northern or Welsh clocks, such as Hampson of Wrexham, Wainwright of Northampton, Deacon of Barton (Leicestershire), Roberts of Llanfair and Jonas Barber II of Winster (Westmorland). As noted earlier, the only other Kent makers to number their clocks are Austen of Challock and Thomas Hall of Maidstone.

The only other information is that the Ashford maker John Wimble and his wife Dorothy owned property in the High Street (now number 17) together with the land behind. This was mortgaged to secure a loan of £15 7s 6d in 1741 to John Baldwin. Wimble died in 1741 and the property was subsequently sold in 1742.

John Baldwin was obviously a very competent clockmaker judging from the little evidence we have, and if the numbering of his clocks is to be believed, it does seem surprising that no other work by him has surfaced in recent years. There must be numerous clocks by him in the Faversham area and it is hoped that more will be learned of this intriguing maker.

BALDWIN, J.   **Sittingbourne**   pre-1770

In July 1770 a silver watch by this maker was reported as lost or stolen (see Appendix I). He is probably the same as John Baldwin of Faversham, given the close proximity of the two towns.

| BALLARD & CO | Cranbrook | 1845 |

| BALLARD, HENRY | Cranbrook | 1847-58 |

| BALLARD, ISAAC | Lamberhurst | 1838 |

Clockmaker.

| BALLARD, JAMES | Lamberhurst | 1826-55 |

Longcase clocks with painted dials in oak cases are known by this maker who is probably the brother of Isaac Ballard.

| BALLARD, JOHN | Lamberhurst | 1885 |

| BALLARD, JOHN THOMAS | Staplehurst | 1844-8 |

| BALLARD, JOSEPH | Lamberhurst | 1839-74 |

| BALLARD, WILLIAM | Brompton | 1847-74 |

A silver pair-case watch by this maker who worked in the Brompton area of Gillingham with the London Assay marks of 1847 is known.

| BALLARD, WILLIAM (or FREDERICK W.) | Cranbrook | 1826-66 |

A good painted dial 8-day movement with moon phase work housed in a fine mahogany case was sold at Philips, Folkestone, July 1993.

| BANISTER, A. | Hawkhurst | 1882 |

| BANKS, JOHN | Herne | 1763 |

An advertisement in the *Kentish Post* 27-31 August 1763 states that a: 'Brown-Dun Gelding was found between Sturry and Hearne ... by John Banks, Clocksmith of Hearne ... if the owner don't soon apply, he will be sold to pay charges etc'. Nothing else known.

| BARHAM, GEORGE | Hawkhurst | 1837-74 |

| BARCHAM, ASHER | Tonbridge | 1800 and after |

| BARCHAM, ASHER | Sevenoaks | 1874 |

| BARLING, JOSEPH | Maidstone | 1847-74 |

These dates are given in Loomes, but there are entries in *Kelly's Directory* of 1852 and *Vivish's Directory* of 1872 with addresses at 90 High Street in 1839 and 7 High Street in 1872. In 1866 the 4th Earl of Abergavenny presented a clock made by Barling to Birling Church to replace an earlier movement dating from 1700. There is also a watch in Maidstone Museum.

| BARNARD, BENJAMIN | Canterbury | 1832 |

| BARNARD, FREDERICK | Sheerness | 1826-55 |

Watchmaker.

| BARNARD, J. | Maidstone | 1865 |

| BARNARD, JAMES | Erith | 1851-66 |

BARNARD, JAMES	Sittingbourne	end eighteenth century-1847

A painted dial longcase is known by this maker.

BARNETT, WILLIAM	Sevenoaks	1771

Watchmaker. Bankrupt in 1771.

BARNETT, ISRAEL	Ramsgate	1826-55

BARNSBY, JAMES	Maidstone	1865-74

He is noted in *Vivish's Directory* of 1872 as a watchmaker and jeweller at 7 Upper Stone Street.

BARR, —	Dover	pre-1784

One of his watches was reported stolen in September 1784 (see Appendix I).

BARRELL, GEORGE	Woolwich	1838-74

BARRETT, DANIEL	Chatham	1792-1800CC

These dates presumably refer to the beginning of his apprenticeship and the date he obtained the Freedom of the Clockmakers' Company. A good quality oak and mahogany London-style longcase has been recorded housing an 8-day painted dial with moon phase work in the arch.

BARRETT, THOMAS	Canterbury	pre-1636-died 1662/3

BARRETT, THOMAS Snr	Canterbury	1659?-1663-92

He died pre-1733.

BARRETT, THOMAS Jnr	Canterbury	Free 1726

BARRETT, JOHN	Canterbury	pre-1700, died pre-1733

John Barrett was the son of Thomas Snr. Little is known of the three members of the Barrett family who practised the craft of clockmaking in Canterbury during the second half of the seventeenth and the early eighteenth centuries. The dates given in Loomes tend to conflict with the Canterbury Freemen and Apprentice List, which itself is none too clear with its references to the members of this family.

    Their origins are unknown, although it is believed that there may be a connection with the Barretts of Lewes in Sussex. The earliest mention of them in Canterbury is a note of the granting of the Freedom by Redemption (purchase) to Edward Barrett, a locksmith on 17 April 1616. There are numerous references to other members working as locksmiths and gunsmiths at this period and Thomas Barrett is recorded as entering his apprenticeship to his father, Edward, whether locksmith or gunsmith is not clear, on 11 December 1663. However, the published records give rise to a certain amount of confusion as they also note a Thomas Sone, clockmaker apprenticed to Thomas Barrett, clockmaker on 23 June 1659, gaining his Freedom on 11 December 1663, the same day as Thomas is supposed to have started his own apprenticeship. Either there were two men by the name of Thomas or, more likely, the entry relating to Thomas Sone is a misreading of the records, and should read Thomas, son of Edward, obtaining his Freedom on 11 December 1663 rather than starting his apprenticeship at this date as the record suggests. Another source of confusion is the entry in the parish records of St Nicholas, Ash which notes a Thomas Barrett repairing the clock at a cost of £12 as early as 1636. This entry suggests that there must have been another clockmaker called Thomas Barrett in addition to the man known to have been working in 1663. He is possibly the Thomas Barrett noted in the records in 1659 and an inventory of Thomas Barrett, clockmaker, who died on 10 March 1662/3 is probably the same man.

[Kent Record Office PRC 11/19/19]
An Inventory of the Goods and Chattels of Thomas Barrett late of the Citty of Canterbury Clokmaker disseaced taken and apprized the 10th day of March 1662 by us whose names are heareunto subscribed.   l   s   d

Imps.	His purse and money	03	00	00
It:	His wearing apparell	01	10	00

In the Kitchin

It:	one payer of fier Irons. A Jack. A Spitt. A payer of Cobyerns fier tongs bellows and other smale Irons	01	5	0
It:	tenn pewter dishes A brase Chaffindish 3 pewter potts 3 candlesticks wh. other smale peeces of pewter	01	0	0
It:	two Iron potts 3 brase scillets one smale brass Kettle worming pan with some other smale things	00	13	4
It:	A samle silver Cann & seven spoones	03	15	0
It:	A table. A Cubbert A kneeding trought A box three Joynestooles fower Chayers with some olde books	01	0	0

In the Streete Chamber

It:	A bedstedle with Curtains and vallence one fether bed and bolster a greene rug to blankets and to pillows	04	0	0
It:	A Case of drawers A Chest A trunke A Cubbert & fower stooles	02	5	0
It:	A payer of Anyernes A looking Glase and six smale pictures	00	6	8
It:	three hollon sheets 4 pillow coates A long towell a table cloth & 18 napkins	02	10	0

In the back Chamber

It:	A bedsteddle A flock bed and bolster 1 pillow A coverlett and A blankett	01	16	0
It:	A trunke a Chest a smale table and a payer of Anyernes	00	06	8
It:	five payer of Corse sheetes two dossen of napkins and towels six smal table cloths and some other smale linin	02	00	0

In the Shop

It:	five Cloks fower watches sme brase dyalls and other brase worke for cloks	10	00	0
It:	An Anvile 3 vices with some files and other working tooles	04	00	0

In the Cellor

It:	A parcell of wood 3 old tubs one kettle with other Lumber about the house not before mentioned	00	10	0
		18	12	8
	The some totall is	39	17	8

[signed] Tho: Mayne;
      John Barret

Figs 7/10 and 7/11 The dial and movement of earliest known Kent watch, made by Thomas Barrett of Canterbury about 1670 or earlier. The champlevé dial, with a steel poker hand, depicts Canterbury Cathedral, the castle and the West Gate

Thomas senior was married twice as he is recorded as a widower when he obtained a licence to marry Mary Colton at the cathedral on 7 April 1692. He had two sons, John by his first marriage who was apprenticed to his father and gained his Freedom by Patrimony on 9 January 1704/5. He is recorded as a clocksmith in the Freemen Lists. He also had two sons, John and Edward, neither of whom followed their father's trade but obtained their Freedom as tailors in 1733. The records show that their father had died prior to 1733. The second son, also Thomas, from the second marriage to Mary Colton obtained his Freedom by Patrimony on 28 September 1726.

Though five clocks and four watches are noted in his inventory, the only work known by the Barretts is a fine silver pair-cased watch by Thomas senior in Canterbury Museum (Figs 7/10-11). Unfortunately, the outer case is missing. The champlevé dial is finely engraved and depicts the castle and cathedral. The movement has typical square baluster pillars and is fitted with catgut lines and no hairspring, with a single hand only to the dial and probably dates from the 1670s. This watch is the earliest known timepiece by a Canterbury maker and once again we can only guess at the fate of other clocks and watches made by this family and the rest of the early makers from Canterbury, of whose undoubted skills there is no trace today.

One other reference to John Barrett exists in the Ash parish records of 1708, where he is required to 'yearly and every year come over and view the clock and receive 5s. 0d and also make good what shall be needful'. It is known from the Ash churchwardens accounts that Richard Greenhill of Canterbury received payment for repairing the church clock and the records also show that this particular mechanism was replaced by a new clock in the very early years of the eighteenth century. Unfortunately no date or maker's name appears in the accounts, and it is not clear therefore whether Greenhill supplied a new clock just before his death in 1705 or whether in fact it was actually made by John Barrett soon after he gained his Freedom. The clock in question was subsequently removed to Staple Church in 1789, where it still is today, with its single-hand movement now converted to electric winding (Fig 1/16).

BARRETT, WILLIAM   Ashford   married 1614-62

Loomes records a turret clock by this man who is very possibly another member of the family discussed in the previous entry, although no reference to him has been found in the Canterbury archives other than an application for a licence by William Barrett of Ashford, locksmith, on 14 September 1614 to marry Elizabeth Master, her father Edward Master, a schoolmaster, acting as bondsman. However, he is mentioned in the hearth tax returns for Ashford where he is recorded as being liable for tax on three hearths in 1662. The *Old Book of Wye*, by G. E. Hubbard (1951), provides a little more information in that William Barrett is recorded as providing a new clock for Wye Church in 1638 at a cost of £9. This clock was to replace the movement which had been reinstalled in the church in 1634 at a cost of £2 14s 4d following a fire in 1572 when the steeple was burnt down. It is interesting to note that there are references in the records to a clock in the church as early as 1529 (see also page 15). Barrett's name also appears in the churchwardens accounts for Bethersden in connection with work carried out on the clock in addition to a number of other names, including Richard and John Greenhill of Ashford (pages 20-1).

BARTHOLOMEW   Maidstone   1582

It is not certain whether he was in fact a clockmaker or merely the keeper of the town clock. The Maidstone Chamberlain's Accounts includes the following entry: 'paid unto Bartholomew for keeping the towne cloke — xxs'. This was presumably the same clock as that at the town conduit (Figs 1/1-2) mentioned in the Chamberlain's Accounts for 1621 as being mended by Swinhogg.

BARTLETT, F. J.   Maidstone   1865

Watchmaker.

BARTLETT, SAMUEL J.   Maidstone   1821-55

Watchmaker.

BARTLETT & CO   Maidstone   1851

In *Pigot's Directory* of 1832 Samuel is recorded as having a business at 109 Week Street and the same address in the directory of 1840. By 1848 he is noted in *Bradshaw's Directory* as having moved to 89 Week Strett and is noted in the *Topograph of Maidstone* as a watchmaker and silversmith and the heir to Robert Apps, the celebrated maker from Battle, Sussex. William Apps who was in partnership in King Street, Maidstone with Susanna French is probably a family connection.

BATES, —   Canterbury   nineteenth century

A watch signed 'Bates, The Boro, Canterbury' is known.

BATES, HENRY   Ramsgate   1874

BATES, JOHN	Hythe	1847
BATES, JOHN	Ramsgate	1849
BAURLE, L. & F.	Chatham and Rochester	1874

Watchmakers.

BAYL[E]Y, JOHN	Romney	1795

Watchmaker. Probably the same as the son of Richard Bayly Snr, and of Ashford in 1785.

BAYL[E]Y, RICHARD Snr	Ashford	1752-died 1785

Watchmaker. Maidstone Museum has two watches by this maker (Figs 7/12-14), and a longcase with a brass dial signed 'Richard Bayley, Biddenden' has been recorded.

He advertised in the *Kentish Gazette* 24-7 April 1776:

> LIGHT GOLD,
> WHEREAS it may be inconvenient
> to many People to exchange their LIGHT GUINEAS, &c.
> at Places appointed by Government for that Purpose,
> RICHARD BAYLEY,
> WATCHMAKER and SILVERSMITH, Ashford,
> Begs Leave to inform the Public, he will exchange such
> GUINEAS, HALFS, and QUARTERS, as specified in the
> Royal Proclamation, at so small a Premium as
> ONE PENNY per Guinea Discount.
> GUINEAS, &c. under the Weight, and PORTUGAL Coin,
> bought at 3£ 16s. per Ounce, as usual.

An advertisement in June 1776 also lists the exchange rates that he gave for 'deficient Coin of the Realm'.

In June 1785 he announced that in addition to being a watchmaker, silversmith and hardware man trading in the High Street, he was carrying on the weaving business of the late John Collar, selling flax, hemp, hop-bagging, corn sacks and cheese-cloths, while his daughters also ran a millenery, linen-drapery, hosiery and haberdashery business from the same house. But by December Richard Bayly (as the name was consistently spelled at this time) had died and his wife, Elizabeth, announced that she would be continuing the business of watchmaker and silversmith as well as the clothing business. Robert's sons were stated as being John of Ashford and Richard of Maidstone. 'Bayly, Watch-Maker, Silversmith and Hardwareman, Maidstone' (ie Richard Jnr) stated that he would be helping his mother carry on his late father's business, and he had engaged an experienced workman from London (see Appendix I).

BAYL[E]Y, RICHARD Jnr	Folkestone & Maidstone	1780-5

Watchmaker. Came from Ashford in 1780, when he opened a shop 'on the Court leading to the Beach' at Folkestone (see Appendix I, 11-15 November 1780). He is presumably the son of Richard Bayly Snr, and had moved to Maidstone by 1785. He appears in the Canterbury marriage records as marrying Margaret Tilbey on 2 July 1781 when he was aged 20.

BAYL[E]Y, WILLIAM	Maidstone	1784

In December 1782 'Bayly, Watchmaker and Silversmith, Opposite the Corn Market in the

Figs 7/12, 7/13 and 7/14 A verge watch by Richard Bayly of Ashford in a gilt repoussé case

High Street, Maidstone' advertised that he 'makes, sells and carefully repairs all sorts of WATCHES', but as the christian name is not given it is not known if it was Richard Jnr, who was at Maidstone by 1785, or William, who is listed by Loomes as 1784.

BEAL, RICHARD	Sheerness	1847-55
BEAMONT, CHARLES JOHN	Greenwich	1847-58

Fig 7/15 The superbly engraved 12in square dial of a clock by James Becket of Dover, about 1720. The 8-day movement has inside countwheel striking and five ringed pillars. The unique engraving on the dial centre has the face of the moon around the calendar aperture and human faces at 3 and 9 o'clock. Though the movement is now housed in a mahogany case of the 1770 period, it is probably another example of a clock that was originally in a lacquer case. The lead weights are unusual in having the clockmaker's name cast into the top of each one

BEAN, EDWARD	Kingston	pre-1792-1800

Watchmaker.

BEARD, J.	Tonbridge	c1800-12

A list of clockmakers in the local studies department of Tonbridge Library notes a James Beard, watchmaker in 1812.

BEARD, WILLIAM	Canterbury (from London)	1667-74

Clockmaker. Born 1653 and apprenticed to James Ellis in Oct 1667 and obtained his Freedom in 1674. Nothing else is known. See James Ellis.

BEAVEN, JOHN RICHARD	Woolwich	1838-51

BECKET, JAMES	Dover	1720/1

A walnut longcase clock housing a brass dial movement inscribed 'James Becket Fecit att Dover' has been noted (see also Plate 7/15). Named in the Canterbury records as bondsman to John Watt when he obtained his Freedom on 16 February 1720/1 and noted as a clockmaker of Dover. Possibly the same man as the John Becket[t] apprenticed in London in 1698, qv John Watts.

BECKWITH, WILLIAM	Dartford	1832-9

BEECHING, ELIZABETH	Maidstone	1791-5

Watchmaker.

BEECHING & EDMELL	Maidstone	early nineteenth century

A watch so signed is in Maidstone Museum.

BEECHING, STEPHEN	Maidstone	pre-1791

Mentioned in the *Universal British Directory* of 1791.

BEEDLE, JAMES FRANCIS	Woolwich	1874
BEKELYS (BEKYLLS) NICHOLAS	Canterbury	1485-95

The churchwardens' accounts for St Andrew the Apostle, Canterbury, contain the following:

1485	Item. Paidd to Nicholas Bekelys for makyng of the yerne werk of the clok	XXJd
1494-95	Item Pay'd to Bekylls for makyng of ye clok	VJs
	Item Pay'd for payntyng of ye Dyall	1 js

This suggests that there was already a clock in the church by 1485 and that it was replaced in 1494-5 by a new clock. It seems inconceivable that there was a ten-year delay in making this clock, even though accounts were sometimes paid years in arrears. Nothing else is known of Nicholas Bekylls who was both a blacksmith and clockmaker judging by these entries.

BELL, FRANK GODFREY	Rochester	1839
BELL, FREDERICK & THOMAS	Rochester	1839
BELL, J.	Chatham	1838
BENNETT, ELIZABETH	Lee	1839-55
BENNETT, GEORGE	Greenwich	1802-11

Watchmaker.

BENNETT, G.	Bexleyheath	1855
BENNETT, G. W.	Blackheath	1851
BENNETT, G. W.	Woolwich	1851-5

Almost certainly the same man as the previous entry.

BENNETT, JOHN	Greenwich	1817-24

Watchmaker.

BENNETT, WILLIAM COX	Blackheath	1866
BENZIE, Mrs M. A.	Maidstone	1866
BERRY, JOHN	Canterbury	pre-1753

A watch is noted by this maker in the report of a robbery at Thomas Shindler's shop in 1753, but nothing else is known. He does not appear on the Canterbury Apprentice Lists and presumably came from London.

BIANCHI, B.	Tunbridge Wells	1858

Known as a watch and clockmaker. Maidstone Museum has a watch in its collection.

BING, Mrs	Canterbury	1865

BINOY (BING?), DAVID	Canterbury	late eighteenth/early nineteenth century

A mahogany bracket clock has been recorded. Nothing else is known.

BING, JOHN	Ramsgate	pre-1781-died 1790

A lost watch signed as above was reported in the *Kentish Gazette* in April 1781 (see Appendix I). Nothing else is known other than the entry in Baillie which records his death in 1790 and that he was a freeman of the Clockmakers' Company in London.

BIRCH & MASTERS	Tenterden	1836-47

Watchmakers. A partnership between John Masters and William Birch, both of Tenterden.

BIRCH, WILLIAM	Tenterden	1823-39

Watchmaker. A four-pillar 30-hour movement with a square painted dial in an oak case has been recorded. He also seems to have been in partnership with Thomas Ollive as a number of clocks signed with their joint names have been recorded.

BIRD, WILLIAM	Faversham	1752

From London, watchmaker. This entry is taken from Baillie and no other information, other than the following advertisement which appeared in the *Kentish Post* on 29 January-1 February 1752, has been traced.

> WILLIAM BIRD, Watch-maker, from LONDON,
> Who, for many Years past, has served some of the Principal Masters there.
> Having NOW taken a Shop in Court-Street, opposite to
> Mr. Cobb's, near the Market-Place in FEVERSHAM, Makes,
> Mends, Cleans, and Sells Watches of every kind, either in
> Gold, Silver, or Shagreen Cases, at very Cheap and Reasonable Prices.
> Repeating Clocks, and Watches, and Spring Movements, of every Kind, are likewise carefully
> Cleaned, and Repaired, as Well, and in as Great Perfection, as any where in London.
> All Gentlemen, and others may depend on being
> Honourable, and Faithfully served, by their most obedient Servant. WILLIAM BIRD.
> N.B. He also buys, and gives, the most Money for
> OLD WATCHES, RINGS, and PLATE of every Kind.

It is possible that he is the same William Bird who was apprenticed in London about 1726 and gained his Freedom from the Clockmakers' Company in 1749, subsequently moved to Faversham to work with William Clement Junior.

BISHOP[P], JOHN I	Maidstone	c1650-died 1710

BISHOP[P], JOHN II	Maidstone	born 1680-1733

BISHOP[P], MARTHA	Maidstone	apprenticed 1720-33

The Bishop family are another group of very interesting early makers of whom little is known and, more surprisingly, hardly any of their work survives.

The entry given in Baillie for John Bishop the elder of 1650 has proved difficult to substantiate. The earliest reference found by the author is an entry in the All Saints registers

Fig 7/16 The 12in square dial of a 30-hour clock by John Bishop of Maidstone, with unusual punched decoration, about 1705. The hands are not original

Fig 7/17 An 8-day clock by John Bishop of Maidstone in a red lacquer case, about 1700, with a 12in square dial. The movement has six latched pillars

relating to the christening of two sons named John, one in January 1670 (presumably dying at a very early age) and another John, the son of John and Elizabeth Bishop who was christened on 13 June 1680. The only other entry found in the register is the death in April 1710 of John Bishopp (presumaby the father). There are no entries in the Maidstone Apprentice Lists prior to the John Bishop, locksmith in 1707, and the only other early reference is that of John Bishop on the list of Maidstone Freemen for the year 1690. While he was probably working at an earlier date it is very unlikely that he was working as early as 1650. Is it possible that this date refers to the year of his birth? The other definite date is of the apprenticeship of Martha (a daughter?) to John II in 1720 and she also appears on a document dated 1733 and is noted as a clockmaker.

A lantern clock by John I would appear to date from the 1680 period and three longcase clocks by him are known: a 30-hour (Fig 7/16), an 8-day brass dial with crown-and-cherub spandrels signed 'John Bishopp Maidstone Fecit', and a fine 8-day movement in its original lacquer case, with restored red and gold chinoiserie decoration, dating from about 1700 (Fig 7/17). The movement of this clock is of particularly high quality with six turned and finned pillars, which are latched to the front plate, with the unusual feature of the inside count wheel being positioned immediately in front of the rear plate. The 12in dial is very simple with fine matting to the centre and no ringing around the winding holes, which makes for a beautifully legible dial very much in the tradition of the best London craftsmen of the time. The case is very similar in style to a clock by John Wimble of Ashford also of red lacquer and dating from the same period, although the Bishop movement is of much superior quality.

The only clock seen made by his son John is an arched dial longcase in oak that is definitely of inferior workmanship to those made by his father. John Snr obviously learnt his craft at the hands of a skilled clockmaker and it is regrettable that at the moment we have no idea whom that may have been. The only other Maidstone makers to have made clocks of similar quality are John Greenhill and Edward Cutbush. They were both working during the same period as Bishop, but so far no connection has been found.

BISHOP, JOHN   Deptford   1748
Watchmaker. Baillie notes that he was bankrupt in 1748.

BISHOP, S. R.   Woolwich   1866

BLACKLEY, J. W.   Canterbury   1805

BLACKMORE, JOHN   Rochester   1716
Watchmaker.

BLAXLAND, JOHN   Sittingbourne   1866

BLISS, AMBROSE   Canterbury   Free 1647
Watchmaker. The only record known of this very early maker is that he obtained his Freedom by Redemption (ie by purchase) in Canterbury on 18 August 1647. He is almost certainly the same man as the Ambrose Bliss who gained the Freedom of the Clockmakers' Company in London in 1653 and was known to have worked in Cornhill until 1662.

It is possible that there is a connection with another London maker, Thomas Dennis, who was apprenticed in 1672 to James Wightman, and who may be the same man as the Thomas Dennis who moved to Canterbury in 1684. In 1685 he moved to Maidstone and was practising as a watchmaker at premises in Week Street, subsequently selling the property to Thomas Bliss (a prominent Maidstone citizen who served as Member of Parliament three times for the borough). Thomas Bliss may have been related to Ambrose, although there is no firm evidence to substantiate this at present, nor, unfortunately, has any watch by Ambrose Bliss seemingly survived. After his move to London, the only record of his activities is that he was one of the fourteen signatories to a counter-petition to the Lord Mayor in 1656. Complaints had been made by a large number (thirty-three in total) of members as to the conduct of the Master, Wardens and Assistants of the Clockmakers' Company, in particular that there was no regular meeting place and that although the company had been formed to protect clockmakers from the immigrant 'French' makers it was in fact being run by the self-same French craftsmen. The counter-petition was presented in support of the administration of the company by other rank and file members, of whom Bliss was one.

BLITZ, A.   Deal   nineteenth century
Watchmaker. Watchpaper with address near the India Arms, Beech Street.

BLUNDELL London & Greenwich apprenticed 1678-c1730

This entry is taken from Baillie, but where the reference to Greenwich came from is not clear. He was born about 1665 and apprenticed in December 1678 to George Nau, a watch-case maker, until 1686, but he never applied for the Freedom of the Clockmakers' Company. Both longcase and bracket clocks are known. Brian Loomes (*Early Clockmakers of Great Britain*) notes that a John Blundel worked in Horley, Oxfordshire, but makes no mention of a Greenwich location. Whether they are one and the same merits further investigation.

BODLE Westerham pre-1771

Watchmaker.

Fig 7/18 The dial of a 30-hour birdcage movement by Abraham Body of Cranbrook, with rococo engraving to the silvered dial centre and a square calendar aperture (date ring missing)

Fig 7/19 A 30-hour longcase clock with a birdcage movement by Obadiah Body of Cranbrook in an oak case typical of the majority of Kent clocks of the early/mid-eighteenth century period. This particular clock is signed at Battle

BODLEY, SIDNEY	Ramsgate	1874

BODY, ABRAHAM	Cranbrook	1766-79

Abraham, the son of Obadiah, was probably born in Battle. Whether he was apprenticed to his father or, like him, was trained by either George or Thomas Thatcher of Cranbrook is not definitely known. When he married Mary Nye of Battle in 1765 he was described on the marriage licence as 'of Cranbrook'. Baillie gives a date of 1763 and describes him as a clock and watchmaker of Battle. This date may well be the end of his apprenticeship to his father. There was obviously a close relationship between the Bodys and the Thatchers because Abraham succeeded to the Thatcher business in 1774 and worked in Cranbrook until his death in 1779. Both watches and 30-hour clocks are known to exist (Fig 7/18).

BODY, HENRY	Cranbrook	1859

Loomes gives this date for Henry, but the exact relationship between him and the other Bodys has not been confirmed. He is probably Abraham's son, presumably named after his uncle Henry who was working in Battle until 1818 as a clockmaker in succession to his father, Obadiah.

BODY, OBADIAH	Cranbrook	born 1702-1730
	Battle (Sussex)	1730-died 1767

Clockmaker. Born in 1702, the son of a yeoman farmer from Westfield in Sussex, Obadiah was apprenticed in 1716 to George Thatcher of Cranbrook and after finishing his time continued working in Cranbrook until 1730 when he moved to Battle in Sussex. In 1731 he married Mary Weston, the daughter of the Lewes clockmaker Abraham Weston. A number of clocks by him exist (Fig 7/19), with at least one 30-hour example signed at Cranbrook.

In 1737 he was appointed by the churchwardens to look after the church clock at Battle, which he continued to do until he died in 1767. He obviously gave satisfactory service to the parish because they issued a certificate stating that the parish would look after him in the event of him becoming destitute.

BOOTH & CO	Rochester	1794

See James Booth.

BOOTH, JAMES	Rochester & London	Free 1766, died 1822

The Rochester Freeman List notes that James Booth obtained his Freedom on 15 March 1766, on payment of £20. Very little is known of his working life in Rochester other than that a watch movement signed by him is known. There is also a reference to him in the Council Day Book of 4 May 1771 when the council decided which of three quotations they should accept for replacing the town clock installed at the Corn Exchange. Quotes were also considered from Edward Muddle and George Jeffery, both of Chatham. The contract was awarded to Muddle and the council decided to pay Booth and Jeffery the sum of one guinea for the trouble they had both gone to in preparing and submitting their proposals.

Booth moved to London soon after, as in February 1771 his business premises were advertised for sale with the implication that he would be working for the London maker William Stacey at 186 Fleet Street (see Appendix I). Stacey is recorded until 1784. Booth became a liveryman of the Clockmakers' Company in 1781 after obtaining his Freedom in 1784. He returned to Rochester in 1794 or earlier, when he founded the firm of Booth & Co. Nothing is known of Booth's activities before 1766 and his origins remain a mystery. A suggestion has been made that he was related to Joseph Booth of Sandwich. It is interesting to note that

Baillie also mentions a Joseph Booth, son of James, entering his apprenticeship in London in 1796. The James referred to is presumably the Rochester maker, and maybe the son was named after Joseph Booth of Sandwich, who may have been his grandfather. The firm of Booth & Co may well have been a partnership between father and son. To date, this has not been substantiated and no other references to either the father or the son has been found in the city records other than a note of James's death in 1822.

BOOTH, JOSEPH   London & Rochester (?)   1796

Believed to be the son of James Booth and may have been a partner in Booth & Co. See previous entry.

BOOTH, JOSEPH   Sandwich   1711

Nothing known other than the entry in Loomes. This date refers to him taking Thomas Jenkinson as an apprentice in that year.

BOURN, LODER   Canterbury   1738-88

Watchmaker and goldsmith. In 1738 he was 'At the sign of the Three Golden Candlesticks in Burgate-Street, Canterbury in the Shop where Mr. Robert Anslow the Cutler lately liv'd', which implies that at that date he had only recently moved there (see Appendix I, 31 May-3 June 1738).

BOWEN, DAVID   Canterbury   late eighteenth century-1815

Clock and watchmaker. This entry is taken from Baillie who also mentions a regulator in the Buckley Collection. In *Hilden's Triennial Directory* of 1805 he is stated as working from premises on The Parade. An extraordinarily verbose advertisement, headed with three masonic stamps, was placed by him in *Kentish Gazette* on 4 January 1803 and also a year later:

D. BOWEN,
Clock and Watch-maker, Silversmith, Jeweller, &c. (from London)
Nearly opposite the Union Bank, Parade, Canterbury,

MOST respectfully returns thanks to his friends for the very liberal encouragement he has received during the short period he has resided in Canterbury, ... as a Watch and Clock Maker, in which department he trusts that he has given sufficient proof that he has a complete knowledge of his business, which has been gained by a constant application to it for many years, and also from receiving every possible instruction from men of the first experience in the trade; he therefore flatters himself that he has it in his power to obviate that very general complaint which is alleged against many of the trade, that Clocks and Watches are oftner spoiled by unskilful workmanship, than by the wearer.

All sorts of musical and church clocks, repeating, horizontal, and plain watches made and repaired on the best principles.

He also takes this opportunity of informing the public that he is just returned from London with a large assortment of Jewellery, Plate, Hardware, Cutlery, &c.&c. consisting of silver, plated and gilt tea and coffee urns of various sizes; silver table and tea spoons; candle-sticks, gold chains and seals, wedding and fancy rings; a variety of fashionable beads; elegant tea trays and waiters; pocket pen knives and scissars; trinkets ....

WANTED,
A respectable Youth, as an apprentice to the above Business, where he will have an opportunity of learning every branch, which is seldom the case in the country, and it is much to be lamented that an apprentice, after serving seven years, should not be capable of following his profession

for want of necessary information being given to him during his apprenticeship ....

By January 1808 his advertisments have no mention of clock or watchmaking, and he merely stocks silver, jewellery and watches (see Appendix I).

BOWRA, JAMES   Sevenoaks   born 1685-died 1745

Clockmaker. The Bowra family have lived in the Sevenoaks area from the late seventeenth century until the present day. They originated from Sussex where the family name was spelt Bourer or Boreer. Thomas Bowra, a surgeon, was born in East Grinstead in 1654 and moved to Sevenoaks in 1683, where he is believed to have entered the service of the Duke of Dorset, who as a member of the Sackville family, had vast landholdings in East Sussex and the Sevenoaks area, and while residing at Knole had their ancestral home at Withyham near East Grinstead. Thomas had two sons by his second wife and although he was only 35 years old when he died in 1690, had made arrangements for James, born in 1685, to be trained as a clockmaker. Unfortunately we do not know with whom he served his apprenticeship. It may have been in Maidstone, as there is believed to be a family connection with the Gill family of clockmakers. Another member of the family, William, married Hannah Gill, the sister of William Gill, in 1706. However there is no trace of James's name in the Maidstone lists.

No record of James's marriage to Ann has been found, but the baptisms of the following children are recorded: Ann in 1708, John in 1714 and Elizabeth in 1717. His wife died in 1719 and he subsequently remarried, having three more children including Thomas in 1720 and William in 1724, both of whom became clockmakers in Sevenoaks, presumably learning the craft from their father, who died in 1745. A number of clocks, both 30-hour and 8-day longcases, are known to exist. All of those seen have been housed in oak cases (7/20). The youngest son William was also employed by the 3rd Duke of Dorset as a gamekeeper, and in addition was a well-known cricketer, playing for the Duke's eleven in a famous match against a London eleven on the Vine, in which game he distinguished himself so much that the Duke was heard to exclaim 'Bravo, my little Bowra'. Small stature must have been a family characteristic as another James was rejected for service in the militia in 1803 because he was undersize!

While his younger sons followed their father's trade as clockmakers the eldest son, John, had become a land surveyor and map-maker of some repute. However, before that he had been heavily involved in the smuggling of brandy and tea and was one of the main organizers of the illegal trade in Kent in the 1730s and was well-known to the customs officers. They seized a quantity of tea and arms in one raid on his house at Groombridge and a report filed by an informer states that he had handled some 3,000lb of goods a week during the winter of 1737. Smuggling was an extremely lucrative business and Bowra spent £500 having a house built at Groombridge. Although he was known to be actively involved, at no time was he arrested by the authorities. Is it possible that friends and associates in high places, or even the Duke himself, were protecting him?

BOWRA, GEORGE   Sevenoaks   1835-51

BOWRA, THOMAS   Sevenoaks   born 1720-73

The son of James Bowra (see above) working in Sevenoaks until at least 1773. Watches and longcase clocks with arched painted dials are known.

BOWRA, WILLIAM   Sevenoaks   born 1724-80

The younger son of James, a clockmaker, gamekeeper and also a member of the Duke of

Fig 7/20 (left) An 8-day arch-dial clock with a five-pillar movement, by James Bowra of Sevenoaks, about 1730, in an oak case

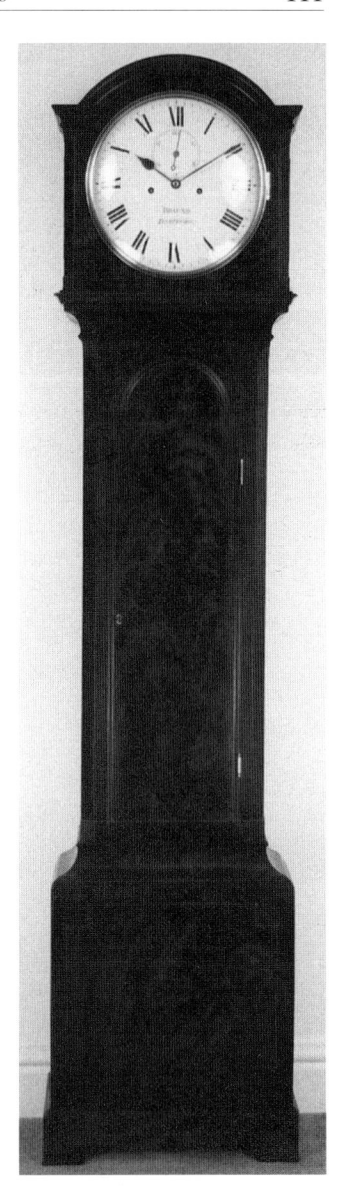

Fig 7/21 (right) A London-style mahogany longcase clock, only 6ft 4in tall, signed 'Braund, Darford'. The round white dial is 12in diameter

Dorset's cricket team. During the 1740s the game of cricket was very well established in Kent and the Duke of Dorset was one of the leading patrons of the game. A number of clocks survive signed 'Will. Bowra 7 Oaks', and an account rendered by William Bowra, clocksmith dated 1753 addressed to the 'Duck of Dosset' is in existence. Spelling was obviously not one of his strong points! A small tavern clock in a black case with traces of original lacquer, which is now hanging in the National Trust shop at Knole may also have been made by him.

BOWRA, WILLIAM Snr   **Sevenoaks**   **married 1789-1802**

It is difficult to unravel the later Bowra family tree because of the common use of the names

William and James, but William is likely to be the son of Thomas. We know that he married in 1789 and the clocks that exist with all-over silver dials with the signature of 'Will. Bowra 7 Oaks' are probably by him.

BOWRA, WILLIAM Jnr　　　　　Sevenoaks　　　　　1823-51
Son of the previous entry.

BOWRA, WILLIAM (E.)　　　　　Sevenoaks　　　　　1823-66
These last two entries are taken from Loomes and nothing else is known, other than that by the middle of the century the family business had degenerated into a small watch-repairing concern, based in the premises now occupied by Warrens, the opticians.

BOWRA, —　　　　　St Mary Cray　　　　　1839
Possibly a small village branch of the business.

BOXER, JOHN Snr　　　　　Folkestone　　　　　1804
Watchmaker. According to Loomes, John, the son of Michael Boxer, was working in London from 1779 until 1804 at which time he obtained the Freedom of Folkestone. Although a watchmaker he also held the position of Postmaster, acted as an auctioneer and librarian!

BOXER, JOHN Jnr　　　　　Folkestone　　　　　born 1779-died 1852

BOXER, MICHAEL　　　　　Folkestone　　　　　pre-1780-died 1810
Watchmaker.

BOXER, THOMAS　　　　　Sandwich　　　　　1832
Watchmaker. Appears in *Pigot's Directory* of 1832.

BOYCE, THOMAS　　　　　Gravesend　　　　　1830

BOYS, LAURENCE　　　　　Deal　　　　　c1720-37
Clock and watchmaker. 'Near the Post-Office in Deal' (see Appendix I, 6 April 1737).

BRABY, J.　　　　　Tunbridge Wells　　　　　1866

BRADBOURNE, JOHN　　　　　Westerham　　　　　1744-64
Clock and watchmaker. Baillie lists him as a watchmaker in 1744 and in Loomes he is noted as working at least until 1764.

BRADSHAW, HENRY CHARLES　　　　　Wye　　　　　1832
Watchmaker.

BRADSHAW, THOMAS　　　　　Canterbury　　　　　c1790-1804
Watchmaker. Very little is known of this maker other than the entry in Loomes and a note in *Hildens Triennial Directory* of 1805 in which he is described as a watch and clockmaker trading from premises in St Peters Street. There is no trace of him on the Canterbury Apprentice Lists but he is probably the father of Henry Bradshaw of Wye.

BRADSTREET, ROBERT　　　　　Rochester　　　　　born 1737-died 1797
Watchmaker. Watch No 450 reported lost in April 1771 (see Appendix I).

BRADY, DANIEL HENRY	Brompton	1847-88

BRATTLE, THOMAS	Brenchley	late eighteenth century

A 30-hour movement with brass dial and minute work in an oak case has been recorded. Nothing else known.

BRAUND, WILLIAM	Dartford	1823-55

A watch and a longcase clock (Fig 7/21) are recorded by this maker.

BRECKLEY, JOHN	Canterbury	1784

Watchmaker.

BRENERTON or BRENSTON, S.	Northfleet	1847-74

BRICE, JOHN	Wingham	married 1708/9

The Canterbury marriage licences record John Brice giving notice of his marriage to Ann Cullen on 24 February 1708/9, and they are both described as living in Wingham. There are references to three other members of the family in the cathedral records and in each case they are described as blacksmiths. On 14 January 1611, a John Brice of Staple was licenced to marry Margaret Philpot of Preston. Subsequently, another John, described as a blacksmith and widower, of Ash, gave notice of his intention to remarry on 4 June 1642, with a Henry Champney, locksmith of Canterbury acting as bondsman. Whether these two entries refer to the same person is not certain, but there is another record of a Simon Bryce of Faversham, a blacksmith also, marrying Margaret Stonehouse of Chillenden in 1641.

It is not definitely known whether any of the above mentioned were actually employed in making clocks, although the two trades tended to overlap in the seventeenth century. However, there is an interesting reference in the Ash churchwardens accounts to 'Symon Brice'. In 1652 he was paid 7s 6d for 'ironworks about the bell and clock'. Whether Simon Brice was actually involved in making the church clock is not known, but there are references to a clock in the church as early as 1633. John Brice is also mentioned in the Ash records in 1672 as carrying out 'iron work about the clock'. The Ash clock was replaced by another in 1708, and there is a possiblity that John, who is believed to be Simon's son, made the replacement mechanism. Wingham Church also has an early birdcage movement with end-to-end trains supposedly dating from the mid-seventeenth century. If this date is correct it may well be the work of Simon Brice, although there is no reference to it in the churchwardens accounts. Unfortunately there is a long gap in the accounts from the 1660s until 1719. From 1719 until 1728 there are various entries referring to regular accounts from John Brice but it does not stipulate exactly what tasks were carried out, other than one reference to repairs to the fourth-bell clapper in 1728.

In 1720 there is a note of 10s 6d being paid for the 'cleaning and mending ye clock', but it gives no indication to whom the money was paid. Interestingly enough, when the clock did need work carried out the accounts refer to John Gardner being paid in 1726, the sum of £2 15s 0d for 'altering, repairing and cleaning clock in ye church'. This almost certainly refers to Gardner of Sandwich.

BRICE, JOHN	Sandwich	apprenticed 1723, died 1755

Clockmaker. Although a record of his birth has not been traced, most probably he is the son of John Brice of Wingham, but it is not certain with whom he served his apprenticeship. He

Fig 7/22 An 8-day longcase brass dial with a five-pillar movement, by John Brice of Sandwich, about 1755

received his Freedom by Redemption on payment of £10 in December 1740 according to the Sandwich Corporation records. A longcase clock is known (Fig 7/22)

The register of St Mary's Sandwich records the birth of William in 1741 to John and Mary Brice, and John (Jnr) on 15 September 1745. On 6 February 1752 he took his son William, then aged 11, apprentice for a term of 7 years.

BRICE, SYMON   Wingham   1641-52 and later
See entry for John Brice of Wingham.

BRICE, WILLIAM   Sandwich   apprenticed 1752-died 1803
The son of John and Mary Brice. Born in 1741 and apprenticed to his father on 6 February 1752. Both clocks and watches are recorded by this maker. A number of longcase clocks by him are recorded and at least two survive, one an 8-day brass and silver dial with an added arch in a pagoda topped case, which was probably originally decorated with lacquer. The other is a five-pillar 8-day 12in movement of good quality in a simple but elegant oak case.

BRINKLEY, WILLIAM   Canterbury   apprenticed 1756-1768
Watchmaker. William Brinkley's father (also William) was a blacksmith and apprenticed his son on 5 June 1756 to Thomas Carrington of London. He was subsequently turned over to Thomas Gray, a clothworker, and after 8 years gained his Freedom on March 3 1766 on payment of £4 to the Drapers' Company. He moved to Canterbury in 1768

BRISTOW, RICHARD   Maidstone   1874
Noted in *Vivish's Directory* 1872 as trading at 11 Sandling Road.

BROCK, — Clockmaker.	Lewisham	c1830

BROCK, — Clockmaker.	Woolwich	nineteenth century

BROCK, JOHN	Lewisham	1839

These references are taken from Loomes and may possibly be the same man. John Brock is known to have made chronometers, and is probably the same John recorded at a London address in 1844-51.

BRODIE, HUGH	Canterbury	1781

Watchmaker. On 14 June 1781 he obtained a licence to marry Mary Robinson and is noted as a watchmaker, aged 23, of St Mary Bredin.

BROMFIELD, — Clocksmith.	Chatham	1707

BROMLEY, E. H.	Bexleyheath	1851-5
BROMLEY, HENRY	Bexleyheath	1847-74
BROMLEY, T.	Woolwich	1851-5
BROMLEY, THOMAS P.	Erith	1866-74
BROMLEY, WILLIAM	Bexleyheath	1832-9
BROOKE, THOMAS	Maidstone	1604

He is mentioned in the marriage licence of William Brooke, yeoman, dated 11 August 1604, as a bondsman and noted as a clocksmith. Nothing else is known.

BROOKS, ELI	Sevenoaks	1847-55

BROOKSTED, JOHN	London	born 1657-c1678
	Tonbridge	c1678-c1695

Born about 1657, he was apprenticed in London to John White until 1678, but did not apply for his Freedom. He is believed to have moved to Tonbridge sometime after 1678. A hooded wall clock with a 30-hour birdcage movement having a very fine chapter ring and cherub spandrels, signed on a cartouche on the dial centre has been inspected. This clock dates from the 1690 period. Nothing else is known.

BROWN, AARON Watchmaker.	Erith	1769-76

BROWN, —	Sevenoaks	1793

Watchmaker. Noted in Loomes as being situated 'next door to John Osbourn'. An 8-day movement with moon-phase work in a black and gold lacquered pagoda-topped case, signed on a cartouche has been recorded.

BROWN, THOMAS   Maidstone   1840
In *Pigot's Directory* of 1840 his place of business is noted as being in St Faiths Street.

BRUSHFIELD, GEORGE   Ashford   1814
Watchmaker.

BUCKLAND, WILLIAM FRIEND   Maidstone   1830-9
Clockmaker. He is listed as a clockmaker and whitesmith in the 1830 *Topograph of Maidstone*, trading from Middle Row.

BUCKLEY, GEORGE   Canterbury   1823

BUCKLEY, JOHN   Canterbury   Free 1782-1805
Clock and watchmaker. His father, Thomas, was a haberdasher and he received his Freedom by Patrimony in 1782. He is also recorded in 1784 as marrying Sarah Court, when he was aged 24. The window tax records of 1788 note him paying tax on a house in the Parish of St Andrew with seventeen windows (a substantial property).

On 27-31 July 1782 he advertised in the *Kentish Gazette* that he had 'taken the late Shop of Mr. Thomas Shindler, Behind the Butter-market: Where the Haberdashery and Millinery Business is carried on, as usual' and solicited a continuance of his watch and clockmaking business at his shop opposite Butchery-lane, Burgate-street. In August 1785 he moved to 'a more commodious Shop, late Mrs. PARKER's, at the corner of Mercery Lane, on the Parade' and he advertised a wide range of goods for sale at this shop in October 1787, April 1788 and October 1791 (see Appendix I).

He must have been a fairly prolific maker as the 1-5 May 1784 issue of the *Kentish Gazette* reported the loss of watch No 102, and No 186 was reported lost in Jan 1786. Both Canterbury and Maidstone Museums have watches by this maker in their collections. Baillie records him as retiring in 1805. A 10in silver dial wall clock with tapered plates and a fine 8-day five-pillar arch-dial with strike/silent work, housed in a mahogany case have also been recorded (Fig 2/9).

BUCKLEY, THOMAS   Canterbury   1784
Watchmaker

BULL, —   Rochester   pre-1753
A watch has been recorded.

BULL, JOHN   Ashford   1845-55

BURCH, ROBERT   Maidstone   1848-52
In *Bagshaw's Directory* of 1848 he is noted as working at 20 Stone Street but he then moved to Bank Street which is the working address in *Kelly's Directory* of 1852.

BURCH, WILLIAM   Maidstone   1795-1813
Watchmaker. A number of watches have been recorded by this maker and his son William Jnr and seven of them are in Maidstone Museum (Figs 7/23, 7/24 and 7/27).

BURCH, WILLIAM Jnr   Maidstone   1815-66
Watchmaker. Baillie records William Jnr being apprenticed in London in 1813 whereas Loomes

Figs 7/23 and 7/24 Two verge watch movements signed William Burch, Maidstone. The one on the left is number 497

records an apprenticeship date of 1815. No other information is known other than the existence of a silver verge watch dated 1823 and a reference to him in the *Topograph of Maistone* of 1839 as a watchmaker. In both *Pigot's Directory* of 1832 and *Bagshaw's Directory* of 1848 his address is given as 69 Bank Street.

BURCH, WILLIAM	Tenterden	1823

Probably the same man as the preceding entry.

BUREGAR, S.	Greenwich	1851
BURTON, —	Eastry	c1820

Clockmaker.

BURTON, FREDERICK	Lewisham	1851
BURTON, HENRY	Penge	1866
BURTON, S. & SON	Lewisham	1851
BURTON, JOHN	Cranbrook	1795

Watchmaker.

BUTLER, W.	Chatham	1851-5
BUTTERLEY, STEPHEN	Dartford	pre-1728-died 1759

The earliest record of this maker is a note in the churchwardens accounts of Holy Trinity that in 1728 he was paid £2s 10s 'for ye clock as pr Bill'. There are various entries relating to him in the accounts, eg in 1746 his bill 'for mending the church clock' was £1. 4s. 6d.

Dartford Museum has a 30-hour movement with a single hand, but with calendar work in a good oak case by this maker in their collection. The mask spandrels and style of case are typical of the 1720-30 period. A fine quality 8-day arch-dial movement in a good lacquer case with a caddy top is also known. The date of 1759, taken from S. K. Keys's *Further Historical Notes of Dartford*, is believed to be the year of his death.

BUTTERLEY, WILLIAM Dartford 1791
Clockmaker. Presumably the son of Stephen, but nothing else is known about him.

CADWELL, CHARLES Maidstone apprenticed 1800
The only record of this maker is that he was apprenticed in 1800 to William Walmesley of Maidstone.

CACKETT, THOMAS Cranbrook 1748, retired 1774
On 7-10 December 1748 the *Kentish Post* reported: 'We hear from Cranbrook that last Saturday night about seven o'clock Mr Thomas Cackett a watchmaker of that place was, near his own house, knocked down, and robb'd of a Guinea and two watches, by persons unknown; and that he lies dangerously ill of the bruises he recieved.' The *Goudhurst Coronation Yearbook* (1937) records much local information and mentions a longcase clock by Cackett which was owned by a Mr Spice of Bell Farm, Goudhurst. The photograph accompanying the article shows a typical 1740s period 30-hour single-handed movement in an oak case with a caddy top, very similar in style to those made by the Thatchers of Cranbrook at the same period. A lantern clock similar in style to the examples by Thatcher and Kingsnorth is also known, and watches have been recorded. Baillie lists a maker of this name at Canterbury in 1748, but this is almost certainly from the newspaper report above, and he may never worked in Canterbury.

CAIRNS, JAMES Deal 1866-74

CAMFIELD, THOMAS Tunbridge Wells 1823-47

CAN..., J. Maidstone early nineteenth century
Watch and clockmaker.
In the British Museum is a torn watchpaper with an address in Week Street, Maidstone.

CANEY, WILLIAM Tunbridge Wells 1839

CANNON, JOHN Greenwich 1874

CANNON, WILLIAM Canterbury 1676
The entry in Loomes is the only record; he does not appear in the Canterbury Apprentice List.

CARPENTER, JAMES Dover 1866-74

CARTER, JONATHAN Faversham 1874

CARTER, JONATHAN Herne Bay 1874
Probably the same man as Carter of Faversham.

CARYER, JESSE Sittingbourne 1874

CASPER, LEWIS Chatham 1832-51

CASS, JOHN	Woolwich	1823
CATCHPOOL, WILLIAM Watchmaker.	Dover	Free 1801
CATTANIO, JOSEPH Barometers have been recorded.	Folkestone	1858-74
CAVE, JOHN R.	Deal	1847-51
CAVE, RICHARD Watchmaker.	Dover	Free 1826-51
CAVE, WILLIAM	Eastry	1832-51
CAVE, WILLIAM JOHN & SON	Dover	1855-89
CAVE, WILLIAM RICHARD	Deal	1874
CAVE, W. J.	Dover	1855

CHALKLIN [CHALKLEN], JOHN   Canterbury   married 1757-died 1766

Watchmaker. Baillie records him in 1761 and as 'an ingenious artificer'. The Canterbury records show that he married Margaret Ward at St Pauls Canterbury on 27 June 1757; his bondsmen were Philip Abbott, gentleman, and Valentine Picard, a perukemaker. He is believed to have come from London, obtaining the Freedom of Canterbury by Redemption on 19 July 1757. On 11 January 1759 he took Richard Cramp as an apprentice who gained his Freedom on 30 Sept 1766, the year of Chalklin's death.

In October 1761 he announced that he would be attending fortnightly on Mondays at the shop of Richard Elgar, cabinetmaker, at Folkestone to do business there, while in September 1762 he also traded at Bridge, no doubt to counter competition from William Nash of Bridge who announced that he was going to trade in Canterbury every Saturday (see Appendix I).

A silver watch, No 17 signed 'John Chalklin' was reported lost in August 1766 and watch No 68 in December 1761. The *Kentish Post* of 21-25 September 1765 announced:

> The watches to shew the three-hundreth part of a minute, were not invented by Mr. Arnold, but by Mr. Chalklin of Canterbury, who hath compleated two of that sort; the first in the year 1754, and the second in the year 1763, which Mr. Chalklin shew'd to Mr. Arnold, soon after it was finished.

This item was the first general news item after the advertisements, which usually implies that it was 'London' news. It suggests that, in common with other watchmakers of the period, notably, John Arnold and Thomas Mudge, that Chalklin was endeavouring to improve the accuracy of watches. It would appear that Arnold had claimed to be the first to make a watch with the precision mentioned and had either borrowed, or stolen the idea from Chalklin without giving him due credit. This intriguing reference is worthy of further investigation. Unfortunately, the watches made by Chalklin do not seem to have survived.

CHALKLIN [CHALKLEN], Mrs Margaret   Canterbury   1766-died 1774

After John Chalklin's death in 1766 his widow carried on the business. Their son John was apprenticed in London in 1774, but there is no record of him gaining his Freedom or returning to Canterbury. Margaret Chalklin obviously prospered in Canterbury and ran a successful business as is evidenced by the auction notices placed in the *Kentish Gazette* on 9-12 February 1774 after her death, with detailed references, not only to stock (including a 'Great Variety of exceeding good Gold and Silver Repeaters; likewise Plain and Chase Gold Watches, Horizontal, Seconds, and plain Silver and metal Watches; several very good Eight Day, Table, Spring, and Thirty-Hour Clocks') and household goods, but also to extensive properties in Canterbury (see Appendix I). Although the advertisement alludes to Mrs. Chalklin being a clock and watchmaker, there is no proof that she had trained and worked herself. It is more likely that the day-to-day running of the business was carried on by Richard Cramp, who had been apprenticed to her husband. The property, advertised for sale in the *Kentish Gazette* on 23-26 February 1774 included: 'A House, in the Occupation of Mr. Mark Giles, Butcher. The House adjoining, late Mrs. Chalklen's, deceased. [and] The Crown Alehouse' all in the High Street, Canterbury.

CHALKLIN [CHALKLEN], WILLIAM   Canterbury   Free 1763-87
Known to be the brother of John, he obtained his Freedom by Redemption on 8 November 1763 and is noted as a watchmaker. After the death of his sister-in-law an advertisement in the *Kentish Gazette* on 2-5 March 1774 assured the public that 'As he never presumed to [solicit] the Custom of one Individual who dealt with his late Sister, Mrs. CHALKLEN, in Highstreet, during her life Time, and whose Business is now given up; he hopes those Ladies and Gentlemen, who favoured her with their Orders, will now lay their future Commands on her Brother'. In November 1775 he moved from King's Bridge to near the Shakespear's Head in Butchery Lane (see Appendix I).

Two good clocks by him are recorded, one an 8-day silvered dial longcase in a fine green lacquered case (Colour Plate VII), the movement with subsidiary date and seconds indicators, was sold at Sothebys in 1992. The other clock is a very fine, small and rare tavern clock in a teardrop lacquer case of a particularly unusual style (Colour Plate VIII). Only one other similarly cased tavern clock exists, by Thomas Fordham of Braintree (Essex). While the distance between Braintree and Canterbury is considerable it seems likely that the two cases were made by the same man, possibly in London, or to a design supplied from London. The case is decorated with finely preserved lacquer in a typical chinoiserie design, the sides having the usual chrysanthemum sprays. The 8-day timepiece movement has tapered plates and four knopped pillars with a five-wheel train and an anchor escapement. The 18lb oval lead weight is suspended on the original brass pulley. The whole clock is 47in tall.

CHAMBERS, EDWARD   Canterbury from London   born c1656-78
Very little is known of this maker but he is almost certainly the same Chambers as the one apprenticed in London to Evan Jones in October 1670. He is recorded as a watchmaker and gained his Freedom by Patrimony. His father Edmund was a milliner in Canterbury.

CHAPMAN, D.   Hythe   1805
Watchmaker. Baillie records his marriage in 1805.

CHAPMAN, DANIEL   Chatham   late eighteenth century-1845
Baillie gives the date for this man as late eighteenth century, whereas Loomes notes a date of 1826-45. They are possibly one and the same. A watch has been noted and an 8-day arch-dialled brass and silver movement in a pagoda-topped case is also recorded.

Colour Plate VII An 8-day longcase clock by William Chalklen of Canterbury, about 1770, with a single-sheet silvered brass dial in a green lacquer case

Colour Plate VIII A tavern clock by William Chalklen of Canterbury in a lacquer case, about 1770. The 8-day timepiece movement has tapered plates, four knopped pillars and a five-wheel train

CHAPMAN, EDWIN	Chatham	**1832-55**
CHAPMAN, THOMAS	Gravesend	**1823-32**
CHARLTON, JOHN	Woolwich	**1838-47**
CHEESE, THOMAS	Canterbury	**died 1779**

Baillie records this man as working in Milton, a small hamlet some two miles from Canterbury, and that his widow, Sarah, continued the business after his death in 1779. Nothing else known. A good quality small 30-hour wall clock with alarm work has been noted.

CHEN, EDWARD   Canterbury   **apprenticed 1680**

The apprentice lists show him entering his apprenticeship on 20 March to Richard Greenhill of Canterbury, but no other details are known. He does not appear to have finished his indentures.

CHIDWICK, WILLIAM   Dover   **1832**

CHITTENDEN, ISAAC   Yalding   **1806**

Watchmaker. A watch by him is in Maidstone Museum, and a 30-hour painted dial movement in an oak case has been recorded.

CHITTENDEN, WILLIAM   Cranbrook   **1773-6**

Clock and watchmaker. Baillie records him as working in Cranbrook from 1773, but in March 1776 there was to let at Cranbrook: 'A House and Shop, well situated for Trade, in the Watch and Clock Business, with the Household Furniture and Tools, suitable for every Convenience in that Branch of Business, and now or late in the Occupation of Mr. William Chittenden. For further Particulars enquire of Mr. George Thatcher, Brazier, at Cranbrook' (see Appendix I).

CLARIS, W.   Canterbury   **1789**

Cabinet and clockcase maker at 15 St Dunstan's Street (see Appendix I, 14-17 April 1789). No clock cases have so far been identifed as being made by him.

CLEMENTS, WILLIAM   Faversham   **1751**

An advertisement in the *Kentish Post* in June 1751 notes that William Clement Junior from London, clockmaker, took a shop in Preston Street and employed William Bird also from London as a watch-finisher. A watch reported lost in July 1775 contained a watch paper from 'Wm. CLEMENTS, Watch maker, Faversham' (see Appendix I). Whether there is any connection with the famous seventeenth-century maker of the same name is not known, but it merits investigation. A high quality three-train longcase clock is known (Fig 7/25).

CLUER, OBADIAH   Lewisham   **apprenticed 1682, CC1709**

Britten records Cluer's apprenticeship to Henry Evans in 1682 and notes a longcase clock with a rocking ship in the arch, the ship named *The Royal Ann*.

COCKINGS, RICHARD   Deal   **1874**

COCKLE, THOMAS   Deptford   **1866-74**

COCKLING, WILLIAM   Sandwich   **born 1725, married 1752-55**

Clockmaker. Little is known of this maker other than he was born in Sandwich and baptized

Fig 7/25 A fine three-train longcase clock by William Clement of Faversham, chiming on eight bells, in a mahogany case, about 1765

Fig 7/26 A bracket clock by Richard Collins of Margate, in a fruitwood case that would have originally been ebonised, about 1770s

at St Clements on 2 May 1725. In July 1748 his first wife Ann died and he subsequently married Mary Goldfinch on 16 June 1752. A longcase clock has been recorded.

| COGGER, WILLIAM | Maidstone | 1838 |

In the *Topograph of Maidstone* (1839) his place of business is given as Gabriels Hill.

| CIGILL, H. W. | Chatham | 1855-66 |

| COLEMAN, HENRY S. | Sandwich | born 1810-67 |

Kelly's Directory of 1862 gives an address in Market Street.

| COLEMAN, WILLIAM | Greenwich | 1805-8 |

| COLLINS, RICHARD | Margate | 1780-1827 |

Baillie and Loomes note this man as a watchmaker but a longcase clock and a good bracket clock have appeared on the market in the 1990s. The 8-day square dialled longcase in oak has a very finely engraved silvered centre depicting Margate Harbour and a full-rigged ship.

The bracket clock (Fig 7/26) has particularly fine spandrels and the rear plate has superb scrolling foliage decoration typical of the 1770s period. The case of fruitwood would have orignally been ebonized. Advertisements in the *Kentish Gazette* on February 1780 reporting watch No 590 lost and 23 April 1783 reporting a stolen watch signed 'Collins Margate No. 810' suggests that he was a prolific watchmaker if the numbering can be believed.

| COLYER, ALLEN | Dover | 1845-47 |

| COLYER, ALLEN | Ramsgate | 1874 |

Presumably the same man as previous entry.

| COOPER, JOHN | Sevenoaks | 1845 |

| COPSEY, SAMUEL | Gravesend | 1839 |

| CORNELL, ALFRED | Tonbridge | 1845-96 |

| CORNELL, GEORGE | Maidstone | 1874 |

In *Vivish's Directory* of 1872 their place of business was noted as 74 Bank Street. It is interesting to note that this firm is still in existence today as jewellers in Maidstone and Canterbury.

| CORNISH, RICHARD | Westerham | early eighteenth century |

Baillie notes a longcase clock by this man. Nothing else known.

| CORNISH, ROBERT | Dartford | 1721-died 1762 |

Baillie records this maker as working prior to 1777 and notes that he was a watchmaker. However, research has shown that he was active at a much earlier date, and there is also a possibility that there were two makers of the same name, father and son, in the eighteenth century. A local historian has suggested that there may be a family connection with three other members of the Cornish family who were working in London in the mid to late seventeenth century, notably Michael Cornish who was a Freeman of the Clockmakers' Company in 1661-93. As yet, this has not been confirmed and it must remain only a possibility. A family connection with Robert Cornish of East Grinstead, who is recorded as working on the church clock at Battle, Sussex in 1686-7, is also possible. He is noted in the church records as receiving £1 10s 0d for work carried out to the clock, in all probability supplying the additional wheelwork on the external dial. A theory has been propounded that he was the father of

Robert Cornish of Dartford and this possible connection is worthy of further investigation.

The earliest reference to the Cornish family in Dartford is in the parish registers of Holy Trinity which note that Elizabeth, daughter of Robert and Mary Cornish, was baptized on 25 August 1721. After the death of Mary in the following December, Robert married again and his third child and only son, Robert, was baptized in October 1733. His second wife, Anne, died in December 1736 and his third wife, Barbara, who had no children, died in 1741. Robert Cornish lived until 1762 and his burial is recorded on 3 November.

His son Robert married on September 30 1759 to Mary Bird of Dartford, and their eldest son, also Robert, was baptized in August 1763. Mary Cornish died in February 1768 and on 10 June 1771 Robert married his second wife, Sarah Arnold. She died in November 1788 followed by her husband in October 1797.

No entry appears in the marriage registers for the grandson but, on January 29 1795, William, son of Robert and Lydia Cornish was baptized at Holy Trinity. The last record is in February 1802, when their daughter Jane was baptized.

An early 30-hour birdcage clock with crown-and-cherub spandrels, fitted with a hoop and spurs for wall mounting, which has been recorded dating from the 1720 period, is obviously the work of the elder Robert Cornish. As far is known. only two other clocks signed Robert Cornish have survived and the style of the 8-day example with arch-dial and engraved silver centre typical of the 1760-70 period leads one to believe that it may be the work of his son. This particular clock at one time was in the possession of the Bull Hotel in Dartford. Unfortunately there is no evidence to suggest that Robert Jnr was trained as a clockmaker.

CORRINGHAM, R.	Canterbury	1858-66
CORRINGHAM, RICHARD	Strood	1874

Possibly the same man.

COTTAM, JOSEPH	Maidstone	1694-9

He obtained the Freedom of Maidstone in 1694 and is recorded as living in the High Street in 1699. In 1697 he took John Hughes as an apprentice and William Gilbert in 1698. A longcase clock signed 'Joseph Cottam Maidstone fecit' is known.

COUCH, GEORGE	Gravesend	1847
COUTTS, JOHN FRANCIS	Woolwich	1874
COVENEY, ZEBULON	Dover	apprenticed 1815
CRAMP, RICHARD	Canterbury	apprenticed 1759-99

Clockmaker. He entered his apprenticeship with John Chalklin of Canterbury on 11 January 1759 and gained his Freedom on 30 September 1766, the year of Chalklin's death. After gaining his Freedom he worked as a journeyman in London as is made plain by the following advertisement which he placed in the *Kentish Gazette* on 17 August 1768, after the opening of a shop in Burgate.

RICHARD CRAMP,
WATCH and CLOCKMAKER,
In Burgate Street, CANTERBURY,
MAKES and repairs Repeating, Horizontal and Common CLOCKS and WATCHES. He hopes

Fig 7/27 Verge watch movements by (left to right): William Cranbrook, Deal (balance cock missing); William Flint, Ashford, No 10887; William Burch, Maidstone, No 339

that the Knowledge of the business which he acquired from his late Master Mr. JOHN CHALKLEN, and the Improvements which he has since made in the Service of an eminent workman in London, will recommend him to the Favor of the Public, which will ever be gratefully acknowledged by their very humble Servant,   RICHARD CRAMP.

By January 1775 he was at a shop on the Parade, as he announced his move from there to the Golden Cup, opposite the Red lion in the High Street, but within sixteen months he was bankrupt and his stock, including 'a very handsome Eight-day Chiming Clock (by Mr. Chalklen) in a Mahogany Case, inlaid with Brass and Brass Ornaments', was sold off (see Appendix I, May 1776). However, he still appears in the city records as late as 1799 taking on apprentices (see Dowsett and Mummery).

CRAMPTON, EDWARD   Canterbury   1796
Recorded as a watchmaker from Kennington Cross, London. He obtained the Freedom of Canterbury on 24 May 1796 by his marriage to Elizabeth Bunyar. Nothing else is known.

CRANBROOK, STEPHEN   Dover   1791-5
Watchmaker. A good 8-day brass and silver archdial movement with a finely engraved centre with the unusual feature of a three-quarter-size pendulum was sold at Olympia in 1992. This clock was in a good quality mahogany case typical of Goulden of Canterbury and almost certainly came from his workshop.

CRANBROOK, WILLIAM   Deal   1823-8
Probably the son of Stephen Cranbrook. Watch known (Fig 7/27).

CRAYTHORN, JOHN   Maidstone   pre-1747
A watch and clock by this maker have been recorded.

CRITTENDEN, JOHN CHAPMAN   Sittingbourne   pre-1796
A watch has been recorded.

CRITTENDEN, JOHN   Chatham   1838-47

Fig 7/28 The dial of the turret clock made by Francis Crow in 1800 for the Guildhall, Faversham

Fig 7/29 An 11in square painted dial by Francis Crow of Faversham, about 1790

CRITTENDEN, GEORGE	Chatham	1838-47
CROSBEY, WILLIAM	Dover	pre-1778

A watch has been recorded by this man.

CROUCH, DANIEL	Dover	1847-55
CROW, EDWARD	Faversham	c1800-45

Watch in Maidstone Museum. The son of Francis Crow, also recorded in directories in 1823-45.

CROW, FRANCIS　　　　　　　　Faversham　　　　　apprenticed 1770-1803

Clock and watchmaker. Baillie gives his dates as 1780-95, however he is believed to have been working as late as 1803. Loomes notes a Crow working in 1802 and this probably refers to Francis. He was apprenticed in 1770 to another Faversham maker, John Vidion. In July 1778 he opened a shop near the Ship Inn in West Street and in October 1780 advertised: 'that he makes for Sale TIME-KEEPERS on the most simple and best-approved Principles; they go eight Days — shew Seconds, Minutes, and Hours. — They have a compound Pendulum on a new Construction, whereby the Inequalities of Expansion and Contraction are greatly removed.' (See Appendix I.) He installed the clock in Faversham Guildhall in 1800 (Fig 7/28).

A number of longcase clocks are known, all with painted dials (Fig 7/29). Those inspected do not have a false-plate between the dial and movement. This is quite common on Kentish clocks of the early painted dial period. Canterbury Museum has a wooden dialled wall clock in a 'Norwich'-type case in their collection.

CROW, FRANCIS — Margate — 1800-23
Probably the same maker as previous entry.

CRUNDWELL, BENJAMIN — Edenbridge — 1874

CRUNDWELL, JOHN — Edenbridge — 1829-55

CRUNDWELL, JOSEPH — Hadlow — 1855-74

CRUNDWELL, S. — Yalding — 1866

CRUNDWELL, SAMUEL — Tunbridge Wells — 1847-74

CRUNDWELL, STEPHEN — Tonbridge — 1823-66

CRUNDWELL, WILLIAM — Edenbridge — 1838

CUMMING, JOSEPH — Sandwich — 1858-62
Watchmaker. Traded at an address in New Street, *Kelly's Directory* 1862.

CURRIE, ARCHIBALD JOHN — Maidstone — 1826-40
Watchmaker. *Pigot's Directory* of 1832 notes his place of business as 96 Week Sreet. Two 8-day painted-dial movements housed in good quality mahogany cases have been recorded.

CUTBUSH, CHARLES — Maidstone — apprenticed 1734-59
Clocksmith. See Chapter 4 for details of the Cutbush family.

CUTBUSH, EDWARD — Maidstone — 1652
Locksmith and clockmaker.

CUTBUSH, EDWARD Jnr — Maidstone — 1699-1702
Locksmith and clockmaker. (Possibly the same man as above.) Clocks known (Figs 4/1-2).

CUTBUSH, GEORGE — Maidstone — apprenticed 1724

CUTBUSH, JOHN I — Maidstone — married 1680
Clockmaker.

CUTBUSH, JOHN II — Maidstone — 1722-34?
Clockmaker.

CUTBUSH, JOHN III — Maidstone — apprenticed 1765

CUTBUSH, JOHN IV — Maidstone — apprenticed 1772
(Possibly the same man, but ending his apprenticeship in 1772, rather than starting it then.)

CUTBUSH, RICHARD	Maidstone	apprenticed 1761-post 1779
CUTBUSH, ROBERT I Locksmith.	Maidstone	apprenticed 1652
CUTBUSH, ROBERT II	Maidstone	apprenticed 1702-72

Clockmaker. Two longcase clocks known: about 1720-30 (Fig 4/4) and about 1760 (Fig 4/3).

CUTBUSH, THOMAS I Locksmith.	Maidstone	died 1678
CUTBUSH, THOMAS II Locksmith.	Maidstone	1702
CUTBUSH, WILLIAM Clock and watchmaker.	Maidstone	apprenticed 1727-91
DALE, W. F.	Lee	1866-74
DALGETY, ALEXANDER	Deptford	1847
DAMPER, WILLIAM	Tunbridge Wells	1838-55
DAMPER, Mrs J. Widow of William.	Tunbridge Wells	1866
DANIEL [DANELL], JAMES	New Romney	1779-died 1788

The above dates are taken from Baillie, but Loomes records a lantern clock by this maker and suggests that he may have been working in the early eighteenth century. In June 1779 he advertised for a journeyman clockmaker 'Who is a good Hand at new Work' (see Appendix I).

DANIEL [DANELL], JAMES	Sutton Valence	eighteenth century

A longcase clock about 1775 has been recorded. Possibly the same as the New Romney man.

DANELL, JAMES See previous entries.	Lenham	eighteenth century
DANN, R. B.	Greenwich	1874
DANN, W.	Maidstone	1826-66
DARLEY, F. H.	Folkestone	1866
DAVIES, JAMES	Blackheath	1874
DAVIS, JOHN	Ashford	1730

An Advertisement in the *Kentish Post* announced that 'JOHN DAVIS, late from Stroud, is Removed to ASHFORD' (see Appendix I, 27-30 May 1730). As no maker of this name is recorded for Stroud, Gloucestershire, Strood is probably meant. He was primarily a retailer, selling a wide range of goods, though he does state: 'at the same Place, Watches are carefully Clean'd and Mended, and Jewellers Work performed by Me'.

DAWES, WILLIAM M.	Gravesend	1823-7

DAY, JOHN	Deal	1832-45

Loomes gives these dates for this maker, but it is likely that he was working earlier. An 8-day silvered dial movement in a good mahogany case with an arched top in typical 1790-1800 style has been recorded.

DEACON, ARTHUR WILLIAM SCRIPPS	Sandwich	1862-1909

DEALE, THOMAS	Ashford	pre-1680-died 1687

Clockmaker. See Chapter 5.

DEAN, JAMES	West Malling	1874

DENDY, JOHN	Canterbury	apprenticed 1753, Free 1760

Clockmaker.

DELACOUR, GEORGE	Chatham	1823-55

DENNIS, GEORGE	Maidstone	pre-1670-84

Clockmaker. On 26 July 1670, George Dennis, aged 27, described as a clockmaker was granted a licence to marry Ann Joanes of Thurnham.

DENNIS, THOMAS	Canterbury & Maidstone	apprenticed 1672, free 1684-5

Watchmaker.

DENNIS, THOMAS Jnr	Maidstone	apprenticed 1698

Little is known of these three makers, who are believed to be members of the same family. Thomas Dennis was apprenticed in London to James Wightman in 1672 and after serving his time then moved to Canterbury and gained the Freedom of the city by Redemption in 1684, although there is no record of him having worked in Canterbury. It appears that he moved almost immediately to Maidstone where he also appears on the list of Freemen in 1684. George Dennis also appears on the Freemen list for that year. According to the records they both seem to have been involved in selling a property in Week Street, Maidstone to Thomas Bliss in 1685. At that date Thomas Dennis is described as a watchmaker. George and Thomas are believed to be the sons of George Dennis, a locksmith, who when aged 26 was granted a licence to marry Elizabeth Greenhill aged 19, the daughter of John and Elizabeth Greenhill on 20 February 1632/3. The records show that John was dead at this date.

The Maidstone records note Thomas Dennis Jnr, son of George Dennis, starting his apprenticeship to David Webb, a blacksmith, on 10 May 1698. Unfortunately no watch or clock by the Dennis family appears to have survived and nothing else is known.

DIXON, WILLIAM	Minster	c1770

An arched silvered dial with the name engraved on a cartouche is known but no other information has been recorded.

DOBELL, JESSE (& E.)	Canterbury	1851-74

The 'E' Dobell referred to in Loomes entry is presumably Ebenezer who was subsequently in business on his own in Hastings (Sussex) in 1855-78.

DOBSON, —   Maidstone   early nineteenth century

A longcase clock with a painted dial has been recorded and an enamelled-dial wall clock has been noted in the Bull Inn at Barming.

DODD, JOHN   Faversham   pre-1693-1700

Clocksmith. John Dodd was a member of a well-known family of gunsmiths who were descended from yeoman farming stock in the village of Stockbury, on the edge of the North Downs between Maidstone and Sittingbourne. The Greenhill family, to whom they were related by marriage, also came from this village.

In 1635 William Dodd married Elizabeth Greenhill, the daughter of Richard Greenhill of Stockbury. In 1668, William's son John, a gunsmith aged 24, obtaining a licence to marry Ann Nower of Ashford on 30 March, with Solomon Woodfall, a silkier of Canterbury acting as Bondsman. The next entry found in the marriage records is of his son John, a clocksmith, marrying Sarah Audley of Wickhambreux on 3 October 1693 when he was aged 22.

It is not certain where John served his apprenticeship, there is no trace of him on any apprentice list, but it seems reasonable to suppose that he learnt his trade under the tutelage of his relative Richard Greenhill of Ashford. Of course it may be that he was trained by his father. During the seventeenth century the gunsmith and clockmaking trades did overlap, as in the case of the Barrett family in Canterbury. There is the added complication that Richard Greenhill died in 1687 and as it is unlikely that John Dodd was apprenticed before he was fourteen years of age, ie in 1685, he would only have served two years before his master's death. If he was apprenticed to Richard he must have finished his training under another clockmaker, either John Greenhill, the eldest son who also worked in Ashford, or possibly the younger son Richard, who was working in Canterbury in 1676. However he does not appear among the names of Richard's apprentices.

As far as is known only one clock made by him has survived and that is a fine large lantern clock illustrated in Sir George White's book *The English Lantern Clock*. Although this clock appears to date from the 1690s, judging by the style of dial and Dodd's known working dates, it does bear a striking resemblance in all other respects to the two superb large lantern clocks made by Richard Greenhill of Ashford, also illustrated in White's book. The Dodd example is similarly fitted with jacks dressed in Jacobean costume striking the quarters. These three clocks are unique and as far as is known no others exist of this type. In addition, the dial engraving is almost the same as other Greenhill clocks. Is it possible that this clock was a product of Richard Greenhill and finished off by Dodd after the former's death? Sadly, we shall never know for certain. John Dodd's clock is now on display in the Gotesborgs Historiska Museum in Sweden. (See Figs 6/1-2, page 68.)

DOLD, ALEXANDER   Maidstone   1847-51

Bagshaw's Directory of 1848 notes an address in Astley Street. Both wooden dial and enamel dial clocks have been recorded.

DOLD, P.   Maidstone   1855

Presumably related to Alexander Dold.

DORER, L. & P.   Greenwich   1866

DORER, P.   Greenwich   1874

DORRER, EAGEN & CO	Dover	1845
DOVE, E. C.	Rochester	1874
DOW, ALEXANDER	Bromley	1845
DOWN, JOHN G.	Wrotham	1847-55
DOWSETT, CHARLES	Margate	married 1799-26

Watchmaker. Baillie notes this maker as marrying in 1799 and being in the Clockmakers' Company. The Canterbury lists note his apprenticeship to Richard Cramp, watchmaker, on 26 July 1792. He presumably moved to Margate soon after completing his indentures. A bracket clock has been recorded. In January 1808 he advertised 'Improved Mathematical Spectacles' (see Appendix I).

DRAYSON, DOUGLAS	Sandwich	1862-91
DUNKLEY, JOHN	Greenwich	1847
DUNN, SAMUEL EDWARD	Elham	c1820, died 1874

A wooden-dialled 'Norwich'-type wall clock by this maker is known (Fig 7/30). After his death in 1874 the business was carried on by his widow, Mrs Esther Dunn.

DUNSTALL, JAMES	Gravesend	1874
EALAND, B. C.	Dartford	early nineteenth century

A watch has been recorded by this maker.

EASTES, WILLIAM	Sandgate	1845
EDGECUMBE, EDWIN	Dartford	1866-74
EKINS, F. G.	Greenwich	pre-1855

He is almost certainly the same man who was apprenticed on 26 July 1792 to Thomas Parker of Canterbury.

ELDRIDGE, A.	Tunbridge Wells	1866
ELLIOTT, ALFRED	Greenwich	1847-55
ELLIOTT, EDWARD	Lenham	1817-47

Watchmaker. Numerous watches and longcase clocks with painted dials have been recorded by this maker.

ELLIOTT, GEORGE	Ashford	1866
ELLIOTT, JAMES	Greenwich	1720

Watchmaker.

ELLIOTT, JOHN (& SON)	Ashford	1802-51

The *Kentish Gazette* January 1803 records him in business as a watchmaker, silversmith, sta-

Fig 7/30 Mahogany cased 'Norwich'-style wall clock by S. E. Dunn of Elham, about 1820

Fig 7/32 right) Early eighteenth-century long-case clock by Thomas Elliott of Greenwich, in a black and gold lacquer case, about 1715 (see also Colour Plate I)

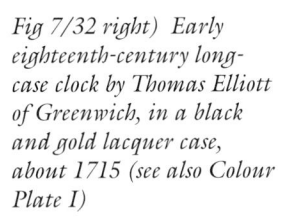

Fig 7/31 Verge watch No 8009 by Elliott of Ashford

tioner and bookseller trading from premises opposite the White Hart Inn, High Street, Ashford. Verge watch known (Fig 7/31).

ELLIOTT, THOMAS — Greenwich — 1720

Watchmaker. The entries relating to Thomas and James Elliott are taken from Loomes, but little else is known of these two early eighteenth-century Greenwich makers, other than the record that James insured his 'goods and merchandize in his dwelling in Church Street, Greenwich' on 23 March 1720 with the Sun Fire Insurance Office. It is assumed that they were related but it is not known where and when they were apprenticed. The London Apprentice Lists note a Thomas Elliott being apprenticed on 24 May 1721 to Valentine Jones for a term of seven years, but it is not certain that this is the same man. Of the two longcase clocks known by Thomas Elliott, one example in a black and gold lacquer case seems to be much earlier than 1730 and is dated nearer to 1715-20. The five-pillar movement is very well made with good engraving to the 11½in dial, which is an unusual size. While the case has been restored the trunk has its original decoration (Colour Plate I and Fig 7/32).

There is a possibility that they were related to Henry Elliott who is believed to have worked in Rotherhithe in the late seventeenth century. The two locations are very near to each other.

ELLIS, GOODMAN — Rochester — c1599

Known to have supplied a new clock platform and dial for All Saints, Lydd in 1599, 1600 and 1601.

ELLIS, JAMES — Canterbury & London — 1658-died 1680

Ellis was apprenticed in London in July 1658 to John Frowd and subsequently transferred to Benjamin Bell. Whether there was any family connection in Canterbury is not clear, but he obtained Freedom by Redemption on 27 August 1666, more than a year before he gained the Freedom of the London Clockmakers' Company in October 1667. He took three apprentices in London: William Beard, October 1667; Charles Annott, September 1673, and Aucher Gate in July 1674. What is extraordinary is that Beard and Gate appear on the Canterbury lists also in 1666/7 and 5 August 1674 respectively. None of the three applied for their Freedom in London but both Beard and Annott obtained the Freedom of Canterbury, the former in 1674 the later in 1680, by which time Ellis had died.

'Aucher' is likely to be a misspelling in the London records for Arthur, and although he is noted as being apprenticed to Ellis until 1681 he does not appear on either the London or Canterbury list of Freemen, so his later life remains a mystery.

The reasons for the double apprenticeships are also a mystery and one can only assume that Ellis operated from workshops in both London and Canterbury. As far as can be ascertained there is no record of any surviving clocks and watches by Ellis and his three apprentices.

He is also recorded as taking Robert Hayman (Heyman) apprentice in Canterbury on 11 September 1678. In view of Ellis's death in 1680 it must be assumed that he continued his indentures with another maker. He is recorded as being free in the city in 1722.

ELLIS, JAMES — Canterbury — 1764

Watchmaker. This entry appears in Baillie and may possibly refer to a son of the earlier James Ellis, but there is no other mention of him in the city records.

ELLIS, JAMES — Dover — 1731

Watchmaker. Possibly the same as the previous entry. The only reference to him appears in the

Kentish Post 10-13 November 1731, recording his arrest for passing counterfeit French coins (see Appendix I).

ELVIS, BENJAMIN	Wingham	1866-74
EMANUEL, EMANUEL	Chatham	1832
EMANUEL, EMANUEL	Deal	1797-1826
EMANUEL, LEVI	Canterbury	1781

A watch by him reported lost in September 1781, to be returned to him at St Dunstan's, Canterbury (see Appendix).

EMBERSON, JAMES Marden late eighteenth/early nineteenth century

A painted dial clock in an oak case has been recorded and there is a watch by this maker in the Canterbury Museum collection.

ENGEHAM, JOHN Canterbury apprenticed 1771

The apprentice lists note 1771 as the year of his apprenticeship to James Warren but nothing else is known. He may well be the same man as the maker from Yalding.

ENGEHAM, JOHN Yalding Free 1789

Other than this entry in Loomes nothing else is known of this maker. Probably the same as the previous maker.

ERIC, SAMUEL Gravesend 1756

Watchmaker.

ETCHER, WILLIAM	Canterbury	1845
EVANS, JAMES WILLIAM	Lewisham	1874
EVELEIGH, JAMES	Gravesend	1839
EVENS, ROBERT	Halstead	c1720

Baillie gives this date for Robert Evans and suggests that he is the maker of the same name who was apprenticed in London in 1706 to Henry Adeane junior. A lantern clock is illustrated in Britten and a good 8-day movement in a fine mahogany case has been seen. The movement has good quality mask-and-dolphin spandrels with the maker's signature on a boss in the arch, typical of the 1730-40 period. Halstead is a small hamlet between Orpington and Sevenoaks, but no other other reference to this maker has been traced.

EVENS, ROBERT Halstead 1795

Baillie records a date of 1795 and states that he was a watchmaker. Nothing else is known but he is assumed to be the son of the earlier Robert Evens.

FAGG, — Sandwich 1753-c1770

Little is known of this maker, who is believed to have been working until at least the 1770 period. Loomes gives a date of 1753 and records him as a clockmaker but he has not been traced on any apprentice list. At least two large tavern clocks have been recorded by him. One is hanging in the old Court House and Guildhall at Sandwich. The other is reputed to have been in the County Hotel in Canterbury and is possibly the clock that can be seen in the reception area. Unfortunately the dial and case have no signature or original decoration, and as the author has been unable to inspect the movement it cannot be stated with certainty that this particular clock is by Fagg. Mr Mason, a member of the family who formerly owned the hotel, states that at one time there were at least three tavern clocks on the premises, one of which was definitely by Fagg. He is unable to recollect who the others were made by.

FAGG, ALFRED   Margate   1866-74

FAGG, JACOB   Folkestone   1823

FAGG, JACOB   Ramsgate   1826-8
Probably the same man as Fagg of Folkestone.

FAGG, JOHN   Margate   1800-51
Baillie records him as about 1800 and he is almost certainly related to Fagg of Sandwich.

FAIREY, JAMES   Maidstone   1840
Pigot's Directory of 1840 gives his address as 44 Earl Street.

FARLEY, THOMAS   Faversham   1778-1802

FARMER, R. W.   Hollingbourne   nineteenth century
An early nineteenth-century watchpaper, which states that he was a 'Watch & Clock Manufacturer' (Fig 7/103) is in the Maidstone Museum, but nothing else is known of him.

FEHRENBACH, B.   Ramsgate   1845

FEHRENBACH, EMILIAN   Dover   1839-51

FEHRENBACH, F.   Dover   1851

FEHRENBACH, F. & O.   Dover   1855

FEILD [FIELD], JAMES   Canterbury   1690-died 1710/11
This maker does not appear on any Canterbury Apprentice List and the only record known is an inventory taken after his death on 6 February 1710/11, where he is noted as a watchmaker. Probate was granted to his widow Anne on 9 February 1710/11.

> An Inventory of [all] and singular the Goods Chattles and Credits of James Feild late of the Citty of Canterbury Watchmaker Deceased ...

	li	s	d
Impris. the deceaseds Weareing apparell both Linnen and wollen and ready money the Sume of	2	10	0
Item Seaven Watches being very old wanting setting upp and being of little value	3	0	0
Itm. two Dyalls	0	13	6
Itm. the Deceaseds Working Tooles	0	5	0
Itm. Books	0	5	0
Itm. an old clock	0	10	0

Itm. in the Lodgeing Camber one Fether-Bedde Bedsteddle Curtaines & Vallance Matt and Cord one Rugg one blankett two Boulsters 2 Pillows a [sic] old Chest and old Boxes & 5 Chaires Andirons Firepan and Tongs and a Little Earthen Ware on the Mantle one old Lookeinglass a table and Pictures in all   3 0 0

Itm, in the Closett in a Chest four pairs of Sheetes & an old one Earthen Ware and some Fould Linnen and six Glass Bottles in all   0 10 0

In the Lower Roome

Itm. two tables a Kneading troff & Safe   0 3 0
Itm. for Chaires 6 Candlesticks   0 1 0
Itm. one pre. of Bellows one pre. of Andirons Fire shovell and tongs in all   0 1 0
Itm. Gridiron and tryvett & Firepann   0 1 0
Itm. three Skilletts a Warmeingpan and two Iron Potts in all   0 3 0
Itm. Six Small Pewter Dishes   0 5 0
Itm. Six Cupps and Drinkeing Potts   0 0 6
Itm. a Gill and two Spitts   0 7 0
Old Lumber and things forgott   0 1 0

Sume Totall   11 15 0
[should total   11 16 0]

Appraized by us   Tho Corpe [?Cape]
                       Solaman Randalph

In 1690 he was a witness to the will of Thomas Shindler I's mother, and by 1705 he appears to have had a business, and possibly a personal, relationship with Elizabeth Shindler, Thomas I's widow (see Shindler Family). He may be the same man as the James Field, citizen and waxchandler of London (and member of the Waxchandlers' Company) who was never admitted to the Clockmakers' Company, but took an apprentice through them in July 1672.

FENN, CHARLES	Boughton	1874
FENN, CHARLES	Whitstable	1866-74
FENN, G.	Lewisham	1851
FENN, J. & SON	Greenwich	1851-74
FENN, JOHN	Boughton	1874
FERRALL, EDWARD	Cranbrook	1686-1706

In Loomes he is given as working between these dates and is noted as repairing the church clock. Nothing else known.

FIELD, BENJAMIN	Tunbridge Wells	1823-51
FIELD, BENJAMIN Jnr	Tunbridge Wells	1851
FIELD, JAMES, see FEILD, JAMES		
FIELDING, A.	Greenwich	1855
FIELDING, AUGUSTIN	Canterbury	1838-59

Fig 7/33 (left) A 30-hour white dial signed 'William Flint, Ashford', about 1800

Fig 7/34 (Left) A timepiece/alarm by Flint of Charing, about 1770

Fig 7/35 (right) An 8-day mahogany longcase clock by Flint of Charing, about 1775

FIELDING, GEORGE	Greenwich	1839-51
FITCH, GEORGE	Gravesend	1839-45
FLEMING, RICHARD	Sevenoaks	1793
FLINT, WILLIAM Snr	Ashford	born 1757-died 1813

Baillie records him as working at the end of the eighteenth century, but it is now known that he was active at a much earlier date. The Ashford records note that he was born in 1757 and was married to Elizabeth Burch at St Mary's Church Ashford on 8 April 1780. She was, very probably, a member of the Burch family of watchmakers from Tenterden and Maidstone.

A number of watches by him have been recorded in recent years (Fig 7/27, page 126), together with 30-hour and 8-day longcase clocks in oak cases, invariably with painted dials (Fig 7/33). An interesting late tavern clock in a small teardrop case decorated with typical chinoiserie designs dating from about 1790 was for sale in 1996. He died in 1813 and was buried at Charing. This has given rise to much confusion, as Flint of Charing was also buried in the same churchyard in 1793 and many people have assumed that they were one and the same person. Although they may very well have been related, it has not been possible to ascertain the exact connection. They were definitely not father and son or brothers.

FLINT, WILLIAM Jnr   Ashford   born1787-died 1871

William, the son of the previous entry was born in 1787 and died in 1871. Presumably trained by his father, like him he also made watches and 30-hour painted dial clocks. The Maidstone Museum collection has the turret clock from Wittersham Church in store and while this has a setting dial signed by Flint the actual clock was almost certainly made by Moore of Clerkenwell.

FLINT, WILLIAM   Charing   born 1733, died 1793

As noted above this maker is often confused with the man from Ashford. The date of death given in Baillie, 1795, is incorrect. His gravestone in Charing churchyard states that he died in 1793 leaving a son, Thomas, and two daughters. Numerous examples of his work have survived, many of them small hooded wall-clocks, some with alarm work and many without cases.

FOOT, R.   Faversham   c1820-55

A 30-hour painted dial movement with arabic numerals housed in an oak case has been recorded.

FORSTER, G.   Sittingbourne   1802-13

Watch and clockmaker.

FORSTER, JOHN   Sheerness   1790-1823

This man is recorded as a clockmaker by Baillie and may have served his apprenticeship in London. Loomes also notes a Forster, Sheerness, about 1780-90. This is almost certainly the same maker. Both 30-hour and 8-day clocks in oak and mahogany cases have been noted and two bracket clocks with painted dials in arched-topped mahogany cases are also known. Both of these movements have backplates engraved with floral motifs around the edges.

FORSTER, WILLIAM   Gravesend   1845-7

FORSTER, WILLIAM HENRY   Sheerness   1832-8

FORSTER, WILLIAM HENRY   Gravesend   1838-66

These two makers are presumed to be the same.

FOSTER, HENRY   Ashford   1858-74

FOSTER, HENRY   Margate   1874

Colour Plate IX An 8-day striking and repeating bracket clock by Gabriel Fowkes of Dartford, in a mahogany case, about 1770

FOSTER, WILLIAM	Lydd	1845

FOWKES, GABRIEL	Lewisham	early eighteenth century-1759

Baillie records a longcase clock by this maker and the dates given are taken from that source. He also mentions a son, Gabriel, being apprenticed in London in 1759. He is almost certainly the maker of that name in the following entry.

FOWKES, GABRIEL	Dartford	apprenticed 1759-1808

Baillie notes a date of 1791 for this maker, but there is a reference to him in the Dartford records as late as 1808. He is known to be the son of the previous entry and was apprenticed in London in 1759, moving to Dartford after finishing his indentures, presumably about 1766. It is possible that his father Gabriel had also moved to Dartford at an earlier date, as there is a reference to him in there as early as 1750, but this has not been substantiated.

A number of longcase clocks signed by Gabriel Fowkes are known and with one exception they are all housed in pagoda topped cases decorated with lacquer. All of the movements are of good quality with five pillars and fine engraving to the dial centres in typical rococo fashion. A bracket clock dating from the 1780s is also known (Colour Plate IX). He was also the maker of a fine shield-dial tavern clock that hung in the Royal Victoria and Bull Hotel, Dartford, until February 1997. This clock, which had been in this old coaching inn on the main route from London to Dover for over 220 years, was sold by the brewers when they gutted the hotel's interior during its transformation to a yuppies' gin-palace.

Fowkes played an active part in the life of the Dartford, acting as churchwarden of Holy Trinity, from 1798 until 1800, and in 1799 was appointed feoffee of the local Grammar School. He looked after the church clock from 1775 to 1780 and the following entries are taken from the churchwardens accounts:

1775	Jan. 9th	To Mr. Fowkes for looking after the clock ...	18s. 6d
1776	June 28th	To Gabriel Fowkes, as by bill and receipt ...	1. 1. 0
1777	June seventeenth	To Gabl. Fowkes in repairing and oiling the church clock ...	4. 0
1779	Jan eighteenth	To Gabriel Fowkes, as by bill and receipt ...	12. 6
	Sept. 24th	To Gabriel Fowkes, as by bill ...	14. 6

These accounts continue until 1780, but from 1781 until 1806 he was paid an annual salary of £6, paid half yearly, for winding the clock and general maintenance. His name also appears in the accounts between the years of 1786 and 1805 as one of the parish signatories.

The most recent clock by this maker to appear on the market was another lacquered example with a fine five-pillar movement, sold at Phillips, Folkestone in 1993. The dial of this clock was also engraved in rococo style and the high standard of workmanship shown, particularly the finish of the spandrels and the engraving is typical of this maker's clocks.

FOWLE, EDWARD	Westerham	1838-died 1879

FOWLE, HUMPFREY	Westerham	1769

Watchmaker. A 30-hour birdcage movement in an oak case has been recorded.

FOWLE, JOHN	Sevenoaks	1802

An 8-day longcase clock in a mahogany case was sold at Olympia in 1992. The painted dial had fine decoration and was signed John Fowle, Westerham. It is not certain whether he moved the business from Sevenoaks, but it is likely as Edward is believed to be his son.

FOWLE, THOMAS — Canterbury — Free 1609

Clockmaker. Nothing is known of this maker other than the fact that he obtained the Freedom of Canterbury by Redemption (ie purchase) on 14 February 1608/9. There is no record of him having worked in Canterbury and no surviving clock or watch made by him is known.

Another early maker named Edward Fowle, Free of the Clockmakers' Company in London in 1670, may be related and it is thought that the late seventeenth/early eighteenth century East Grinstead maker, Thomas Fowle, is also connected. Other members of the family carried on the craft of clockmaking in the East Grinstead and Westerham area during the eighteenth century and it is likely that they too were descended from this very early maker.

FRANCIS, W. — Forest Hill — 1874

FRANCKE, RICHARD — Canterbury — apprenticed 1700

Noted as entering his apprenticeship to Robert Hayman, clockmaker, on 13 September 1700. Nothing else is known.

FRANKS, E. — Crayford — c1790

Watchmaker.

FRENCH, A. W. — Maidstone — 1812-72

In the *Topograph of Maidstone* published in 1830 the address is noted as 5 King Street as clockmakers. Also appears in *Vivish's Directory* of 1872. *Bagshaw's Directory* of 1848 records William Apps being in partnership with Susanna French from the same address and other records show the business carrying on until 1874.

FRENCH, JOHN — Yalding — 1832-53

A watchpaper is known (Fig 7/103).

FRENCH, RICHARD VIGOR — Maidstone — c1800-52

Baillie gives this date but he also appears in *Pigot's Directory* of 1832 at 77 King Street and at the same premises in 1852 in *Kelly's Directory*.

FRENCH, R. V. & Son — Maidstone — 1852 onwards

FRENCH, STEPHEN — Maidstone — 1847

Bagshaw's Directory of 1848 notes an address in Paradise Row.

FRENCH, SUSANNA & WILLIAM APPS — Maidstone — 1848-74

See A. W. French above.

FUCHTER, FIDEL — Maidstone — 1874

Vivish's Directory of 1872 gives an address at 15 Stone Street.

FULWELL, THOMAS — Deal — 1838-55

GAMBIER, GEORGE — Canterbury — 1847-51

Bagshaw's Directory of 1847 records an address in Stour Street.

GARDNER, JOHN — Lydd — 1826-8

GARDNER, JOHN Sandwich pre-1726-52

Clockmaker. Loomes gives 1747-52 for this maker, but it is believed that he was working earlier in the century. As noted in the entry for John Brice, Gardner is mentioned in the Ash churchwardens accounts as receiving £2 15s 0d for 'altering, repairing and cleaning clock in ye church' in 1726, so his working dates must predate this. No other definite information is known, but he is believed to be the son of Joseph Gardner. A good quality 8-day movement with five pillars, inside locking plate striking, finely-engraved 11in dial and chapter ring, and with crown-and-cherub spandrels typical of the 1700-10 period has been inspected.

GARDNER, JOSEPH Sandwich 1691

Clockmaker. Nothing else known but believed to be the father of John Gardner.

GARDNER, WILLIAM Sandwich 1733-died 1758

Nothing known other than his marriage to Hannah Knapp at St Peters and he is likely to be the son of John Gardner, although, to date, this has not been confirmed. A 30-hour birdcage movement in an oak case has been recorded.

GATE, ARTHUR Canterbury 1674-81

See entry for James Ellis.

GATES, THOMAS Faversham late seventeenth century

A lantern clock is noted by this maker but nothing else is known.

GATWARD, JOSEPH Tonbridge 1802

Loomes gives a date of 1802 for this man but the Tonbridge local records mention a Joseph Gatward coming from London at the turn of the century. Is it possible that he is the same man who is recorded in London in 1790 in Baillie? The only other reference to him is his signature as a witness on a conveyance of a property in Bank Street in 1801. A longcase clock has been recorded.

GATWARD, JOSHUA Sevenoaks 1784

Baillie records this maker as obtaining his Freedom from the Clockmakers' Compnay in 1784 and that he was a watchmaker.

GEE, JOHN Canterbury apprenticed 1680/1

Apprenticed to Richard Greenhill of Canterbury 20 March 1680/1. Nothing else known and he does not appear in the list of Freemen.

GELHALTER, LEON Sevenoaks nineteenth century

Known to have worked from premises on St John's Hill.

GIBBETT, — Hadlow c1750

A longcase clock has been recorded. Nothing else known.

GILBERT, JEFFERY New Romney born 1814-74

GILBERT, THOMAS Hythe early nineteenth century

Bagshaw's Directory of 1848 notes an address at 31 Mill Street and a watch is known signed by him. An 8-day painted dial clock by him has been recorded.

Fig 7/36 The dial of a fine 8-day, five-pillar clock by William Gill of Maidstone, about 1715-20

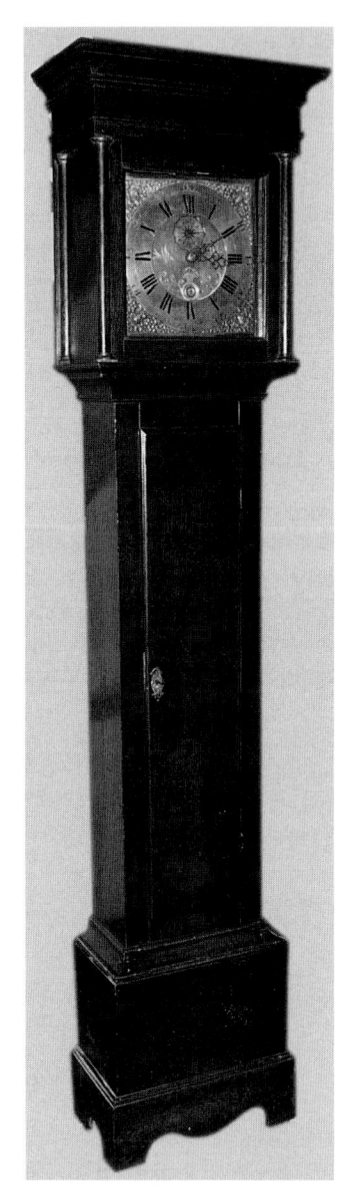

Fig 7/37 Oak longcase clock by John Godden of West Malling, about 1750, in its original condition

Fig 7/38 The 11in square dial of the John Godden clock

| GILES, JOHN SCOTT | Maidstone | 1845-55 |

| GILES, NICHOLAS | Maidstone | 1807-40 |

In the *Topograph of Maidstone* he is noted as a watchmaker with an address at 125 Stone Street. Also appears in *Pigot's Directory* of 1840 at the same address. A watch signed by him is in the Maidstone Museum collection.

| GILL, GEORGE | Maidstone | apprenticed 1737 |

The son of Thomas Gill, a barber, was apprenticed in 1737 to William Gill, who was one of Maidstone's most well-known clockmakers of the eighteenth century. There is no further mention of George and it must be assumed that he did not finish his apprenticeship.

| GILL, WILLIAM | Maidstone | apprenticed 1704-70 |

One of Maidstone's most celebrated and competent clockmakers, entering his apprenticeship to Walter Harris in 1704. He is noted on the Maidstone Freeman lists as obtaining his Freedom in 1710 and supplied the clock to Rainham Church in 1713. He was also responsible for the turret clock in Leeds Church; it has a finely engraved setting dial with engraving very similar to that found on early eighteenth-century lantern clock dials (Figs 1/20-21). A lantern clock of about 1710 with a floral engraved dial centre and signed 'William Gill, Maidstone, Fecit' was sold at Christies, South Kensington in 1997.

A number of fine longcase clocks by this maker are known, all of them exhibiting a high standard of workmanship and one particularly fine example is pictured in Fig 7/36.

In 1727 he took John Hughes as an apprentice and, as has been noted above, George Gill, who may have been his nephew, was apprenticed to him in 1737. The churchwardens accounts for All Saints in Maidstone for the year 1737 record 'William Gill for altering chimes £6 6s 2d'.

| GILLOTT, WILLIAM | Tonbridge | 1845-7 |

| GLADDISH, THOMAS | Yalding | c1720 |

A 30-hour birdcage movement with a 10in dial with two hands, contained in an oak case has been recorded, but nothing else is known.

| GLADDISH, WILLIAM | Yalding | 1788-94 |

These dates are taken from Baillie in which he is noted as a clock and watchmaker. He is possibly the son of Thomas Gladdish, but nothing else is known.

| GOATLEY, DANIEL | Canterbury | 1823-8 |

| GOATLEY, — | Ramsgate | c1790 |

Baillie records this maker as being free of the Clockmakers' Company at this date. He is possibly the same man as the Canterbury maker but nothing else is known at present.

| GODDARD, HENRY | Tenterden | c1730-67 |
| | Dover | after 1767 |

Known to have been working in Tenterden in the middle of the eighteenth century but moved to a shop in Snargate Street, Dover in 1767, when he was described as a watchmaker and silversmith. Goddard's shop in Tenterden was taken over by Owen Jackson of Cranbrook. (See Appendix I, 23 December 1767 and 30 December 1768.)

A good quality 30-hour clock in an oak case is known. The finely engraved brass and silvered dial is signed 'Tenterden'. A silver watch with steel chain and a Turk's Head silver seal, signed 'Henry Goddard, Dover' was reported as lost on Romney Marsh (26 April 1783), while a similarly signed gold watch No 1395 was reported lost in June 1776 (see Appendix).

GODDEN, GEORGE	Tonbridge	1847
GODDEN, HENRY	Town Malling	1823-47
GODDEN, JOHN	Town Malling	1784

The dates for this man and Henry Godden, who are thought to be father and son, are taken from Baillie and Loomes; other than this virtually nothing else is known. At least three silver-dialled 30-hour longcase clocks signed 'Godden, Town Malling' have been recorded. All of these have been in oak cases and the style of engraving on the dials is typical of the late eighteenth century so they are more than likely to be the work of John Godden. A 30-hour clock signed 'John Godden, West Malling' is also known (Figs 7/37-38).

In October 1772 an 'almost new' silver watch with a 'China Face, Makers Name John Godden' was reported stolen (see Appendix I), but as no place was given it is not certain whether it was made by this man or the following one.

GODDEN, JOHN	Wingham	c1750

Clockmaker. Whether there is any connection with the Godden family of West Malling (sometimes known as Town Malling in the seventeenth and eighteenth centuries) is not certain. At least one very fine 8-day arch-dialled movement in a good mahogany pagoda-topped case has been inspected. The all-over silver dial is superbly engraved with floral motifs and swags in a typical rococo fashion with the name signed in the arch, and is very much in the style of the 1770s (Fig 7/39).

GODDEN, THOMAS	Canterbury	1832
GODDEN, THOMAS	Hythe	1784-1824
GOFFE, THOMAS	Maidstone	1709

The only record is a note of his apprenticeship to John Hughes of Maidstone in 1709.

GOODCHILD, RICHARD	Canterbury	1756

Watchmaker. A watch was reported stolen from his shop in St Dunstan's Street in February 1756 (see Appendix).

GOODCHILD, RICHARD	Harbledown	1767

Watchmaker. Amost certainly the same man as the previous entry. Harbledown is a village 1½ miles from Canterbury and he must have been working or living there in August 1767, when a found watch could be claimed from him (see Appendix I). Nothing else is known.

GOODLAD, RICHARD	Canterbury & London	1677-1716

Little is known of Goodlad, who was the first of Richard Greenhill's apprentices. His place or date of birth has not been traced and there is no record of him ever marrying during his time in Canterbury. He may have come from Maidstone as there are references to members of the Goodlad family in the registers of All Saints.

Fig 7/39 The superbly engraved dial of an 8-day, five-pillar clock by John Godden of Wingham, about 1770

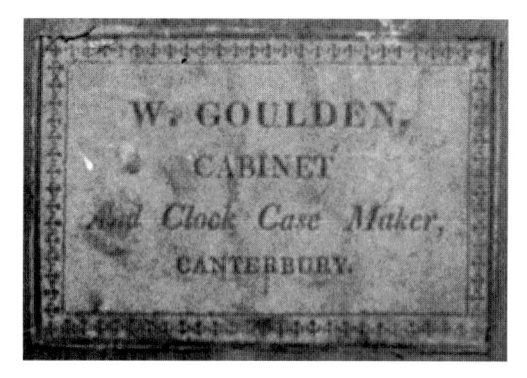

Fig 7/40 The trade label of William Goulden, clockcase maker, found in many of the East Kent cases of the late eighteenth/early nineteenth centuries. A round version is sometimes found

He began his apprenticeship on 17 February 1677/8. After completing his training there is no record of him until July 1688 when his name appears in the London records as promising to join the Clockmakers' Company at the next court. He was admitted as a Free Brother in September 1689, described as a 'Great Clockmaker'. This is the term used to describe makers who specialized in turret clocks, which is exactly the type of work he would have been trained in with Greenhill, who is known to have made large turret clocks.

After being admitted to the company he paid no more quarterage (fees) and it must be assumed that he returned to Canterbury, possibly working for Greenhill as a journeyman clockmaker. He did not apply for the Freedom of Canterbury until 13 April 1705, some five weeks after the death of his master Richard, who was buried on 5 March 1705

The only other reference to him appears in the churchwardens accounts for St Martins, Herne for 1715/16: 'Feb. 4th. Paid Goodlad for cleaning the church clock 7s. 0d'. This clock had been supplied to the church by Greenhill in 1677/8 at a cost of £14 and presumably Goodlad had been involved in its manufacture.

GORRUM, WILLIAM   Canterbury   pre-1788
Baillie records a watch signed by this maker.

GOULDEN, WILLIAM   Canterbury   born 1782-1822
William, the son of Samuel Goulden, shopkeeper, was apprenticed to Charles Lepine (*qv*), cabinetmaker in St Peter's Street near the Kings Bridge and who owned ten houses nearby. After completing his apprenticeship William worked as a journeyman for Lepine, sebsequently becoming a Freeman of the city, and married his master's daughter Hannah. After Lepine's death in 1822 they inherited the business. At least one case, now in the possession of a

relative, housing a clock by Robbins is known to have been made by William.

GOULDEN, WILLIAM   Canterbury   pre-1797-early nineteenth century
Samuel Goulden, father of the previous William, had a brother William, who had a grocery business at 18 Church Street, The Borough. His son, also William, was apprenticed to James Duryez, a turner and woodworker, gaining his Freedom in 1797. He set up in business at 16 Church Street, the Borough, and appears in various directories as a spinning-wheel maker, turner and chairmaker (*Hilden's Triennial Directory*, 1805). His trade label (Fig 7/40), sometimes styling him as 'clock-case maker and chairmaker' can be found in many East Kent cases of the early nineteenth century. Hence there were two William Gouldens, who were cousins, making clock cases in Canterbury at the same period.

GRAHAM, THOMAS   Bellingham   1785-1827

GRAHAM, WILLIAM   Bellingham   1785
Baillie notes longcase clocks by this maker and he is assumed to be related to Thomas.

GRAY, STEPHEN   Canterbury   apprenticed 1682
Apprenticed to Charles Annott in Canterbury on 9 June 1682. Nothing else is known.

GREENE, WILLIAM   Maidstone   late eighteenth century
A longcase clock by this hitherto unknown maker was sold at auction in 1996 by Olivers of Sudbury. The 8-day movement, housed in a walnut case typical of the 1770 period, appears to be of a high quality, with a brass dial and silvered chapter ring, date aperture, subsidiary dials for strike/silent and fast/slow set within the top spandrels. The arch contains a Father Time in front of a recess painted with a moon and stars. The movement and case bear a strong resemblance to work by the Cutbush family and it is possible that Greene was the owner of the clock, rather than the maker.

GREENHILL, JOHN I   Maidstone   Free 1607, died pre-1632
Smith ?

GREENHILL, JOHN II   Maidstone   born 1608, Free 1636
Locksmith.

GREENHILL, JOHN III   Maidstone   born 1655, died 1712
Gunsmith and clockmaker. Married Alice Harris, sister or aunt of Walter Harris Snr, clockmaker, in 1680.

GREENHILL, JOHN   Ashford   born 1644, died 1706
Clockmaker. Moved to Ashford from Maidstone after his marriage in 1642/3.

GREENHILL, RICHARD I   Ashford   born 1616, died 1687
Clockmaker.

GREENHILL, RICHARD II   Canterbury   born 1648, Free 1676, died 1705/6

GREENHILL, SAMUEL   Canterbury   born 1684, bankrupt 1723
Son of Richard II, apprenticed to his father in 1699. The Sun Fire Insurance Office has a

policy dated 22 November 1716 for the house only of Samuel Greenhill, watchmaker, in Winecheap Street.
See Chapter 6 for details of the Greenhill family.

| GREENLAND, JOHN | Ashford | 1675 |

Clockmaker. This entry is taken from Loomes and almost certainly is the same man as John Greenhill of Ashford. In the marriage records the name is sometimes written Greenhill, Grennell or Greenland. The latter is a misspelling of the family name.

| GREENWOOD, JOHN | Canterbury | 1832-40 |

Watchmaker. Watches signed by this maker have been recorded and examples of his work are held by both Maidstone and Canterbury Museums.

| GREENWOOD, JOHN | Dover | 1823-8 |

| GREENWOOD, JOHN Snr | Rochester | 1793-95 |

Watchmaker, but a clock is also known (Plate 7/41). Loomes notes an address 'opposite the rump in High St'. He obtained the Freedom of Rochester on 25 March 1793 on payment of £20.

| GREENWOOD, JOHN Jnr | Rochester | 1815 |

The son of John Senior, he was presumably trained by his father, obtaining his Freedom on 6 November 1815 by Patrimony. Nothing else is known although it is possible that he is the same John Greenwood who appears in Dover and Canterbury during 1823-1840

| GREENWOOD, WILLIAM | Rochester | 1823-47 |

Believed to be the younger son of John Snr and is particularly notable for looking after the town clock during the early nineteenth century.

| GREGSBY, EDMUND | Sittingbourne | c1673-c1685 |

A lantern clock formerly in the Ilbert Collection is now in the British Museum.

| GREMELS & SCHWAR(Z) | Greenwich | 1874 |

| GRUNWALD, J. | Dover | nineteenth century |

A watch signed by this maker has been recorded.

| GULLVEN, THOMAS | Horsmonden | early eighteenth century |

A 30-hour birdcage movement with a 10in brass dial with crown-and-cherub spandrels in an oak case about 1720 has been recorded. Another clock by this unknown maker was reported in the Bull Inn, Barming in 1965

| HAIZMAN, — | Canterbury | 1865 |

See Heitzman of Canterbury.

| HALL, KENNETT | Dover | 1838-66 |

| HALL, THOMAS | Maidstone | 1761-8 |

Loomes notes this maker working in the early years of the eighteenth century but very little information has come to light. His name appears in the Maidstone Poll Books for 1761-8 and

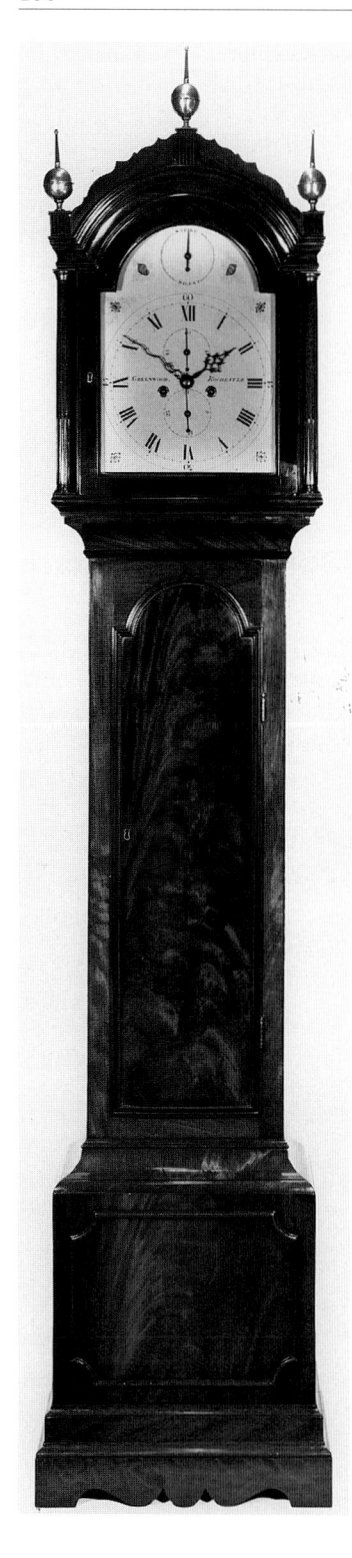

Fig 7/42 An 8-day clock by Thomas Hall of Maidstone, in an oak and mahogany case with holly inlays, about 1765. The rococo-style dial is signed on a boss in the arch and numbered No 77

Fig 7/41 A fine 8-day mahogany longcase clock by Greenwood of Rochester, 7ft 8½in tall. The painted dial has gilt flower decoration on a tracery background, about 1795

the Maidstone Museum have the dial of a 30-hour clock in their collection.

Another fine clock made by him is shown in Fig 7/42. The engraving on the dial is finely executed and the whole clock exudes quality. The movement with five pillars is very well made and is signed on a boss in the arch together with the number 77. Thomas Hall is the only Kent maker other than Austen of Challock and Baldwin of Faversham to number his clocks, and the number of this particular example leads one to suppose that there must be others in existence. The oak case of this clock is also interesting in that it bears no resemblance to the case style used by the other Maidstone makers who were his contemporaries. The only similarity is to a clock of the same period by Mercer of Hythe.

HANDS, FREDERICK ADOLPHUS　Ramsgate　　　　　　　　　1866-74

HANKINS, JACOB　　　　　　　　Farningham　　　　　　　　　1826-45

HANSELL, HENRY　　　　　　　　Chatham　　　　　　　　　　1839

HARDEMAN, EDWIN SAMUEL　Canterbury　　　　　　　　　1838-55
See Samuel Hardeman. *Bagshaw's Directory* of 1847 records a business address at 2 St Georges Street.

HARDEMAN, SAMUEL　　　　　　Bridge and Canterbury　　　1794-1839
Prior to 1794 Hardeman had been taken into partnership by William Nash of Bridge who was working between 1762 and 1794, the year he died. Following Nash's death Hardeman moved the business to Canterbury where it continued to be run after his death by his son Edwin Samuel. A year-going clock signed 'Hardeman & Son, Bridge' is shown in Figs 7/43-45

Numerous examples of his work have survived, both bracket and longcase clocks, invariably with painted dials, sometimes signed with dual signatures and others by Hardeman alone. The latter, which usually date from after the time of William Nash's death, often bear Canterbury as the place of working.

HARDEMAN, WILLIAM HENRY　Bridge　　　　　　　　　　　1848-74
Believed to be the younger son of Samuel Hardeman.

HARDEN, CHARLES　　　　　　　Hythe　　　　　　　　　　　1793-1861
Loomes notes Harden's date of birth as 1793 and that he was known as a watchmaker. Nothing else is known.

HARDING, E. H.　　　　　　　　　Canterbury　　　　　　　　　1865

HARFLETE, CORNELIUS　　　　　Sandwich　　　　　　　　　　pre-1747

HARLAND, CHRISTOPHER　　　Ramsgate　　　　　　　　　　1858

HARLAND, E.　　　　　　　　　　Ramsgate　　　　　　　　　　1855
Known to have made chronometers.

HARRIS, CHARLES　　　　　　　Tunbridge Wells　　　　　　　1874

HARRIS, JAMES　　　　　　　　　Maidstone　　　　　　　　　　pre-1754
A watch signed by this maker was reported as being lost or stolen, and having 'a most remark-

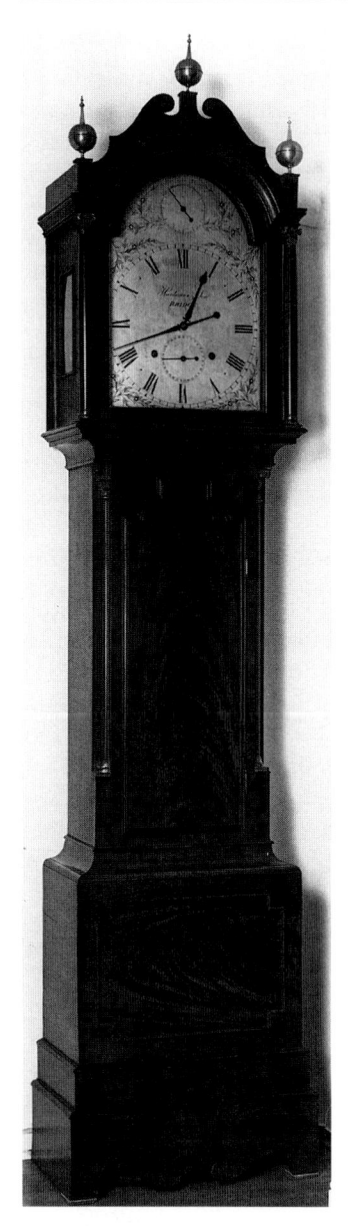

Fig 7/43 A year-going striking clock by Hardeman & Son, Bridge, about 1810. The massive case — 9ft 2in tall, including the finial — has a substantial mahogany carcase and fine veneers, with cresting on the top of the hood and stands on a double plinth

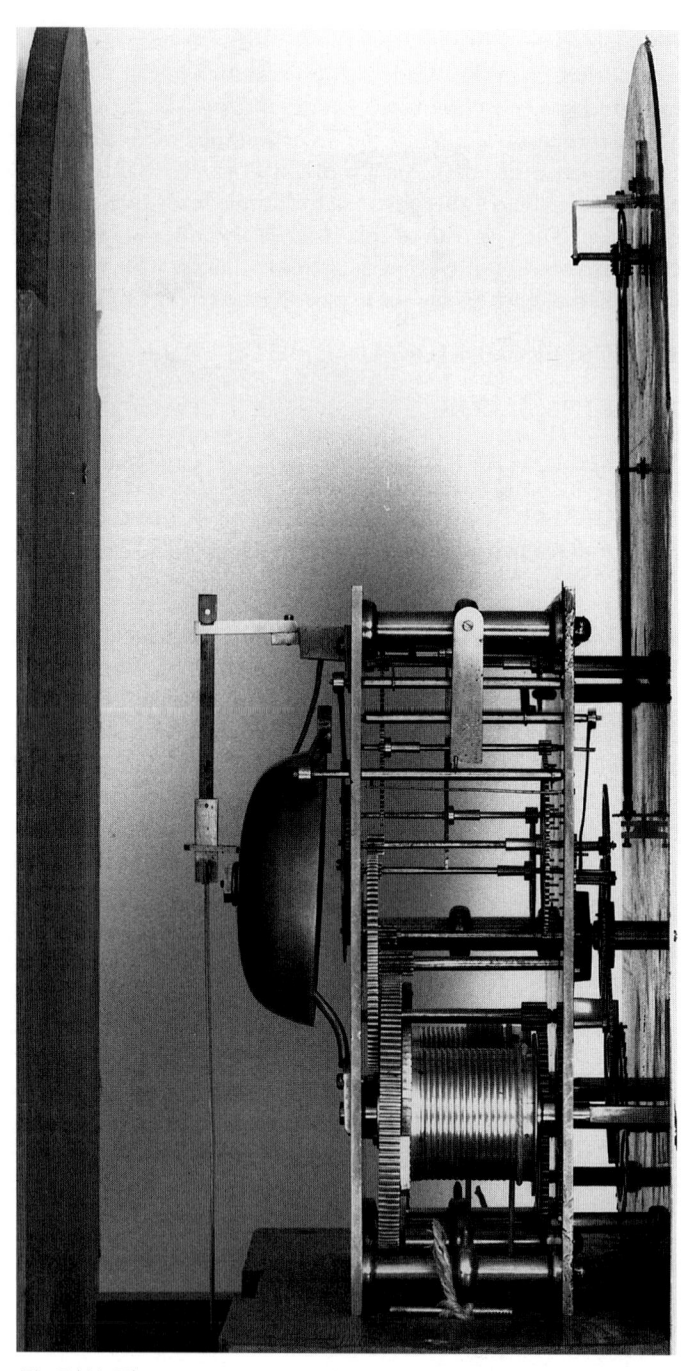

Fig 7/44 The movement of Hardeman's year-going clock with a six-wheel going train and a five-wheel striking train. The top end of the vertical arbor behind the dial operates a month indicator in the arch. The seven-pillar movement has plates 9¼in wide and 11in tall

Fig 7/45 The 16in wide silvered brass dial of Hardeman's year clock, with a month dial in the arch. The minute hand is counterbalanced and there are no minute numbers

Figs 7/46, 7/47 and 7/48 A 30-hour oak longcase clock by James Harris of Westerham, about 1770. The birdcage movement has iron corner posts

able Dial Plate, for instead of the usual Letters round it, are the twelve Following, viz. ITRADMEKSHAM' (see Appendix I, 8-11 May 1754); the significance of this is not known. Maidstone Museum has the dial only of a 30-hour clock signed 'James Harris, Town Malling', presumably the same man.

HARRIS, JAMES   Westerham   pre-1769

Baillie gives this date for Harris but does not record a Christian name. However, it is likely to be James as this name appears on a silvered-dial 30-hour longcase clock typical of the 1760-70 period (Figs 7/46-48). It is also probable that he is the same man as the James Harris of Maidstone. A watch has also been recorded.

HARRIS, JOHN   Sevenoaks   1855-66

HARRIS, JOHN   Sutton Valence   1847-51

HARRIS, JOHN   Deptford   1791
Watchmaker.

HARRIS, JOSEPH   Maidstone   pre-1743-70

The entry in Baillie records a watch and a turret clock by this maker. Maidstone Museum has an address plate dated 1770 engraved by the famous eighteenth-century pictorial and brass engraver William Woollett, who was born in Maidstone and is known to have worked for many of the local goldsmiths and silversmiths (Fig7/49). As Woollett engraved this plate for Harris there is every reason to suppose that he may have carried out engraving on many of the fine brass dials which exist on Maidstone clocks of the period. The plate reads:

> Joseph Harris — Watchmaker — Goldsmith from London — at the Dial and Cup —
> just above ye Upper Court House Maidstone —
> makes mends and sells all sorts of gold and silver watches at the lowest prices
> Also gives the most money for secondhand plate, jewels and watches.

A watch by Joseph Harris, No 3109, was reported lost in August 1768 (see Appendix I). There is also a possible connection with Christopher Harris, the London Maker who was working from 1782 to 1823. A watchpaper of his was found in a watch by Joseph Harris in which Christopher is described as a goldsmith and silversmith with a date of 1794.

HARRIS, SAMUEL   Gillingham   1830

HARRIS, STEPHEN   Tonbridge   born 1692-1755

Stephen, the son of William and Rose Harris was christened on 12 September 1692, and presumably was apprenticed to his father, a clocksmith by trade, although no indenture survives. There are various references to him in the town rate books between1728 and 1755. Unfortunately, the earlier books are missing so the exact date that he started working is not certain.

A number of clocks, both 30-hour and 8-day longcases, have survived and a particularly fine example is shown in Figs 7/51-52. The case is well-made of oak with walnut cross-banding to the trunk door and base, with a flat top to the arched hood. The good quality 8-day movement has five ringed pillars and rack-striking. The dial has good engraving with spandrels typical of the 1715-20 period and features the sun on a boss in the arch above well-

Fig 7/49 Trade card of Joseph Harris of Maidstone, engraved by William Woollett in 1770

Fig 7/50 An early eighteenth-century lantern clock by Stephen Harris of Tunbridge, about 1715-20

made dolphin spandrels. A 30-hour wall clock in a lacquered hooded case is shown in Colour Plate IV, and a lantern clock in Fig 7/50.

HARRIS, WALTER Snr **Canterbury** **apprenticed 1693**
The Canterbury Apprentice Lists note that Walter Harris was apprenticed to Richard Greenhill of Canterbury on 29 April 1693, but there is no record of him having obtained his Freedom after serving his time with Greenhill. However the Freemen Lists record Walter Harris Senior, Gentleman of Maidstone, receiving the Freedom of Canterbury by gift on the same day as Walter Jnr was apprenticed.

Although there is no firm evidence, he is believed to be the same Walter Harris who is known to have been working in Maidstone in 1704. The Canterbury records are confusing as they suggest that there was a Walter Harris Jnr entering his apprenticeship to Greenhill on 14 May 1700. It is possible that this is a mis-reading of the records and that this date refers to Harris obtaining his Freedom. To confuse the matter further his name appears as an appraiser on the inventory taken on 27 March 1706 after the death of Richard Greenhill. If the records are correct then this signature is that of Walter Harris Jnr.

Fig 7/51 A fine 8-day five-pillar longcase clock by Stephen Harris of Tonbridge, about 1715-20. The oak case is cross-banded with walnut.

Fig 7/52 The hood and dial of the Stephen Harris clock. Originally there would have been corner frets on the hood above the arch

HARRIS, WALTER Jnr   Canterbury   apprenticed 1700
Apprenticed 14 May 1700. See previous entry.

HARRIS, WALTER   Maidstone   1704
The Maidstone records note William Gill apprenticed in 1704 to Walter Harris, clockmaker. Nothing else is known, although he is believed to be the same man as Walter Harris Snr of Canterbury.

There may also be another connection with the Greenhill family. In John Greenhill's will dated 1711 made in Maidstone he leaves a bequest to his daughter Elizabeth Harris, possibly married to Walter Harris. The two families are known to be related as John Greenhill, the gunsmith, was married to Alice Harris in 1680. However, the exact relationship between the Harris and Greenhill families remains a mystery.

HARRIS, WILLIAM   Tonbridge   1684-1725

Neither Baillie or Loomes records this maker who worked as a clocksmith in Tonbridge from at least as early as 1684. There are other references to him in the town records in 1719, 1722 and 1725 and the baptism of his son Stephen is recorded in the parish registers as noted in the earlier entry.

Although he and his son are the earliest known Tonbridge makers (Figs 7/53-54), they do not appear to have had anything to do with the supply or upkeep of the two town clocks, ie the church clock and the four-sided dial clock on the former market house. Their names do not appear in the churchwardens' accounts who employed the services of a local blacksmith, George Summerton, to repair the church clock.

Figs 7/53-541 A single-handed longcase clock by William Harris of Tonbridge in an oak case, about 1715. The 30-hour birdcage movement has brass corner posts

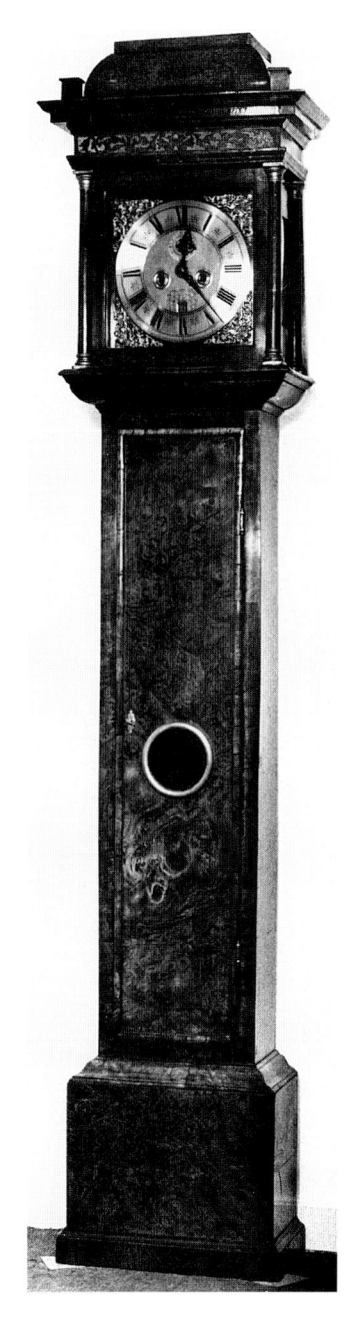

Figs 7/55-56 A longcase clock signed 'John Heydon, Rotherhith' in a burr elm case with a caddy top, 7ft 2in tall, about 1695-1700. The movement pillars show the fins typical of the period

HART, H.	Woolwich	1847
HART, JAMES	Canterbury	married 1752-61
HASCOCK, JOHN	Tunbridge Wells	1874

Figs 7/57 The London-style 11in square dial of the 'John Heydon, Rotherhith' clock, with ringed winding holes, engraved decoration around the date aperture and crown-and-cherub spandrels, about 1695-1700

HAYCOCK, SILAS HENRY　　　　　Ramsgate　　　　　　　　　　1874-94

HAYDEN, JOHN　　　　　　　　　Deptford　　　　　　early eighteenth century

Britten records a longcase clock by this man but nothing else is known of him. Brian Loomes also records a fine 8-day movement with an 11in dial in a burr-elm case with convex moulding to the hood, typical of the 1700 period (Fig 7/55-57). The chapter ring is signed 'John Haydon, Rotherhith,' but it must be assumed that they are one and the same, as the two places are adjacent to each other. It is possible that he is the same man as the John Hayden noted by Loomes as a watchmaker working in Croydon (Surrey), but not heard of after 1681 when he promised to become a Brother in the Clockmakers' Company at the next Court.

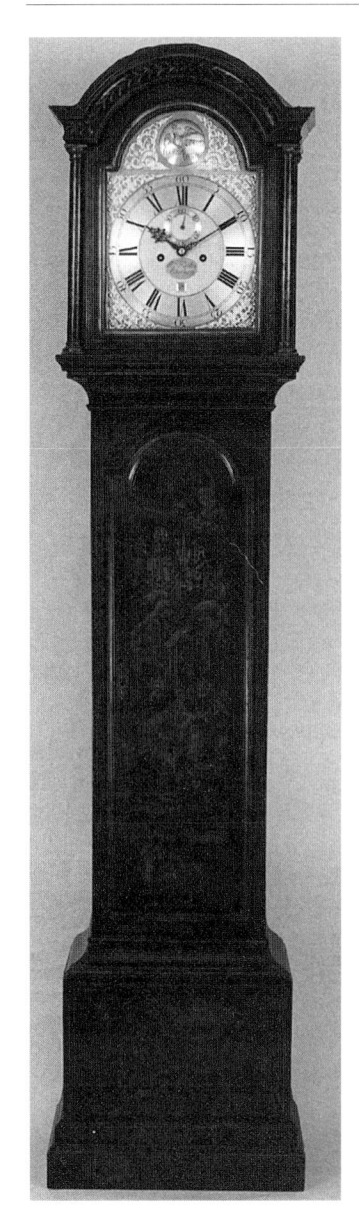

Fig 7/58-59 An 8-day five-pillar longcase clock by William Hayler of Chatham in a green lacquer case 7ft 3in tall, about 1765

HAYLER, BENJAMIN Chatham pre-1769-1800
Noted by Loomes as a watchmaker.

HAYLER, WILLIAM Chatham c1765-1851
Believed to be the son of Benjamin Hayler. A watchpaper exists advertising his business and giving an address at 59 High Street, near Chest Arms, Chatham. Two 8-day longcase clocks, both with brass dials in lacquered cases, are known (Figs 7/58-61). In view of the long working dates possibly two men of this name, maybe a son.

Fig 7/60 A lacquer longcase clock by William Hayler of Chatham, about 1775

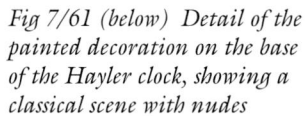

Fig 7/61 (below) Detail of the painted decoration on the base of the Hayler clock, showing a classical scene with nudes

Fig 7/62 (right) An 8-day four-pillar longcase clock by Benjamin Heeley of Deptford, about 1735. The 8ft 1in tall green lacquer case, in fine original condition, has chinoiserie decoration in red and brown with gold leaf on raised gesso

HAYMAN [HEYMAN], ROBERT	Canterbury	Free 1722

The Canterbury Freemen Lists note this man as a watchmaker being granted his Freedom in 1722. Recent research has discovered the following note in the City Accounts (9CC/FA29, f216): 'Item paid the xjth of June 1685 to Robert Heyman of this City, Watchmaker being a young beginner in his said Trade £5 [being the charitable gift of Henry Robinson Esq, being part of the interest on £100]'. No watch made by him appears to have survived. Possibly the same man as the one apprenticed to James Ellis on 11 September 1678, but this seems too early for him to be Free in 1722, so there is the possibility of two men of this name.

HAYWARD, E.	Deal	1866

HAYWARD, EDWARD	Ashford	1847-74

HAYWARD, EDWARD	Folkestone	1866-74

Possibly the same man as Hayward of Ashford.

HAYWARD, WILLIAM	Edenbridge	1858-66

HEELEY, BENJAMIN	Deptford	pre-1747

Both longcase clocks (Fig 7/62) and watches have been recorded by this maker. He is thought to be the brother of Joseph Heeley.

HEELEY, JOSEPH Snr	Deptford?	c1715

A 10in dial with 30-hour four-pillar movement dating from the 1715 period, signed 'Joseph Heeley Fecit' has been recorded. Because of the style of the dial and movement it is believed to be by an earlier member of the family, possibly the father of Benjamin and Joseph.

HEELEY, JOSEPH	Deptford	1747

Believed to be the brother of Benjamin and also known as a watchmaker. A fine and unusual dial clock, thought to be made by this maker is illustrated in R. Rose's *English Dial Clocks*. The silvered brass dial with outside numeral markings is mounted within a very well carved mahogany case in the cartel style and dates from the 1760s. The movement has tapered plates and early style pillars with a replaced verge escapement.

HEDGES, JOHN & SON	Maidstone	1832

Pigot's Directory of 1832 notes an address at 6 Union Street.

HEITZMAN, —	Faversham	1858

HEITZMAN, ANTHONY	Rochester	1838-74

HEITZMAN, CHARLES	Canterbury	1832-47

Working at an address in Church Street in 1847.

HEITZMAN, CHARLES	Maidstone	1839

The *Topograph of Maidstone* published in 1839 records this maker as a 'Dutch clockmaker'.

HEITZMAN, GEORGE	Maidstone	1832-47

Advertised in *Pigot's Directory* 1832 as making German clocks from 72 Week Street.

HEITZMAN, JOHN	Canterbury	1840
HEITZMAN, M. G. & B.	Canterbury	1838-40

Canterbury Museum has an 8-day arch-dial painted with pastoral scenes, housed in a good quality mahogany case, almost certainly from the Goulden workshop.

HEITZMAN, M. & J.	Canterbury	1847-51

Bagshaw's Directory records a business address in Westgate Street

HENDRICK, CHARLES	Dartford	1851-5
HERMAN, JOSEPH	Dartford	1874
HEYDON, THOMAS	Farnborough	c1780-1800

Loomes records him as a clockmaker. Nothing else has been discovered and it is not certain whether the Farnborough is the Kent village of that name, or if he was a Hampshire maker.

HEYMAN JOHN	Canterbury	Free 1714

Recorded as obtaining his Freedom in 1714 as a watchmaker, but nothing else is known.

HEYMAN, ROBERT, see HAYMAN, ROBERT

HICK, CHARLE F.	Rochester	1845
HICKMOTT, CHARLES	Maidstone	1865
HIGGINS, CHARLES	Canterbury	1865-74
HIGGINS, JOHN SIMMS	Canterbury	1859-74
HIGGINSON, SAMUEL	Chatham	1725

Nothing else is known of this clockmaker other than a record of his bankruptcy in 1725.

HILL, JOHN	Lydd	1847-55
HILLS, W. & G.	Rochester	1874
HILLS, WILLIAM	Rochester	1847-74
HISLOP, WILLIAM	Tunbridge Wells	1874
HOAD, HENRY	West Malling	1858-74
HOBBS, SAMUEL	Canterbury	Free 1742/3

The son of John Hobbs, a linen draper, he obtained his Freedom by Patrimony on 5 Jan 1742/3 as a watchmaker. There is no indication where and to whom he was apprenticed.

HODGES, JOHN & SON	Maidstone	1839
HODGESKYNNE, —	Maidstone	c1578

The Maidstone Chamberlain's Accounts for 1578 records:

> 1578 item payd to Hodgeskynne for keepinge of the town clocke   XVs

It is not known whether Hodgeskynne was actually a clocksmith or merely responsible for maintaining or winding the town clock, which probably is the same clock as the one installed at the conduit in the middle of Maidstone. Nothing else is known of this man.

| **HOGBEN, THOMAS** | Smarden | c1740-92 |

| **HOMERSHAM, JOHN** | Canterbury | 1838 |

A silver lever watch with the assay marks for 1835 has been recorded.

| **HOPE, W. D.** | Sevenoaks | 1874 |

| **HOPKINS, HENRY** | Deptford | pre-1780 |

| **HOPKINS, HENRY** | Deptford | 1802-24 |

Baillie records watches by these two men, who may be same person. A good quality mahogany longcase clock was sold at Christies in 1997.

| **HOPKINS, WILLIAM** | Tenterden | c1760-70 |

Two fine 30-hour movements, both with silvered dial centres with good engraving, mask-and-rococo pattern spandrels and with two hands, housed in elegant oak cases have been inspected. Both clocks are typical of the 1760-70 period but, to date, no other information concerning this maker has been discovered. However, there is a possibility that he is the same man as the maker with a similar name, apprenticed in London in 1742 and known to have obtained his Freedom in 1751, who is not heard of in London after 1759.

| **HOPLEY, GEORGE HENRY** | Dover | 1851 |

| **HOPLEY, W. F.** | Dover | 1855 |

| **HOPLEY, WILLIAM** | Dover | Free 1818-55 |

| **HORMER, J.** | Tonbridge | 1802 |

Watchmaker.

| **HORNE, J.** | Canterbury | 1784-1828 |

Loomes records this maker as marrying in 1784.

| **HOSMER, —** | Tonbridge | Free 1790 |

Baillie notes this man as gaining his Freedom in 1790 but there are no references to him in the records in Tonbridge Library Local Studies Department.

| **HOW, JAMES** | Bromley | 1802-74 |

Baillie records James How as working until 1824 as a watchmaker, but Loomes extends this date to 1874.

| **HOW, MARTHA, THOMAS & JOHN** | Bromley | 1847-66 |

| **HOW, PETER** | Eltham | 1832-39 |

| **HOW, PETER** | Woolwich | 1847-66 |

Probably the same man as previous entry.

HOW, SILAS SAMUEL	Tonbridge	1826-66
HUBBLE, DANIEL	East Malling	1847-74
HUGGETT, FREDERICK & THOMAS	Ramsgate	1832-47

HUGGETT, JOHN Hawkhurst 1832-9
Clockmaker.

HUGGETT, THOMAS Lamberhurst
 late eighteenth/early nineteenth century
A 30-hour painted dial movement in an oak case has been recorded. The movement has the unusual feature of a vertical fly operating the striking mechanism. This feature is peculiar to clocks made by Robert Apps of Battle, Sussex.

HUGHES, JOHN Maidstone apprenticed 1697-1709
Little is known of this maker who is named as a clocksmith in the 1709 Apprentice Lists when he took Thomas Goffe as his apprentice. He was the son of a cordwainer, Thomas Hughes, and was apprenticed to another little-known Maidstone maker, Joseph Cottam, in 1697.

HUGHES, JOHN Jnr Maidstone apprenticed 1727
According to the Maidstone Apprentice Lists John, presumably the son of the earlier John Hughes, was apprenticed to William Gill in 1727. A clock signed by John Hughes has been recorded, but it is not certain whether it is the work of the father or of John junior.

HUGHES, RICHARD Canterbury 1784
The only reference there is to this maker is in Baillie where he is described as a watchmaker.

HUGHES, WILLIAM Maidstone Free 1732
He is recorded as being the eldest son of John Hughes and obtained his Freedom in 1732. However, there is no record of with whom he served his apprenticeship and it must be assumed that his father trained him as a clocksmith.

HUKINS, G. N. Hawkhurst 1851

HUKINS, GEORGE HOPPER Tenterden born 1824, died 1910
Loomes notes that this maker died in Guernsey. Nothing else known.

HUKINS, JAMES Tenterden born 1800, died 1882
Presumably related to the previous entry.

HUMPHFREYS, WILLIAM	Deal	1832-40
HUNT, CHARLES	Headcorn	1826-66
HUNTER, THOMAS	Ramsgate	1823-51
HUNTER, WILLIAM D.	Ramsgate	1855

HURST, EDWARD Margate c1715-20
A good 8-day brass dialled longcase clock with a sunburst in the arch and dolphin and mask-

head spandrels has been recorded, but nothing else is known of this hitherto unrecorded maker.

HURT, ARTHUR Ashford married 1704-died 1741

Arthur Hurt was the son-in-law of John Greenhill of Ashford, and is recorded in the parish registers as marrying Lydia, the youngest daughter, on 7 December 1704. In 1707 they had a daughter Mary who was christened on 13 May. The only other reference traced is a note of his death in 1741 when he is described as a clocksmith and his shop was taken over by Abraham Webb of Wye (see Appendix 24-27 June 1741). As far is known no clocks signed by him are recorded other than the lantern clock noted in Brian Loomes's *Early Clockmakers of Great Britain*, where he is recorded as Arthur Hurst. This is a mis-spelling.

HUXLEY, THOMAS	Tunbridge Wells	1832
HYMAN, PHILIP	Chatham	1866-74
IGGLESDEN, JOHN	Chatham	1823-32

A watch has been recorded by this maker who may be the same man as the following entry. *Pigot's Directory* of 1832 gives an address at 97 High Street.

IGGLESDEN, JOHN Dover Free 1818

Loomes records this maker obtaining his Freedom in 1818 and notes that he was a watchmaker. He is believed to be the maker of the same name working in Chatham in the 1820s.

IGGLESDEN, G.R.	Dover	1866-74
ILLMAN, CHARLES & SON	Greenwich	1939-74
INGRAM, J.	Bexley	pre-1733

A watch signed by this maker has been recorded.

INTROSS, A. Strood c1800

Bracket clock known (Fig 7/63).

ISAACS, JOHN	Chatham	1839-47
ISAACS, JOHN	Woolwich	1817-24
ISAACS, M.	Woolwich	1817-24
JACKMAN, GEORGE	Tunbridge Wells	1855-74
JACKSON, JOHN	Tenterden	1796

Son of Owen Jackson; they were in partnership together in 1796.

JACKSON, J.	Rochester	1866
JACKSON, JOHN	Gravesend	1832
JACKSON, JOHN	Brompton	1826-8

Fig 7/63 A tavern clock by Owen Jackson of Cranbrook, about 1765, in a lacquer case with chinoiserie decoration

Fig 7/64 A Regency-period bracket clock signed 'A. Intross, Strood, Kent'

JACKSON, OWEN	Cranbrook	c1760-8
	Tenterden	1768-1803

Little is known of the background of this very interesting clockmaker who is recorded in Baillie as working from 1767 until 1783. However he is thought to have been working at least as early as 1760. His origins are unclear, but there may be a relationship with Richard Jackson of Maidstone, but as yet a definite connection has not been made.

Recent research has shown him to be the son of John Jackson, gunsmith of Cranbrook, with a brother Thomas, with whom he appears to have had a serious disagreement after the death of their father some time prior to March 1766. Both sons are known to have been gunsmiths, but it is not clear who taught Owen the craft of clockmaking, a trade which his brother does not seem to have practised.

On 22-25 January 1766 he placed the following notice in the *Kentish Post* announcing that he was carrying on his father's business, but at new premises in Cranbrook.

> Whereas Mr. John Jackson, Gun-maker, late of Cranbrook, being deceased, his Son, Owen Jackson, begs leave to inform Gentlemen and others, that he carries on the Gun-business in all its Branches, opposite the Crane in the said Town; where they may be supplied with any Sort of Guns or Pistols mounted in Silver, Steel, or Brass, according to the best Improvements and newest Fashions.
> N.B. Watches, Rings, and Plate, as usual, at the lowest prices.

This advertisement was followed by another in the 1-5 March issue placed by his brother Thomas refuting the notice placed by Owen.

> Whereas it hath been industriously reported, that the Busness at the old Shop, late John Jackson, Gun-maker at Cranbrook, was stopt, and the Trade carried to another Shop; therefore, as such Report tends greatly to my Prejudice, I Thomas Jackson, Son and Administrator of the said John Jackson, deceased, beg leave to inform all Gentlemen and others, that I carry on the Business which my Father did, in all its Branches, at the said Shop, known by the Sign of the Golden-Gun, next Door to the George and Bull Inns in Cranbrook aforesaid, where, as well as at my Shop in Maidstone, all Persons may depend on being served with the best Goods, on the cheapest Terms, and that their Favours will be gratefully acknowledged.
>
> Maidstone.   By their much obliged, And most obedient Servant,
> Feb. 20, 1766.   THO. JACKSON.
> N.B. Wanted as an Apprentice, an active ingenious Boy, not exceeding fourteen Years of Age. For Particulars, enquire as above.

The implication is that it must have been difficult for the two brothers to trade in the same town, for in 30 December-2 January 1768 Owen placed another advertisement in the *Kentish Post* stating that it was his intention to move to Tenterden, taking over Henry Goddard's old shop after the latter's move to Dover in 1767.

> OWEN JACKSON
> Of Cranbrook in Kent; Watch and Gun-maker, and Silver-smith
> BEGS Leave to inform Gentlemen, Ladies, and others, that he has taken the House and Shop late Mr. Goddards at Tenterden, in the said County, where he intends to carry on the above Trades in all their various Branches. Those Gentlemen that please to honour him with their Commands, in any of the above Particulars, may depend on having them duly and faithfully executed, and the Favour gratefully acknowledged, by their most humble and

obedient Servant, Owen Jackson.
N.B. As I cannot remove immediately to Tenterden, I propose opening Shop on New-Year's Day, and attending weekly on Fridays, for the Conveniency of Gentlemen and others, who may have Occasion to deal in any of the above-mentioned Articles, and may depend on their being used well. Cranbrook in Kent, December the 9th 1767.

In December 1771 Owen Jackson, by then at Tenterden, advertised that he would be attending at the Bull Inn at Cranbrook once a fortnight and on alternate weeks at Rye to see his customers (see Appendix I).

How long the estrangement between the brothers lasted is not known, but Owen was forced to place another notice in the 18-21 June 1783 issue of the *Kentish Gazette*:

Tenterden, June 18, 1783.
WHEREAS a Report has been industriously spread in and about the [Environs] of Tenterden, that I have declined the WATCH TRADE, I respectfully take this Method of informing my Customers in particular, and the Public in general, that such Report is void of the least Foundation in Truth; and that I carry on my several Branches of Trade, as usual, at the old Shop, opposite the Prison in Tenterden; where all Orders will be thankfully received and carefully executed, with the utmost Dispatch, By their most obedient Servant,

OWEN JACKSON.

Neat NEW SILVER WATCHES, at Three Guineas each, equal to any at the Price sold in this Kingdom.
Likewise all Sorts of NEW CLOCKs, and OLD WORK, repaired on easy Terms.

In 1796 he is known to have been in partnership with his son, John, and the only other reference that has been traced is a notice in the *Kentish Gazette* on 18 August 1803, stating that the business was to be sold due to his retirement after 35 years.

Two very fine, almost identical, shield-dial tavern clocks exist, one signed Cranbrook (Fig 7/63), the other Tenterden. They both date from the 1775-85 period and are typical of many of the tavern clocks made by Kentish makers, which were formerly found in the coaching inns on the main roads from London to the coastal ports.

JACKSON, RICHARD   Maidstone   c1760
Loomes notes this maker as working during the 1760s, but the only clock seen suggests an earlier date. The 30-hour movement of birdcage construction is housed in an oak case and the 11¼in brass dial is fitted with early cherub spandrels. Nothing else is recorded.

JACKSON, —   Ashford   died 1790
It is known that this man was a watchmaker and died in 1790. Unfortunately there is no record of his christian name and nothing else is known.

JACKSON, WILLIAM   Maidstone   1796-c1800
Baillie records this maker as being Free of the Clockmakers' Company in 1800 and that he was primarily a watchmaker.

JACKSON, WILLIAM   Maidstone   1796-1823

JACKSON, WILLIAM RICHARD RADLEY   Maidstone   1823-29/32?
The previous three entries are taken from Baillie and Loomes. Whether there is any connection between them is not clear, or if they are one and the same. The only other information is

that *Pigot's Directory* of 1832 notes a William Jackson working at 39 Gabriels Hill.

JAGGER, HANNAH	Canterbury	1823

JAMES, THOMAS	Woolwich	1847

JEFFERYS, EDWARD	Chatham	married 1789-1824

Edward, the son of George Jefferys, was presumably trained by his father and also worked in Chatham after he finished his training. He was married in 1789 and a number of watches signed by him have been recorded. One, a silver pair-case example, has the assay marks for 1824 so presumably he was working at least until this date.

JEFFERYS, GEORGE	Chatham	apprenticed 1748-83

George Jefferys is another of the very able clockmakers originating from the Maidstone area during the second half of the eighteenth century. He was the son of William Jefferys, the carpenter, to whom John Cutbush apprenticed his son John in 1722. George was apprenticed in 1748 to another member of the Cutbush family, Robert, a good maker in his own right. There is no record of him petitioning for his Freedom in Maidstone and it is assumed that after finishing his apprenticeship he moved to Chatham sometime after 1755.

A number of good longcase clocks (Fig 7/66) and watches, including a repeating example, have been recorded and a particularly fine and very rare dial clock by him is illustrated in Fig 7/65. The 10in engraved silvered brass dial is housed in a salt-box case with a finely-turned wooden bezel. The substantial verge movement has tapered plates with five heavy pillars and an offset barrel. English dial clocks with a diameter of less than 12in are very rare and this is a very good example of the type.

In 1770 he submitted a quote for the replacement of the Rochester Corn Exchange clock in competition with Muddle of Chatham and James Booth. The contract was finally awarded to Edward Muddle and Jefferys received one guinea as compensation for costs incurred in drawing up his plans for the new clock. See entry for Edward Muddle.

JEFFERYS, JAMES	Canterbury	1823-65

A silver pair-case watch and a longcase clock signed 'Jeffrey, Canterbury' in a Goulden-style case have been recorded (Fig 7/67). Noted as working at 3 Parade in 1847.

JEFFERYS, JAMES	Wingham	1847-55

Probably the same man as Jefferys of Canterbury.

JEFFERYS, WILLIAM R.	Ashford	1874

JENKINS, GEORGE	Hythe	1845

JENKINS, J. & M.	Rochester	1784

Noted in Baillie as watchmakers, and presumably a partnership involving Mason Jenkins referred to in a previous entry.

JENKINS, MASON	Rochester	pre-1787, died 1802

JENKINS, THOMAS	Rochester	1787-died 1789

Watchmaker (see Appendix I, November 1787). His effects advertised for sale in April 1789.

Fig 7/65 A rare English dial timepiece by George Jefferys of Chatham, with a 10in diameter engraved brass dial, about 1765. The verge movement has tapered plates and five heavy pillars

Fig 7/66 A fine 8-day five pillar longcase clock with a silvered brass dial by George Jefferys of Chatham, about 1765. The mahogany case has a pagoda top to the hood

Fig 7/67 An 8-day painted dial signed 'Jefferys, Canterbury' in a mahogany case in the Goulden style. About 1830

JENKINSON, J. Sandwich 1770
Britten records a maker by this name working about 1770, but no other reference seems to exist. Possibility he has been confused with Thomas Jenkinson.

JENKINSON, S. Sandwich c1760
The Victoria and Albert Museum have a watch in their collection signed by this maker. Nothing else is known.

JENKINSON, THOMAS Sandwich born 1696, died 1755
Thomas, one of Sandwich's best-known makers, was born in 1696 the eldest son of John Jenkinson, a grocer, who had married Elizabeth Jennings in 1694. In 1711 he was apprenticed to Joseph Booth of Sandwich, his mother standing bond for him in the sum of £15 as his father had died in 1703. After finishing his training as a clockmaker he petitioned for his Freedom, which was granted in 1719. He is also in the Sandwich Poll Book for 1734.

The Canterbury marriage licences record him as applying for a licence to marry Sarah Bing of Ash on 11 January 1723. She was probably related to Daniel Bing who was working between 1727 and 1757 in the town of Sandwich. Numerous examples of his clocks and

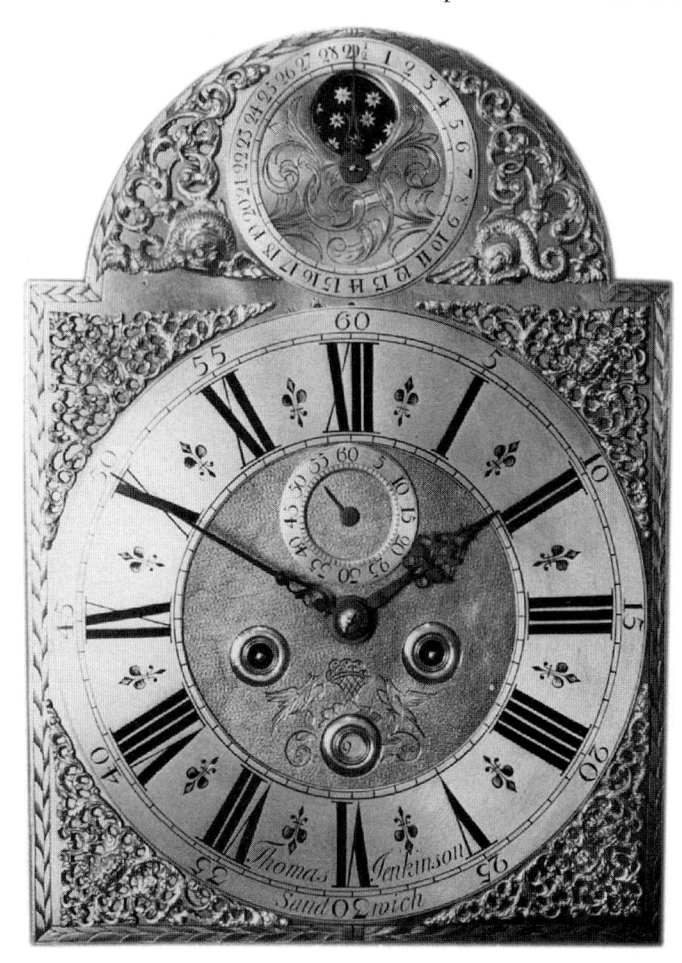

Fig 7/68 A very attractive brass dial by Thomas Jenkinson of Sandwich, about 1720. The winding holes and the circular calendar aperture are ringed and there is the typical engraving of flowers and a basket of fruit around the calendar. The arch has a 'penny' moon set in acanthus engraving and a lunar date indicator, while the dial and arch are surrounded by herringbone engraving. The green lacquer case is shown in Colour Plate X

Fig 7/69 An oak-cased 8-day clock by Thomas Jenkinson of Sandwich, about 1730

Fig 7/70 The the arch of the dial has a silvered name boss with a broad ring of herringbone engraving around it

watches have survived, the clocks in particular are invariably of good quality with very attractive dials, usually engraved with a herring-bone border. At least two in lacquer cases are known (Colour Plate X and Fig 7/68) and others in walnut and oak (Figs 7/69-70). A miniature lantern clock by him is featured in Sir George White's book *English Lantern Clocks*. He died in August 1755 (see Appendix I) and a memorial tablet to him is above the north door of St Peters Church in Sandwich.

JOBSON, FREDERICK WILLIAM	Marden	1866-74
JOHNSON, B.	Greenwich	1851
JOHNSON, CHARLES	Canterbury	1667-82

Recorded in the Canterbury lists as a smith, obtaining his Freedom by marriage to Susan, daughter of Samuel Bond, cordwainer deceased, in 1682. Thomas Ventisan is noted as being apprenticed to him, when he is described as a clocksmith. Nothing else is known.

JOHNSON, JEREMIAH	Deptford	apprenticed 1770-1808

Watches signed by this maker have been recorded.

Fig 7/71 11in square dial by James Jordan, with crown-and-cherub spandrels, ringed winding holes and engraved decoration around the square calendar aperture. The high quality movement has five pillars

Fig 7/72 (left) An early eighteenth-century longcase clock in an ebonised case by James Jordan of Chatham. The hood has gilded pillars and the low caddy top often seen on Kent clocks of the 1705 period. The dial is shown in Fig 7/71

Fig 7/73 (right) A month-going longcase clock with a 12in square dial by James Jordan of Chatham, about 1705. The 6ft 11in tall fruitwood and walnut arabesque marquetry case is decorated with strapwork, scrolls and cherubs. The movement has five pillars, five wheels in each train and an outside countwheel

JORDAN, JAMES Chatham c1705-40

Loomes gives a date of about 1740 for this maker of whom very little is known. Judging by the style of two of the longcase clocks that have been inspected it is suggested that he was working at a much earlier date in the eighteenth century. Fig 7/73 shows a fine 12in diall five-pillar month clock housed in a good arabesque marquetry case. The movement has five wheels in each train and outside countwheel striking with cherub-and-crown spandrels, dating from no later than 1720 and possibly as early as 1705 from the style of the case. The clock shown in Figs 7/71-72 has an 11in dial and five-pillar 8-day movement, once again of very high quality, but dating possibly from 1710-15, judging by the engraving around the calendar aperture and the slightly wider minute section of the chapter ring. The ebonized pine case has the later feature of a concave moulding below the hood.

Numbers of other clocks have also been recorded, some in lacquer cases, and all exhibiting the same high standard of workmanship, comparable to the best London work of the period. Unfortunately, there appears to be no record of where or with whom he was trained.

JONES, SAMUEL Sevenoaks 1826-51

JUDGE, T. Canterbury 1866

JUDGE, THOMAS Hythe 1874
Possibly the same man as Judge of Canterbury.

KAYSER, JOSEPH Sheerness 1839-47

KEARLY, R. Brompton c1800
Loomes records this man as a clockmaker but nothing else is known.

KELVEY, ROBERT Dover late eighteenth century
Noted in Loomes as a clockmaker.

KEMP, ALFRED Tunbridge Wells 1851-74

KEMP, C. Groombridge early nineteenth century
A 30-hour 12in square painted dial clock in an oak case has been recorded.

KENDALL, RICHARD Canterbury 1838-40

KENDALL, — Canterbury early nineteenth century
A longcase clock with the dial signed 'Kendall' is known. He is probably the same maker as the previous entry.

KETTERER, CHARLES Greenwich 1847

KETTERER, CRISPIN Greenwich 1847

KEW, WILLIAM Ashford 1874

KIBBLE, RICHARD Greenwich 1847-74

KILLMAN, JOHN Deptford 1847

KIMPTON, BENJAMIN	Eltham	1874

KINGSNORTH, JOHN	Tenterden	apprenticed 1688

Very little is known of this interesting maker and moreover, so far, only two clocks have been recorded by him. The exact date of his arrival in Tenterden has not been discovered, but he is believed to be the the man of this name who was apprenticed in London on 29 September 1688 to Daniel Steevens and subsequently turned over to Thomas Stubbs, a well-known London maker. After serving his seven-year apprenticeship it is assumed that he moved to Kent some time after 1695.

An 8-day movement signed 'Kingsnorth, Tenterden', housed in a lacquer case has been recorded, and a lantern clock in early-eighteenth century style with an anchor escapement was sold at auction in 1993 (Fig 7/74). The dial was decorated in typical fashion with scrolling foliage, a design which was common to a great number of the lantern clocks produced in and around London during the eighteenth century. However, it does bear a striking resemblance to two very similar lantern clocks made by George Thatcher of Cranbrook who was working about 1716 onwards, and is identical in almost every respect.

It is not known who taught Thatcher the craft of clockmaking, but in view of the closeness of the two towns and the overlapping of working dates, is it possible that Kingsnorth was responsible for passing on the skills he had learned in London to the man from Cranbrook?

KINGSNORTH, THOMAS	Tenterden	1714-15

Mentioned in the accounts for St Margarets, Bethersden. Previously unrecorded. May be the son of John Kingsnorth.

Kingston, J. T.	Ramsgate	1855-74

KINLAN, THOMAS	Tunbridge Wells	1866

KIRBY, GEORGE	Canterbury	1851-74

KIRBY, JOHN	Bromley	c1730

Baillie notes this man as a clockmaker working in the early eighteenth century, but nothing else is known.

KIRBY, JOHN	Greenwich	1866-74

KISSAR, SAMUEL	London	apprenticed 1700, CC1712
	Canterbury	1712/13-32

Samuel Kissar was apprenticed in London on 18 July 1700 to Edward Enys, and was subsequently turned over to Thomas Walford for a period of seven years, obtaining his Freedom on 29 September 1712. The Clockmakers' Company records note that he was the 'son of Stephen Kissar, late of Canterbury, Woolcomber deceased'. The Canterbury marriage licences register notes that he gave notice of his intention to marry Mary Parker of St Mary Bredin, Canterbury on 16 January 1712/13. He presumably moved to Canterbury after obtaining the Freedom of the Clockmakers' Company.

He insured his goods and merchandise on 4 May 1719 with the Sun Fire Insurance Company, and subsequently 'removed to the corner of Canterbury Lane in St. George's Street' in 1724. In May 1731 he announced that he was 'to leave off Shop-keeping' near the Mermaid

Fig 7/74 A lantern clock by John Kingsnorth of Tenterden, about 1720

Fig 7/75 (top right) The 14in late-seventeenth-century square dial of a rare three-train wall clock by John Knight of Faversham. The winged cherub head spandrels are of a very unusual design. The six pillar movement is missing the third train and quarter-chiming bells. The hands date from when it was put into a longcase

Fig 7/76 The mahogany case specially-made in the 1770s for the John Knight three-train chiming movement and dial

in St Margaret's Streeet, and offered for sale his stock of linen, haberdashery goods, new clocks and watches , but by June of the next year he had moved to the Crown and Dial in Burgate, where 'all sorts of Watches are made, mended and sold'. In August 1732 he sought to purchase 'Fleece Wooll', yet the next month he was still advertised his watchmaking business, warranting that he would keep his 5-guinea watches in repair for seven years (see Appendix I).

He is another of the numerous early Canterbury makers of whom, as far as is known, no clock or watch signed by him survives today.

KNIGHT, GEORGE   Faversham & Canterbury   Free 1683-96

The Canterbury lists record the granting of the Freedom of the city to this man by his marriage on 12 November 1683 to Elizabeth Ludd, the daughter of Randolph Ludd, a glazier. He is described as a clockmaker of Faversham. References to him appears in two depositions dating from 1696 in the Faversham Borough records to the effect that John Austen, a saddler, had declared himself to have Jacobite sympathies at the Ship Inn in Faversham. The depositions informing on him are signed by William Mitchell, mariner, and George Knight, described as a clocksmith. The Ship Inn is in use as licensed premises to this day. He may also have been the George Knight who signed Thomas Deale's inventory in 1687 and who is named as an executor (see pages 64-6).

KNIGHT, JOHN   Faversham
                         late seventeenth-early eighteenth century

The only clock so far recorded by this hitherto unknown maker is a superb three-train 8-day movement with the unusual feature of a 14in square dial in typical late seventeenth century style with a finely-engraved dial (Fig 7/75). The movement has six pillars, but unfortunately the third barrel and wheelwork for the quarter-striking train is missing, although the clock possesses its three original brass-cased weights. The spandrels are of a most unusual design and the edge of the dial is engraved with a wheatear border.

The movement is now housed in a good mahogany case of the 1770 period (Fig 7/76), which because of the unusual dial size has obviously been purpose-made for the clock, as have the replacement hands, which also date from the same period. Whether the movement was always housed in a longcase is open to question. It is more likely to have been in an appropriately sized wall mounted case. This is the only known example of an early quarter-striking movement by a Kent maker, with the exception of the two lantern clocks by Greenhill and the movement by Dodd of Faversham.

No trace of this maker has been found in either the Faversham or Canterbury records, other than the record of John Knight of Faversham obtaining a licence to marry Mary Fremoult of Westgate, Canterbury on 13 July 1716, but it is not certain that this refers to the maker of this fine clock. There is also a memorial in the churchyard at Faversham to two children and the second wife, Mary, of a John Knight. Mary is noted as dying in 1779. It appears likely that this refers to Mary Fremoult. At the present time there is no evidence to connect John with George Knight of Faversham, although it seems possible. The only other reference to a maker of the same name is in Baillie where a John Knight is recorded as having made a watch in 1684 for Charles II to give to Muley Hamet. As there appears to be no other reference to this maker in the Clockmakers' Company records in London, they may be one and the same, although it is strange that no other clocks or watches made by someone who was obviously a fine craftsman appear to have survived.

Fig 7/77 A shield-dial tavern clock by William Knight, Jnr, West Marden, about 1760. The five-pillar movement has a five-wheel train

KNIGHT, WILLIAM   Marden   c1760

Loomes records him as a clockmaker working during the middle of the eighteenth century, but nothing more about him has been discovered. The only clock recorded by him is the very fine shield dial tavern clock illustrated in Fig 7/77. This clock is signed 'William Knight Jnr. West Marden' and is almost certainly by the man mentioned in Loomes. However, the place-name does cause some confusion because the only West Marden mentioned in modern gazetteers is in West Sussex. Marden itself is a small village in the Weald and while being within a few miles of Staplehurst and Cranbrook, is not directly on the route of the old coach road to the coast. The clock in question does bear a remarkable resemblance to a similar shield tavern clock by Owen Jackson of Cranbrook and the cases and decoration are virtually identical.

    The case is constructed of oak throughout and is the final development of the shield style before it was superseded by the round dial. The access doors retain their original wire catches and the dial is secured to the trunk by unusual wire pegs. The dial is made up of three oak planks within a deep moulded border surmounted by two gilt wood ball-and-spire finials and is finely decorated with gold leaf to the chapter ring and outside five minute markings with an elegantly written signature filling the base of the dial. The brass hands are particularly unusual with their heart tips and scalloped diamond piercing.

    The movement has thin rectangular plates with five knopped pillars and a five-wheel train, with the unusual feature of a plain ungrooved barrel. Because of the clock's similarity to other Kentish examples, particularly the one by Jackson, it must be assumed that this clock and its maker are indeed from Marden in Kent, although the place designation is extremely puzzling.

KOLLSALL, —   Maidstone   1719

This man is noted as a watchmaker in an entry in the Maidstone archives, in which Robert Southgate is accused of being under age at the 'Court meet' in 1719

KNOWLER, SAMUEL   Chilham   c1770

An 8-day brass dialled movement in an oak longcase typical of the 1770 period has been recorded, but nothing else is known.

KUNER & CASEY   Sheerness   1866

KUNER, S.   Canterbury   1865

This man was presumably a member of the partnership in the previous entry.

LADBROOK, DANIEL   Maidstone   1858

LADE, MICHAEL   Canterbury   married 1723-died 1737

Loomes records his marriage in 1723 and that he died in 1737. He is believed to be a member of a well-known family of goldsmiths and silversmiths in Canterbury, one of whom served as Mayor of Canterbury. No work by him has been traced, but he is assumed to be a watchmaker. There are several reports of watches signed by him either stolen or lost in the *Kentish Post*: watch No 16 (August 1750), No 549 (June 1745) and No 2241 July 1760, and two with calendar work on the dial plates (July 1730 and July 1760). See Appendix I.

LAMB, SIMON   London   apprenticed 1669/70-c1677
                                       Rochester   1677-c1700

Baillie notes a longcase clock by this man, who is believed to have been working until 1700 at least. Thought to have been the man apprenticed in February 1669/70 to William Glazier. He is not recorded as a Freeman of the Clockmakers' Company and presumably moved to Rochester after 1676. Noted in the Rochester records as working on the town clock in 1677. A month-going marquetry longcase clock is known.

LAMB, H. T.	Woolwich	1855
LANGLEY, CHARLES	River	1851-74
LAWRENCE, J.	Eltham	1847-74
LEE, I. J.	Woolwich	1851
LEE, L.	Margate	1797

Watchmaker.

LEE, SAMUEL	Dover	1742/3-65

Baillie records a watch signed by this man. He is known to have previously worked in London (see Appendix I, March 1742/3).

LEECH, WILLIAM	Maidstone	early nineteenth century

Pigot's Directory of 1832 gives an address in Week Street.

LEOFFLER & CO	Greenwich	1849
LEOFFLER, PETER	Greenwich	1839
LEMAITRE, JULES	Canterbury	1865

Came from Paris.

LEPINE, CHARLES	Deal	1839-45

Clockcase maker? Brother of Henry Lepine of Canterbury, watchmaker.

LEPINE, CHARLES	Canterbury	died 1822

Clockcase maker. A reference to Charles Lepine in the Canterbury records describes him as a cabinetmaker. Had premises in St Peter's Street near the Kings Bridge and took William Goulden (*qv*) as an apprentice. Goulden married Lepine's daughter and they inherited the business after Lepine's death. in 1822. A trade label of Charles Lepine of Canterbury, a cabinet-maker and clockcase maker, has been found pasted inside a cherrywood case housing a painted dial 8-day clock by Thomas Pegden of Sandwich (Figs 7/78 and 7/98).

LEPINE, HENRY	Canterbury	1823-38

He was the brother of Charles Lepine and was known as a watchmaker.

LE PLASTRIER, ROBERT	Dover	1810-40

He is recorded in Loomes as obtaining his Freedom in 1810 and a fine ebonized balloon-case bracket clock signed by him has been recorded.

LE PLASTRIER, ROBERT LOUIS	Ramsgate	1832

Fig 7/78 The only known trade label for Charles Lepine, clock-case maker of Canterbury, found inside a longcase clock by Thomas Pegden of Sandwhich (Fig 7/98)

Fig 7/79 A lacquer cased round-dial tavern clock by Levey (possibly James & Benjamin Levi) of Dover, about 1800

LE PLASTRIER, WILLIAM Dover Free 1800-2
Noted by Loomes as a watchmaker.

LE PLASTRIER, — Deal early nineteenth century
Watches signed Le Plastrier, Deal have been recorded but it is not certain whether they were made by William or Robert.

LEVEY [LEVI?], EMANUEL Dover Free 1818-1851
Levey worked in premises in Strand Street and is reputed to have made the tavern clock for the Guildhall which was formerly in the Market Square. This clock now hangs in the Maison Dieu. Possibly a member of the Levi family.

LEVI, A.   Folkestone   1799
Recorded in Baillie as marrying in 1799 and known as a clock and watchmaker.

LEVI, BENJAMIN   Canterbury   1783-91
0n 7-11 August 1784 he placed the following advertisement in the *Kentish Gazette*:

BENJAMIN LEVI,
CLOCK and WATCH MAKER, and SILVERSMITH,
In PALACE-STREET, CANTERBURY.

Respectfully acquaints his Friends and the Public in general, that he has opened a Shop in Palace Street, where he has a very neat Assortment of WATCHES of all kinds, both New and Second Hand; likewise, his new-invented Watches that will go two, three, or eight Days without winding up; also Repeating, Horizontal, and Patent Stop Watches. He has likewise a great Variety of Plate, and Plated Goods, both New and Second Hand. He hopes by his strict attention to Business, and the reasonableness of his Charge to merit the Continuance of any Favours that may be conferred on him.
N.B. All sorts of Clocks and Watches cleaned and repaired in the best Manner; Mourning Rings on the shortest
Notice. Gives the utmost Value for old Silver, Gold, &c.

After 1791 he joined his brother James in partnership in Dover, and after the latter's death in February 1803 formed a partnership with Emanuel Levi of Dover, which was dissolved by mutual consent in August of that year (see Appendix I).

LEVI, EMANUEL   Margate   1804
Noted in Baillie as a clock and watchmaker. Before moving to Margate he had been in partnership with Benjamin Levi in Dover for a short period of six months in 1803.

LEVI, JAMES   Dover   born c1753- died 1803
In 1791 formed a partnership fwith his brother Benjamin, when the latter left Canterbury and moved to Dover. An advertisement in the 18 February 1803 issue of the *Kentish Gazette* reported the death of James Levi aged 50 and notes him as senior partner. The partnership was noted as clock and watchmakers. An 8-day painted dial in a fine inlaid mahogany case has been recorded.

LEVI, NOAH   Ramsgate   pre-1789-1802
In September 1789 described himself as a 'Watch and Clock Maker, Working Goldsmith and Jeweller, South Street near the Market Place, Ramsgate', who had 'removed from his late Shop to No. 12, on the other Side of the Way' (see Appendix I).

LEVI, SOLOMON   Canterbury   1793
Recorded in Loomes as working between 1788 and 1793 and one of the members of the thriving Jewish community in Canterbury, many of whom were clock and watchmakers. To date, the family connections between the Levis has not been unravelled, but it must be assumed that they were all related to one another.

LEVIEN, LOUIS W.   Chatham   1845

LINGHAM, H.   Maidstone   1855

Fig 7/80 A mahogany drop-dial wall clock with brass inlay, the 12in diameter white dial signed by W. Loof, Tunbridge Wells, about 1840

Fig 7/81 A mahogany 12in silvered dial wall clock by Thomas Lowe of Dartford, about 1800

LION, ISAAC	Ramsgate	1826-40
LITTLEWOOD, B.	Woolwich	1866-74
LITTLEWOOD, BENJAMIN Jnr	Woolwich	1874
LONDON, GEORGE	Cranbrook	c1760

A 30-hour four-pillar plated movement with two hands and an engraved and matted brass dial with silver chapter ring has been noted. Nothing else is known of this unrecorded maker.

LONG, RICHARD	Deal	1784

Watchmaker.

LONG, WILLIAM	Sandwich	early nineteenth century

A watch signed by this maker has been recorded. He is believed to be the son of Richard Long, also a watchmaker.

LONGHURST, —	Chatham	CC1790

Noted in Baillie as having obtained his Freedom in that year but nothing else is known.

LOOF, EDWARD FRY	Tunbridge Wells	1855-74
LOOF, WILLIAM	Tunbridge Wells	1823-55
LOOF, WILLIAM Jnr	Tunbridge Wells	1845-7

A mahogany drop-dial wall clock by one of these two makers is known (Fig 7/80).

LOWE, THOMAS Dartford 1770-CC1790-1836

Baillie notes a Lowe, christian name unknown, as obtaining his Freedom from the Clockmakers' Company about 1790, but Loomes extends these dates to 1770-1830 and states that he was a watch and clockmaker. He also records Thomas Lowe as working in Dartford between 1823 and 1832. These dates are taken from entries in *Pigot's Directory*. The local records only show one maker of this name and it likely that these references to Lowe all refer to the same man. Keys's *Historical Notes of Dartford* record an address at 15 High Street in 1836. He is known to have made watches, one is illustrated in the 6th edition of *Britten's Old Clocks and Watches*. A fine dial clock dating from about 1800 is also known (Fig 7/81).

LOWMAN, JEREMIAH Ramsgate c1810-32

Noted in Loomes as a clock and watchmaker and stated by Baillie to have originated in London before moving to Ramsgate.

LYNE, — Deptford pre-1745

Baillie records a watch signed by this man, christian name unknown, but nothing else is known.

LYON, ISAAC Ramsgate 1768-1840

Loomes suggests that there may be two watchmakers of this name.

LYON, SIMON Rochester 1866-74

LYON, L. & S. Chatham 1845

Possibly a partnership involving Simon Lyon who worked in Rochester from 1866.

LYONS, JAMES Folkestone 1845

McCLELLAN, — Deptford 1839

McCLELLAN, S. Greenwich 1851-5

McCLENNAN, ALEXANDER JOHN Greenwich 1866-74

McGHAN, R. Greenwich 1851-66

McKELLOW, JOHN Maidstone 1823-died 1862

Pigot's Directory of 1832 notes his address as 56 King Street. He died on 17 November 1862.

McKELLOW, Miss J. Maidstone 1851-5

McKENZIE, ALEXANDER Brompton 1847

McNEIL, WILLIAM Greenwich 1874

NAGNUS, SIMON Chatham 1838

MALLERY, JOHN Chatham 1832

MANLEY, JOHN London pre-1777
  Chatham 1777-1828

Baillie records this makers dates as pre-1782-c1790. He was a Freeman of the Clockmakers'

Company and worked primarily as a watchmaker. Loomes extends these dates to 1777-1828. An advertisement in the *Kentish Gazette* in August 1777 states that he had moved from London and had opened a shop opposite the George in Globe Lane. An apprentice was also required (see Appendix I).

| MANNING, W. H. | Tunbridge Wells | 1874 |

| MANSELL, Mrs MARY | Boughton | 1847 |

| MANWARING, RICHARD | Maidstone | 1838-58 |

In the *Topograph of Maidstone* published in 1839 his address is given as 38 Earl Street and he is noted as a watchmaker, silversmith and jeweller. Other directories in 1858 also give Earl Street as his address, but note that he was a postmaster at this date.

| MARLOW, JOHN | Cranbrook | 1732-47 |

Noted in Loomes as a watchmaker, and a watch signed by him is in Canterbury Museum.

| MARSH, JOHN Snr | Eastry | 1771, died 1788 |

Noted in Loomes as a clockmaker and as dying in 1788. A watch by him was reported lost in June 1786, and in September 1788 William Potts, who was either a journeyman or an apprentice, absconded 'taking with him several of his Master's Tools' (see Appendix I).

| MARSH, JOHN Jnr | Dover | 1825-47 |

The son of John Marsh of Eastry is noted by Loomes as being apprenticed in 1825

| MARSH, CHARLES HOLLANDS | Dover | 1834-66 |

Loomes suggests that Charles is the son of John Jnr and notes that he was apprenticed in 1834.

| MARSHALL, CHARLES | Greenwich | 1839-47 |

| MARTIN, A. | Canterbury | 1846 |
Watchmaker.

| MARTIN, ALFRED | Greenwich | 1866-74 |

| MARTIN, CELESTIN | Sandwich | 1862-82 |

Kelly's Directory of 1862, 1878 and 1882 note a working address in King Street.

| MARTIN, JAMES | Faversham | 1847-55 |

Known as a watchmaker and believed to be the son of John Martin.

| MARTIN, JOHN | Faversham | 1818-23 |

An 8-day arch-dial painted with floral motifs housed in a typical Kentish case has been inspected (Fig 7/82). Inside is pasted the trade label of Goulden of Canterbury, clockcase maker, who was responsible for the majority of the cases made in the area in the early nineteenth century.

| MARTIN, JOHN | Maidstone | 1832-72 |

Pigot's Directory (1832) lists John Martin working in Middle Row, High Street as a jeweller and silversmith, while *Vivish's Directory* (1872) gives an address at 25 Week Street, so if it is the same man he had moved premises.

MASON, CHARLES   Canterbury   1838-51
A watch signed Mason has been recorded and the entries in Loomes to a Charles Mason in 1838-40 and a C. Mason in 1851 probably refer to the same man.

MASON & SON   Canterbury   1874
Presumably a partnership involving the man noted in the previous entry.

MASON, — WILLIAM?   Bexley   pre-1762
Noted in Baillie as working prior to 1762 and a watch has been recorded. An 8-day movement in a red-gold lacquered longcase dating from the George II period was sold at Henry Duke's auction in Dorchester in 1995

MASON & JENKINS   Rochester   1795

MASOTH, CHARLES   Canterbury   1847
Bagshaw's Directory of 1847 records Charles Masoth at 28 St George's St, but it is probably a mis-spelling of Mason.

MASPOLI, MONTI & CO.   Sandwich   1838-40
See also the entry for the Monti family.

MASPOLI, PETER   Sandwich   1845-7

MASPOLI, VITTORE   Sandwich   1841
Recorded as being born in Italy in 1810.

MASTERS, JOHN   Tenterden   born 1818, died 1887

MASTERS, JOHN NEVE   Rye (Sussex)   born 1846, died 1928
                                                                                  Tenterden   1855
Loomes records a maker of this name in both Rye and Tenterden and is probably the same man. Though he would only have been nine years of age at the date given for Tenterden, this is presumably a reference in a rate book or some similar official return.

MAY, ROBERT   Deptford   1828-39
Noted in Baillie as receiving an award from the Royal Society of Arts for a detached escapement, described in *Trans Soc Arts* Vol 46.

MAY, WILLIAM ALFRED   Dover   1874

MAYHEW, JOSEPH   Rochester   1866-74

MEMESS, ROBERT   Woolwich   1826-8
Noted as a watchmaker and probably the same person as the man of that name who is recorded in Baillie as obtaining his Freedom in London in 1817. He is known to have been active until 1844.

Fig 7/83 A verge watch movement, No 5813 by John Mercer of Hythe

Fig 7/82 A small oak longcase clock with Kent cresting, 7ft 0in tall including finial, by Martin of Faversham, about 1810

Fig 7/84 A painted dial with a hunting scene in the arch by Thomas Mercer, Hythe, about 1770

MEPSTEAD, JOHN Ash pre-1788

In the 1788 Window Tax records he paid tax on a house with '6 window lights' as a watchmaker. Known to have supplied Ash Church with its third clock in 1789

MERCER, JOHN Hythe pre-1736-79

An interesting maker about whom little is known and whose origins are unclear. The *Kentish Post* records a watch by him being lost in February 1736 (Appendix I). The Hythe archives note that John Mercer, clocksmith, was a party to a Land Uses Deed in 1738. He is also known to have been the Mayor in 1756, 1768, 1772, 1777 and 1779. He is recorded in Baillie as a maker of bracket and longcase clocks and also watches, and although a number of longcase clocks and watches signed by him have been seen by the author, no bracket clock has been traced, which is unfortunate given the scarcity of early bracket clocks by Kent makers.

A very good early lantern clock with tulip and floral engraving on the dial and with dolphin frets has been recorded. Regrettably, this clock has been fitted with a fusee movement at a later date, but the style of the dial leads one to suppose that he must have been working very early in the eighteenth century. All of the longcase movements seen have been of good quality with five ringed-pillars and housed in oak or walnut cases.

MERCER, JOHN Jnr Hythe born c1738, married 1780-1797

Both Baillie and Loomes record this man, the son of John, as working towards the end of the century, and numerous watches (Fig 7/83) in particular have survived from this period. However he must have been working at a much earlier date, presumably for his father. The Canterbury records note that he obtained a licence to marry Anne Grigsby aged 32 on 10 August 1780, her father, Elias Grigsby, standing as bondsman. On the licence John is described as a bachelor aged 42, which makes his date of birth some time in 1738, which is approximately the earliest year in which we have a reference to his father. His wife's maiden name of Grigsby is interesting as there is a record of an early maker from Sittingbourne by this name in 1673. Is it possible that there is a family connection and that John senior received his training from Edmund Gregsby (Grigsby)?

MERCER, THOMAS Folkestone pre-1751-1805

Baillie gives his dates as a watchmaker as 1760-71, but as his signature appears on a lease assigning premises in Stade Street in 1805 his working life must be extended to at least then. In 31 July 1751 the Folkestone Assembly recorded:

> Upon the petition of Thomas Mercer watchmaker setting forth that he is desirous of being admitted a freeman of this corporation on such terms and conditions as to this house shall seem meet he is admitted and sworn free on payment of ten pounds ten shillings and court fees.

When, in January 1760 he advertised for an apprentice 'to learn the Art of Watch and Clock Making' he was at Folkestone (Appendix I), but some of the longcase clocks by him are signed at Hythe and both brass and painted-dial examples exist (Fig 7/84). His exact relationship to John Mercer Snr is not known, but he may be one of his sons.

MERCER, WILLIAM Folkestone and Hythe born c1730-died 1791

Recorded by Baillie as a clock and watchmaker, but the actual connection with the other members of the family has not definitely been established. The Canterbury records note a marriage licence in his name in which he is described as a watchmaker age 25 from Hastings in

Sussex. On the licence, dated 31 May 1755, he applied to marry Dorothy Mannings age 22 from Hythe. Whether he settled in Hythe immediately after the marriage is not known, but on 17 October 1761 he advertised in the *Kentish Post* that 'he is coming to settle at Folkestone to carry on the several branches of that business [as clock and watchmaker]' (Appendix I).

Whether he remained in business in Folkestone is not clear, but he appears to have returned to Hythe by at least 1775 as he held the office of Mayor in 1775, 1778, 1780, 1782, 1786, 1788 and 1791 (the year he died).

MERCER, WILLIAM Maidstone eighteenth century
Two clocks signed 'Wm. Mercer Maidstone' have been recorded. One, a 30-hour single handed movement in an oak case, was sold at auction in 1963. Whether he is the same man as the Folkestone maker is not certain, but it seems likely.

MERITO, WILLIAM Canterbury 1847
In 1847 recorded in Watling Street.

MERITT, WILLIAM Canterbury 1838-74
Noted in Loomes as a clockmaker. He is possibly the same man as the previous entry. Merito may be a mis-spelling of Meritt.

MESURE, HENRY Gillingham 1848

MESURE, LIONEL Eltham 1845

MESURE, LIONEL Maidstone 1845

MILLER, WILLIAM Hythe 1826

MILLER, WILLIAM FREDERICK Dover 1823

MILLER, WILLIAM FREDERICK Sandgate 1838
Probably the same man as the Dover entry.

MILNE, Mrs ANN Crayford 1874

MILLES, ISAAC Canterbury 1680/1
Numerous members of this family are noted in the Freemen List of Canterbury, working as linen drapers. On 20 March 1680/1 Isaac was apprenticed to Richard Greenhill, but there is no further mention of him and it is assumed that he did not complete his indentures.

MINORS, G. Woolwich 1855-74

MINNES, JAMES HENRY Tunbridge Wells 1874

MOAT, WILLIAM Sandwich 1867

MOLYNEUX, WILLIAM Blackheath 1802-8

MONK, — Tenterden
See Munk.

| MONTI, ANTHONY | Canterbury | 1845-died 1868 |

Recorded in 1847 at 27½ Palace Street. Died in London 7 December 1868, and described as a silversmith.

| MONTI, ANTONIO | Ramsgate | 1874 |

| MONTI, JOHN | Canterbury | 1847 |

82 Northgate.

| MONTI, JOSEPH | Canterbury | 1838 |

Possibly the Joseph Francis Monti, brother of Anthony Monti (*qv*), who was living in St Helier, Jersey, when the latter died in 1868.

| MONTI, MASPOLI & CO. | Sandwich | 1839 |

Recorded as clock and watchmakers.

| MONTI, PETER | Sandwich | born 1817-95 |

Peter Monti was born in Italy in 1817 and is recorded in various directories in which he is described as a watchmaker and clockmaker and also as a jeweller working from addresses in High Street, New Street and Knightrider Street.

The 1861 Census describes him as a jeweller living in the High Street, married to Amelia who was born in Sandwich and with four children, Thomas, Peter, Laura and Joseph. All of the sons followed him into the business and watches and barometers have been recorded. The Monti family were also involved in a partnership with another Italian immigrant, Peter Maspoli.

| MOON, — | Brenchley | 1845 |

| MOON, GEORGE | Tunbridge Wells | 1839 |

| MOORE, JOHN | Maidstone | 1784 |

Recorded in Baillie as a watchmaker, but nothing else is known.

| MORRIS, JAMES W. | Faversham | 1866-74 |

| MORRIS, JOHN | Gravesend and Dover | 1761 |

Noted in Baillie as a bankrupt in 1761 and as coming to Gravesend from Dover.

| MORTON, PETER | Herne Bay | 1838 |

| MOSES, — | Dover | 1801-28 |

Watchmaker.

| MOUNT, WILLIAM HENRY | Ramsgate | 1847-66 |

| MUDDLE, EDWARD | Chatham | born 1707-1772 |

Very few records exist relating to this maker, who is noted in Baillie as working before 1760 and is known to have made both clocks and watches. His origins are unclear, although there may be a relationship with the Muddles of Tonbridge. Brian Loomes, in *Early Clockmakers of Great Britain*, notes that a longcase clock has been seen dating from the 1700/10 period, although it is difficult to believe that he was working at that early date, particularly as he also has a note that Edward was born in 1707 and was the son of Thomas Muddle of Rotherfield.

A green laquer longcase clock (Figs 7/85-86) and a fine tavern clock made by him with a painted scene of an inn interior on the trunk (Fig 7/87) have been recorded, but very little other work by him is known to have survived other than the turret clock in the Corn Exchange at Rochester. As has been already mentioned in the entry for James Booth, he was responsible for making and installing the new clock in 1771.

The minutes of the Rochester City Council record that in 1770 they 'gave instructions to investigate the overhauling of the Town clock' and at a meeting on 4 May 1771 the council met to decide which of the three quotes they should accept: Edward Muddle, James Booth or George Jeffery. The entry in the minutes reads as follows:

> This day proposals having been delivered by Mr. Edward Muddle, Mr. James Booth and Mr. Jeffery, for the taking down, cleaning and thorough repairing the Town Clock and it appearing that Mr. Muddle's proposal is the lowest and upon the best plan it is agreed that the same shall be done by Mr. Muddle in the manner as set forth in his proposal.

It was also 'ordered that the Chamberlain do pay Mr. James Booth and Mr. Jeffery each one guinea for their trouble in drawing and delivering in plans concerning the Town Clock'.

The clock was presumably delivered and installed by March 1772, as payment of £33 0s 2d to Muddle is recorded in the Chamberlain's accounts, together with further payments of £8 5s 0d and £2 10s 0d later the same year. Whether the clock was just overhauled by Muddle or was a completely new mechanism is not clear. It is known that the original clock installed at the Corn Exchange earlier in the eighteenth century had a square dial, whereas Muddle's clock, which can still be seen housed in the ornate early bracket, has a round dial.

The decision to award the contract to Muddle, an outsider from Chatham, may well have upset the Rochester Freeman Booth, because he immediately moved to London and did not return to Rochester until the 1790s.

MUDDLE, NICHOLAS   Tonbridge   pre-1745-75
The son of Thomas Muddle and keeper of the church and town clocks.

MUDDLE, THOMAS   Tonbridge   1739-75
The Tonbridge records mention this maker as living in Southborough since 1739, but little else is known other than that he was employed to look after the clock on the market house for a guinea per annum plus an extra sum for any repairs which may have been needed. In 1745 his son Nicholas took over this task and his name appears many times in the Tonbridge accounts for looking after and repairing both the town clock and the church clock. The last mention of him is repairing the lock on the church door for a fee of one shilling in 1775. Edward Muddle of Chatham may well be related in some way to Thomas Muddle but no definite connection has been found.

MUMMERY, CHARLES   Folkestone   1839

MUMMERY, THOMAS Snr   Dover   Free 1808-55
Recorded in Loomes as gaining his Freedom in 1808. The Canterbury lists note his apprenticeship to Richard Cramp, watchmaker, on 8 April 1799

MUMMERY, THOMAS Jnr   Dover   1814-47
Noted in Loomes as being apprenticed in 1814, presumably to his father.

Fig 7/85-86 A green lacquer longcase clock, 7ft 2½in tall, by Edward Muddle of Chatham, about 1760. The calendar aperture is surrounded by engraved flowers, birds and a basket of fruit, while the arch has a silvered name boss

MUNGHAM, JOHN   Smarden   1704
Noted in the church records as keeping the church clock, but whether he was actually a working clocksmith is not known.

MUNK, JAMES   Tenterden   c1800
A 30-hour painted dial longcase clock with a plated four pillar movement typical of the early

Fig 7/87 A tavern clock by Edward Muddle of Chatham, with a painting of an inn interior on the trunk, about 1770

Fig 7/88 A 12in square 30-hour painted dial signed 'Munk, Tenterden', about 1800. Like most 30-hour painted dials, there is no falseplate to fix it to the movement. Note the fine brass hands and the arabic hour numerals

nineteenth century signed just 'Munk, Tenterden' (Fig 7/88), and another signed 'James Munk, Tenterden' have been noted by this hitherto unrecorded maker.

MUNN, THOMAS	Maidstone	1858
MUNN, WILLIAM	Sheerness	1838-47
MUNN, WILLIAM E.	Sheerness	1851-74

MURDEN & SON	Dover	1874
MURDOCH, J. G.	Ramsgate	1874
NASH, JOHN	Bekesbourne	1769

In partnership with his brother, William, at Bridge. The two brothers dissolved the partnership as a result of what must have been serious differences, although the details are not known. As a result of slanderous accusations by the brothers (see Appendix I, 26 August and 2 September 1769), John placed the following notice in *Kentish Gazette* on 9-12 December 1769:

> JOHN NASH, Clockmaker, at Beaksbourn
> WHEREAS some malicious Reports have been propagated by my Brother, Wm. Nash, Clock maker at Bridge, representing me as imposing on my kind Employers. — in Particular, that I had overcharged Mr. Drayson of Upstreet, for repairing his Clock, and as such Reports have a manifest tendency to prejudice me in my Business, I have taken the Opportunity to lay the said Charge before some reputable Clock makers of Canterbury; who have confirmed the Equity of the same, and will readily attest, if called upon, the Injustice done to
> JOHN NASH.

NASH, WILLIAM	Bridge	pre-1762, died 1794

Although numerous examples of his watches and clocks have survived the author has been unable to trace his exact origins, but the Nash name appears in the Canterbury records a number of times in the Adisham area, a village close to Bridge. However, there is no reference to him in the marriage licence records.

In September 1762 he announced that he would trade every Saturday at Mr Bushel's, the cornfactor, opposite the market at Canterbury, whereupon John Chalklin of Canterbury — no doubt annoyed that William Nash was trying to poach his customers — announced in the same issue that he would be trading at Bridge (see Appendix I). In February 1763 Nash, described as a clocksmith, reported that his thirteen-year-old apprentice, William Knight, was sent to Littlebourne to 'put up' two clocks, whereupon he absconded with a pony.

He was in partnership with his brother John (*qv*), who is recorded at Bekesbourne in 1769. The content of advertisements placed in the *Kentish Gazette* by his brother suggests that the partnership must have broken up in acrimonious circumstances. The competition with Chalklin, the slanderous accusations against his brother and the absconding of his apprentice, suggest that William Nash may not have been the easiest of people to work with.

He advertised for an apprentice in August 1769 and later took Samuel Hardeman as a partner; Hardeman succeeded to and carried on the business after Nash's death in 1794. His earlier longcase clocks usually have very simple silvered dials with little engraving other than his signature (Fig 7/90). His silvered dial movements were always of good quality with four or five pillars. They can be found in both mahogany and oak cases of the pagoda-topped style. Towards the end of his life he began to use painted dials, but with little decoration other than floral scrolls (Fig 7/89). His painted dials, mostly made by Osborne of Birmingham, were usually attached to the movements by false plates. Watches signed by him are of a high quality (Fig 7/91) and a movement by him is in the Ilbert Collection at the British Museum.

A fine painted dial bracket clock in a lacquer case was sold at auction in 1993 and a lacquer-cased clock by William Nash is another superb example of the tavern clocks produced by Kentish makers during the middle of the eighteenth century (Colour Plate XI).

Fig 7/90 A silvered 30-hour single-sheet arch dial, 12in wide, The arch is finely engraved 'William Nash, Bridge'

Fig 7/89 An 8-day early painted-dial longcase clock by William Nash of Bridge, with gilt decoration on the dial, about 1775

Fig 7/91 A verge watch by William Nash of Bridge

NEALE, RICHARD	Hunton	1866-74

A watchpaper known (Fig 7/103).

NEWINGTON, J. O.	Lamberhurst	1845

NEWMAN, JOHN	Charlton	1821
NEWMAN, J.	Chatham	1855
NEWMAN, JOHN	Bexleyheath	1866-74
NEWMAN, JOHN	Belvedere	1874

Almost certain to be the same man as the Bexleyheath entry.

| NICHOLAS, ROBERT | Sheerness | 1839-47 |
| NICHOLLS, HAMMOND | Canterbury | 1789-1810 |

Noted in Baillie and Loomes as a clock and watchmaker. In 1794 his household furniture was offered for sale, but there is no mention of any working stock or tools (see Appendix I). *Hilden's Triennial Directory* (1805) records him as a watchmaker working in the High Street.

NINNES, J. & W.	Tunbridge Wells	1855-66
NINNES & LOOF	Tunbridge Wells	1851
NOAKES, J. F.	Faversham	1874-84
NOAKES, JOHN THOMAS	Tonbridge	1874
NORTON, WILLIAM	Leeds	1826-39
NORWOOD, GEORGE	Dover	1826-33

Recorded in Loomes as obtaining his Freedom in 1826.

NYE, EDWIN	Woolwich	1874
NYE, THOMAS JOHN	Chatham	1847-66
NYE, THOMAS JOHN	Rochester	1832-9
OCLEE, E. A.	Bromley	1874
OCLEE, JAMES	Ramsgate	c1790

Recorded in Baillie as obtaining his Freedom from the Clockmakers' Company in London in 1790.

OCLEE, JOHN	Ramsgate	1823-7
OCLEE, JOHN	Bexleyheath	1832
OLLIVE, SAMUEL	Lewes (Sussex) & Tonbridge	c1760-1794

Baillie gives 1773-94 for this maker and Loomes extends these to at least 1760. Though the relevant records have yet been consulted to confirm this, the author believes that he was the eldest son of Samuel Ollive, tobacconist of Bull House, Lewes (Sussex), and that he was born in the early 1740s in that town. Derek Roberts (*Precision Pendulum Clocks*, 1986, pages 66-7) illustrates a 30-hour clock by him to the design of James Ferguson FRS, having just a three-wheel train and no motionwork. This clock is signed 'Saml Ollive, Lewes', so he must have worked in his native town before moving to Tonbridge. When he moved and with whom he received his training remains a mystery. An advertisement in September 1780 (Appendix I) reads:

> ... he makes CLOCKS on an entire new Construction, from an Invention of his own, which for Utility, Convenience, and Cheapness, exceed any Thing of the Kind ever before offered to the Public. — They go eight Days, shew the Hour and Minutes, and by Means of pulling a Line, which may be conveyed from the Clock to any Room in the House, strike the Hours the same as an Eight-day Clock. Price 2£. 5s. — with a Case, 3£.6s.
> N.B. As the Beating of Clocks is disagreeable to some People, they may (if required) be made to go silent.

In October 1773 he advertised for an apprentice as a clock and watchmaker, and again in August 1778 (see Appendix I). The latter advertisement, in which he describes himself as a gunmaker, also describes his improvement for charging guns with gunpowder.

From 1775 he seems to have taken over the responsibility for the upkeep of the church clock from Nicholas Muddle. The churchwardens' accounts note that he mended a candlestick in 1773 at a cost of 1s 6d and in 1787 he is recorded as repairing the dial. He also provided a sundial for the churchyard in 1791.

In the *British Directory of Merchants and Trades* published in 1784 he is listed as a watchmaker, clockmaker and gunsmith and he also appears in the Tonbridge Poll Book for 1790. As there is no mention of him in *Finch's Directory* of 1803, Baillie's date of 1794 must be assumed to be correct. For many years a fine tavern clock signed by him hung in the entrance hall of the main coaching inn in the town, the Rose & Crown, in the High Street. This clock was stolen from the hotel a few years ago and to date has not been recovered. Numerous clocks and watches by him have been recorded (Fig 7/92) and a particularly interesting watch with a verge escapement in a later silver and tortoiseshell pair-case was sold at Phillips, Folkestone in 1993. The unusual feature of this watch was that the numerals indicating the hours were replaced by the letters spelling the name FEATHERSTONE with the christian name Ann appearing in the middle of the enamel dial.

OLLIVE, THOMAS   Tenterden   born c1752, married 1777
                                 Cranbrook   1788-died1829

Thomas Ollive was aged 77 when he died on 20 June 1829 and from this we can deduce that he was born in 1752, almost certainly in Lewes (Sussex) and that he was Samuel Ollive's younger brother. Their father Samuel carried on business as a tobacconist from the premises known as Bull House, which can still be seen in the High Street, Lewes. The father had inherited the house from his father, the Rev John Ollive who was minister of the adjacent chapel from 1711 to 1740. As with his brother, his apprentice years are a mystery, but he is believed to have married in 1777, possibly in Tenterden where he was known to have been living in that year, moving to Cranbrook in 1778. Baillie records another Ollive, christian name unknown, as working in Lewes in 1780. Whether this is Thomas is not known for certain and the entry may refer to the third brother, but at the present time there is no evidence to suggest that he too was a clockmaker.

During the research into the Ollive family an interesting connection with the famous radical and revolutionary writer Thomas Paine has emerged. Paine arrived in Lewes in 1768 where he held the post of exciseman, responsible for collecting the duties on tobacco and alcohol, and lodged with the Ollives at Bull House. Prior to this he had led a chequered

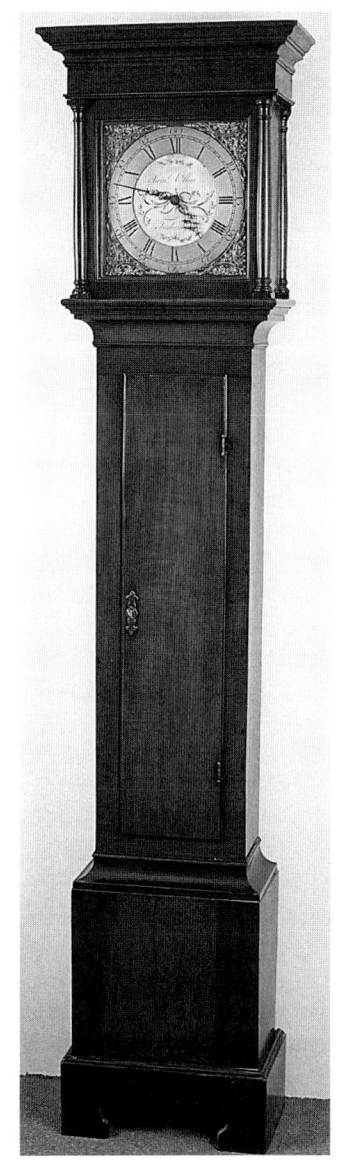

Fig 7/92 A 30-hour longcase clock in a cherrywood case by Samuel Ollive of Tonbridge, about 1770

Fig 7/93 A late eighteenth-century verge watch by Thomas Ollive of Cranbrook

career, being born in Thetford, Norfolk, to Quaker parents and received little or no formal education and was apprenticed to his father, a staymaker, when he was thirteen years old. After learning his trade he joined the crew of a privateer and then worked as a corsetmaker in London for two years, subsequently moving to Dover in 1758. After marrying the daughter of an exciseman he set up in business in Sandwich. The business was a failure and he moved to Margate where his wife died. Following her death he returned to Thetford and applied to join the revenue service in which he served as an exciseman in Lincolnshire at Grantham and then Alford, where he stayed until 1765

Having been dismissed from the service for negligence he once again took up his trade as a staymaker, eventually moving back to London where he worked as an English teacher in a school, while attending lectures at the Royal Society on astronomy and the other sciences. Becoming bored with teaching he applied for reinstatement in the customs service and was given the post of exciseman at Lewes at a salary of £50 per annum.

In July 1769 Samuel Ollive the elder died and Paine joined his widow in the tobacconist business expanding into the selling of groceries. On 26 March 1771, when he was thirty-five years old, he married the daughter Elizabeth, aged twenty-one. During his time in Lewes he had been an active member of a debating club, whose meetings were held at the White Hart, and this marked the beginning of his career as a political writer. In 1772 he wrote a pamphlet urging an increase in wages for members of the revenue service in order to combat the poor conditions that appertained at the time and the consequent corruption in the service owing to the low remuneration they received. Much of his time was spent in London lobbying Members of Parliament and others, and he was subsequently dismissed the service for being absent without leave. His many absences had also caused his grocery business to fail and all his possessions were sold to pay his debts in 1774. On 4 June in the same year he and his wife separated by mutual agreement. In October he left England for America bearing letters of introduction from Benjamin Franklin whom he had met at the Royal Society and who shared a common interest in science, clockmaking and astronomy. In 1776 he published his article entitled 'Common Sense' which formed the basis of the Declaration of Independence, and the rest is history. He returned to London in 1787 where he wrote the *Rights of Man* defending the principles of the French Revolution. On the brink of arrest by the authorities on a charge of sedition he escaped to France where he was elected to the National Convention and subsequently imprisoned, narrowly escaping the guillotine. He subsequently published *The Age of Reason* a treatise against both religious orthodoxy and atheism, and returned to America where he died in 1809 following a truly eventful life.

Soon after their separation his wife Elizabeth went to live at Cranbrook with her clockmaker brother, Thomas. During his time in Cranbrook Thomas Ollive had made the tavern clock with a mahogany case which hangs in the George Hotel. When last seen the case of this clock was in a very poor state and it is understood that at one time the clock was hung on an outside wall of the inn, which would account for its condition. In records dating from 1803 he is described as a clockmaker and silversmith of Cranbrook and in 1805 he took Thomas Oliver Reader into partnerhsip. Owing to the similarity of the two names it has always been assumed that there was some family connection between the two men, but the exact relationship is not known.

Numerous clocks, both with brass and silver dials, have been recorded and the later examples are invariably signed Ollive & Reader. A musical longcase clock signed by both men is noted in Baillie. A watch signed 'Thos. Ollive, Cranbrook' is shown in Fig 7/93.

Following the death of his sister in 1808 Thomas carried on working in partnership with Reader. At some time during this period he also appears to have had a partnership with William Birch of Tenterden. A number of clocks signed jointly by both men have been recorded (Fig 7/94). He died on 20 June 1829 and he, his wife and sister are buried in the churchyard of St Dunstans, Cranbrook, where their gravestones can be seen to the west of the tower. Reader carried on the business after Thomas's death until 1840.

OUSMAR, O.　　　　　　　　　　**West Malling**　　　　　　　　　　**1838**

Fig 7/94 A 12in diameter silvered brass dial wall clock by Ollive & Birch, about 1800. The movement has tapered plates, four pillars and an anchor escapement

Fig 7/95 A mahogany-cased bracket clock by Henry Overall of Dover, about 1780

| **OVERALL, HENRY** | Dover | 1778-95 |

Baillie records him as a watchmaker, but a bracket clock is also known (Fig 7/95). A lacquer longcase clock of about 1780 and a watch signed at Romney recorded by Baillie are probably by the same man.

OVERALL & BRADSHAW	Dover	1802
PADGHAM, JOHN	Bethersden	1855
PAGE, HENRY	Margate	1839
PAGE, JAMES	Dartford	1839-47
PAGE, THOMAS	Dartford	1838-47
PAGE, WALTER	Tunbridge Wells	1874
PAIN(E), MARK	Ash	born c1791-1847
	Sandwich	1845-62

The 1861 Census returns for Sandwich note this man as a watchmaker from Deal and gives an address at King Street, which is the same location as that recorded in *Kelly's Directory* for 1851 and 1855. The Census records his age in 1861 as 70 and states that he was married to Elizabeth aged 71 from Ash. He is noted in *Pigot's Directory* of 1826 as working in Ash. He presumably moved to Sandwich some time after his marriage.

A mahogany longcase clock with a painted dial movement with strike/silent work in the arch has been recorded, signed 'Mark Pain Deal'.

PAIN(E), STEVEN	Sandwich	1858

Recorded as working from the same address in King Street as the previous entry and believed to be his son.

PALMER, H.W.	Margate	pre-1813

The entry in Baillie is wrongly dated as pre-1713. Watches signed by him have been recorded.

PALMER, ROBERT	Ramsgate	1874
PALMER, THOMAS	Woolwich	1845-7
PALMER, WILLIAM	Westerham	1826
PANKHURST, A.	Maidstone	1866
PANKHURST & MUNN	Maidstone	1851
PAPPRILL, JOSEPH	Canterbury	Free 1789

In the Canterbury freemen lists he is noted as a watch-motion maker from London. He married a widow, Elizabeth Leishman, in 1789 and received his Freedom by marriage on 13 October of the same year. After gaining his Freedom he presumably carried on making and supplying watch parts to the numerous watchmakers working in the city at this time, but no other reference to him has been found.

PARKER, AUG^t (AUGUSTUS?)	Milton (Gravesend)	early eighteenth century

A hitherto unrecorded maker. A number of clockmakers named Parker are noted as working in London in the early and late seventeenth century. Whether there is any connection with them is not known.

The only clock seen is a good example of an early eighteenth-century 30-hour bird-cage movement with decorative steelwork and a 10¾in dial, exhibiting many of the characteristics of the 1705 period. The finely-engraved silvered chapter ring has outside minute divisions and is signed 'Augt. Parker, Milton *fecit*'. The dial centre is completely engraved with floral scrolls and foliage reminiscent of the lantern clock dials of the early eighteenth century by Cutbush, Kingsnorth and Thatcher. The spandrels are of the crown-and-cherub type. Both the minute and hour hands are later replacements.

PARKER, JOHN	Greenwich	died 1766

Little is known of this maker other than that he died in 1766 and his widow carried on the business after his death.

PARKER, JASPER OR JOSEPH	Greenwich	pre-1780

PARKER, MARY	Canterbury	1782
PARKHURST, ALEXANDER	Maidstone	1858
PARNELL, THOMAS	Canterbury	apprenticed 1773-1805

Thomas Parnell is noted in the records as a watchmaker and clockmaker and carried on his business by the Westgate. He was the son of Thomas and Mary Parnell and his father was a gingerbreadmaker. In 1773 he was apprenticed to Richard Cramp, watchmaker, and obtained his Freedom by Patrimony on 12 May 1784.

He is recorded as taking Charles Dowsett and Thomas Mummery as apprentices in 1792 and 1799 respectively, but no surviving watches or clocks by him have been traced, though one was reported lost in July 1784 (see Appendix I).

Hilden's Triennial Directory of 1805 gives an address in St Georges Street, which is the other side of the city.

PARQUOT [PARQUAIT], —	Canterbury	pre-1750

A watch by this man was reported lost in February 1749/50 (see Appendix I), but there is no trace of him in the Canterbury records. It is possible that he is the same man as the Peter Parquot who is noted by Baillie as being made bankrupt in London in 1723.

PATRICK, G.	Greenwich	early nineteenth century

Watchmaker.

PATRICK, GEORGE L.	Strood	1866-74
PATRICK, JOHN	Maidstone	1838-47

In *Bagshaw's Directory* of 1848 and *Pigot's Directory* of 1840, his working address is given as 9 Pudding Lane.

PATRICK, JOHN	Strood	1851-5
PATRICK, MARY ANNE	Greenwich	1847
PATRICK, Mrs L.	Strood	1866
PATRICK, MILES	Greenwich	1795

Baillie records a repeating watch movement by this maker.

PATRICK, WILLIAM	Greenwich	1827-39

Recorded in Loomes as a watchmaker.

PATTISON, ROBERT	London	born c1662, apprenticed 1676-83
	Greenwich	after c1683

This little-known maker was born about 1662 and was apprenticed to none other than Thomas Tompion in 1676 until 1683. There is no record of him obtaining the Freedom of the Clockmakers' Company after the completion of his indentures.

However, a longcase clock signed 'Robert Pattison, Greenwich' is known (Fig 7/96) and it must be assumed that he moved there some time after 1683 to set up business on his own. The jurisdiction of the Clockmakers' Company does not seem to have extended as far as Greenwich, which at that time was part of Kent.

Colour Plate X A superb green lacquer longcase clock with a 'penny moon' in the arch, by Thomas Jenkinson of Sandwich, about 1725

Colour Plate XI An 8-day timepiece tavern clock by William Nash of Bridge, with painted decoration on the trunk door, about 1780

| PAUL, JAMES | Gravesend | 1847 |

| PAUL, JAMES | Strood | 1847 |

Probably the same man as the previous entry.

| PAY, JAMES | Hythe | 1866-74 |

| PAYNARD, JAMES | Strood | 1808 |

Loomes records this man as a watchmaker.

| PAYNE, JOHN | Lenham | born 1731-95 |

He is noted in Baillie as being born in 1731 and recorded as both a clock and watchmaker. Nothing else is known, although a number of clocks, both 30-hour and 8-day longcases in oak and mahogany cases, have survived. They are all of good quality, some with painted dials and others with all-over silvered dials (Fig 7/97).

| PAYNE, JOHN | Sevenoaks | late eighteenth century |

At least two very good quality movements with silvered dials signed 'John Payne, Sevenoaks' have been recorded. One of these was a fine late eighteenth-century 30-hour birdcage movement, the other an 8-day example with a silver dial in a London-quality mahogany case. Whether there is any connection between Payne of Lenham and the maker from Sevenoaks has not been established. However, it is possible that they are the same person as the dials and movements bear striking similarities to the clocks bearing the Lenham signature.

| PAYNE, M. | Lenham | late eighteenth century |

An 8-day painted dial clock with date and seconds indicators in a good arched-dial case has been recorded, but nothing else is known of this maker. He is probably related to John Payne.

| PAYNE, T. E. | Tunbridge Wells | 1874 |

| PAYNE, WILLIAM | Deal | 1796 |

Noted in Baillie as a clock and watchmaker. He may be related to Mark Pain of Sandwich who is known to have been born in Deal in 1791.

| PAYNE, WILLIAM | Lenham | died 1803 |

Reported in the *Kentish Gazette* as dying in July 1803 and noted as a watchmaker, leaving a widow and three children. Probably the same man as the maker from Deal.

| PEARSON, WILLIAM | Maidstone | 1826-38 |

| PECKHAM, JOHN | Ashford | 1741 |

Recorded in Baillie as a clock and watchmaker, but nothing else is known other than an advertisement in the *Kentish Post* 12-15 August 1741: 'John Peckham, Clock and watchmaker at the Sign of the Spring Clock in the town of Ashford makes, mends and sells all sorts of repeating and Plain Clocks and Watches at the cheapest rates ....' Loomes records a John Peckham of Lewes, Sussex, apprenticed to Thomas Barrett in 1733. There may very well be a connection and it is possible that the Thomas Barrett of Lewes is related to the Barrett family from Canterbury who were noted gunsmiths and clockmakers in the late seventeenth and early eighteenth centuries.

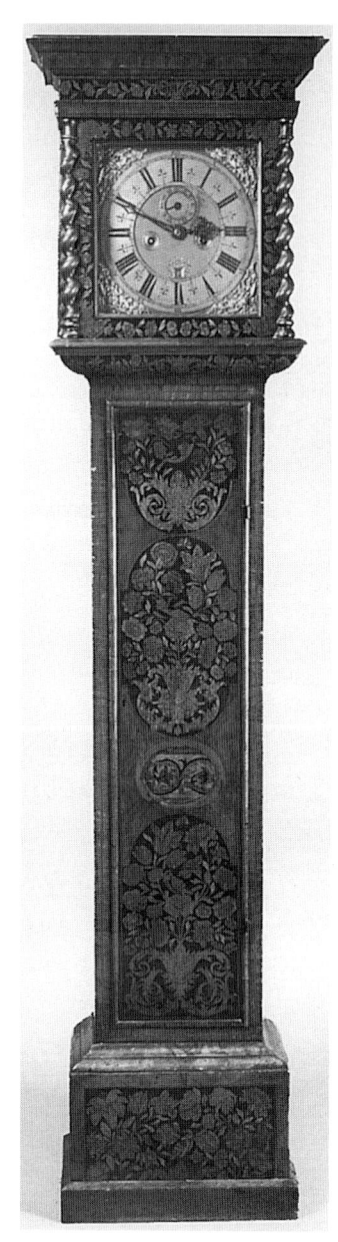

Fig 7/96 A marquetry longcase by Robert Pattison, Greenwich, about 1690, with panels of flowers and birds. The glass lenticle in the trunk door has been replaced by marquetry and the plinth has been shortened

Fig 7/97 An 8-day five-pillar clock by John Payne of Lenham, about 1760. There is good engraving in the dial centre with well-finished rococo-style spandrels. The substantial oak case has a finely pierced fret at the top of the hood

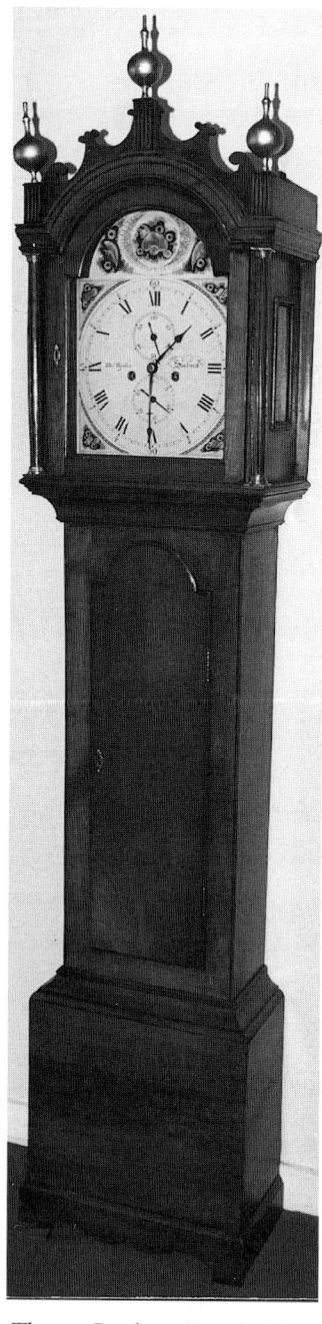

Fig 7/98 (right) An 8-day clock by Thomas Pegden of Sandwich, about 1810. The painted dial has unusual shell decoration in the arch and spandrels. The cherrywood case, with a Charles Lepine trade label, is almost identical to those from the Goulden workshop

| PEGDEN, GEORGE ROBERT | Deal | 1866-74 |

| PEGDEN, THOMAS | Sandwich | 1802-45 |

Finch's Directory (1802) notes him as a watchmaker. *Pigot's Directory* of 1826/7 gives his business address as Strand Street, where he had moved to in 1825/6 as prior to then he had worked in King Street. The last reference to him appears in *Kelly's Directory* of 1845

The Canterbury Museum collection possesses a watch movement, and a fine 8-day painted-dial clock in a very good mahogany case typical of those made by Goulden of Canterbury was sold in Canterbury at auction in 1992. Another similar clock housed in a good cherrywood case was also sold in Canterbury in 1995 (Fig 7/98) and had a trade label pasted inside, advertising Charles Lepine, clockcase maker of Canterbury (Fig 7/78). The style of this case is almost identical to those known to have been made in the Goulden workshop.

| PEGDEN, VINCENT | Deal | 1851-8 |

| PEGDEN, VINCENT | Herne Street | 1826-8 |

| PEMBLE, GILBERT | Eythorne | c1760 |

A 30-hour 12in dial birdcage movement with two hands and with good rococo engraving in an oak case was sold at Phillips Folkestone in 1994. A good five-pillar 8-day movement in an oak case has also been recorded. The dial with fine rococo engraving to the centre surrounding the signature suggests a date of about 1760. Nothing else is known.

| PENNY, ALEXANDER | Dover | 1845 |

| PERCIVAL, JOHN | Woolwich | late eighteenth century-1811 |

These dates are taken from Baillie who also records a clock and watch signed by him.

| PERCIVAL, MARY and THOMAS JAMES | Woolwich | 1817-55 |

Loomes extends Baillie's dates of 1817-24 for Thomas James Percival from 1824 to 1855. Nothing else is known but they and the following man are presumably related to John Percival.

| PERCIVAL, JAMES | Woolwich | 1839 |

| PERDUNE, ROBERT | Margate | 1826-8 |

PEREN, THOMAS — see PERRIN, THOMAS

| PERKINS, RICHARD | Dover | 1726 |

Goldsmith who also made and sold 'good watches' (see Appendix I, 22-26 October 1726).

| PERRIN, THOMAS | Smarden | pre-1705-34 |

The churchwardens' accounts of 1705 record that Thomas Perrin was paid £2 5s for 'looking after the church clock', but the only other record is an 8-day movement with a brass and silver dial in a typical case of the 1730 period seen in the Goudhurst area.

| PERRIN, THOMAS | Smarden | c1760-80 |

Loomes also records a Thomas Peren about 1760-80, which is probably a mis-spelling. There were probably two makers of this name, possibly father and son, but at the present time there is no evidence to confirm this.

PERSE, FREDERICK	**Chatham**	**1874**
PETT, ROBERT	**West Malling**	**1729**

Little is known of this hitherto unrecorded maker who may have been working as early as 1712. The Vestry Book for West Malling Church records a payment of £ 2 17s 0d in November of that year for repairs to the clock following damage sustained during the great storm of 17 November. Unfortunately there is no mention of to whom the payment was made. We do know, however, that Pett was responsible for the upkeep of the church clock as the following appears in the churchwardens' accounts for December 1729: 'Paid Rob Pett as his bill appears … 3. 8. 4.' There is also mention of a payment of 3s 6d to a Thomas Pett in the same accounts. Whether he too was a clockmaker is not known, but the style of the known clocks signed either 'Robert Pett' or just 'Pett' suggest that either Robert worked for a very long period during the eighteenth century or that there were two makers bearing the same name.

The earliest clock to survive is a fine example of a 30-hour birdcage movement with brass pillars and good steelwork, with date and seconds indicators on a very well engraved 10in dial. The spandrels are finely cast with cherubs-and-crown typical of the 1705-1715 period and this clock is probably no later than 1710 (Figs 7/99-100). At least two 8-day examples exist dating from the 1730-40 period, both with arch-dials and seconds and calendar work. One has a sunburst in the arch, the other is signed on a cartouche and has dolphin arch spandrels with mask spandrels surrounding the chapter ring. Both of these clocks are of London quality with five-pillar movements. The latter clock is housed in a good lacquer case and,

Figs 7/99-100 The 10in square dial and movement of a 30-hour clock with date and seconds dials by Robert Pett of Malling, about 1710. The birdcage movement is of high quality with brass pillars and good steelwork, and is housed in a tall, slim oak ase

according to a letter on file at the Maidstone Museum, is in the Spanish Embassy in Lisbon, but there is no explanation as to how it found its way to its present location.

At least two other 30-hour clocks have been recorded, both with silver dials in the 1790s style, together with a small arch-dial 30-hour movement with alarm work typical of the last quarter of the eighteenth century. These three clocks are all signed 'Pett' with no christian name. It is possible that there were indeed two makers, perhaps father and son, but so far this has not been substantiated. Moreover, there is no record of a Robert Pett on any apprentice or freemen list, and the only other makers with this name or similar are William and Samuel Petty. Baillie suggests that Samuel also used the name Pett, but any connection with these two London makers must remain pure conjecture at this stage.

Name	Place	Date
PETT, THOMAS (?)	Malling	mid-late eighteenth century

See previous entry.

PHILCOX, GEORGE (?)	Canterbury	1823
PHILCOX, GEORGE	Chatham	1826-8

Believed to be the same man as the previous entry.

PHILLIP, MYER	Rochester	pre-1767

Recorded by Loomes as a watchmaker.

PHILLIPS, SOLOMON	Canterbury	1832
PIERCE, THOMAS	Ash	1866-74
PIERCE, THOMAS	New Romney	late eighteenth century

Recorded by Loomes as a clockmaker, but nothing else is known.

PIKE, JAMES	Eltham	1805-32

Baillie records this man as working 1805-8 but Loomes extends these dates to 1832 and notes him as a clockmaker.

PIKE, JOHN	Eltham	1820-42

Presumably related to the previous entry.

PIKE, LAWRENCE J.	Eltham	1839
PIKE, RUTH	Eltham	1839-55
PIKE, Mrs SARAH	Eltham	1845
PILKINGTON, J.	Woolwich	1826-39
PIRKIS, G.	Strood	1838-55
PIRKIS, GEORGE	Gravesend	1847

Probably the same man as the previous entry.

PIXLEY, JOHN	Gravesend	early nineteenth century

Baillie records a watch signed by this man.

POLLARD, J.	Canterbury	1855
POLLARD, J.	Wye	1866
POLLARD, JAMES	Eastry	1874
POLLARD, JAMES	Lenham	1838
POLLARD, SAMUEL	Canterbury	1840, died pre-1865

Noted in *Bagshaw's Directory* as being at 5 St Peter's Street.

POLLARD, SAMUEL	Whitstable	1838
POLLARD, T. S.	Gravesend	1855
POPE, JOHN	Canterbury	after 1766

Though listed separately by Baillie, he is almost certainly the same as the watchmaker from Margate.

POPE, JOHN	Margate	pre-1766-9

An advertisement in the *Kentish Post* 1-4 October 1766 stated:

> MARGATE, JOHN POPE, Watch-Maker,
> TAKES this Opportunity of returning his sincere Thanks to his Friends, and the Publick, for their past Favours when in Business with Mrs. Swan: He also begs Leave to acquaint them, that he hath taken a Shop next door to Mr. Covell, Hoyman, in Highstreet, where he hopes for a continuance of their Favours. He intends to Make and Repair all sorts of Clocks and Watches, and warrant them to perform very Correct, on moderate Terms ....

John Pope, noted as 'an eminent watch-maker' was admitted as a Freeman of Margate in 1769, and a watch by him was reported lost in May 1785 (see Appendix I).

PORTER, CHARLES	Dartford	1823
POTTS, WILLIAM	Eastry	1788

Clockmaker, who as either a journeyman or an apprentice to John Marsh, 'left his Work unfinished, taking with him several of his said Masters Tools ... he is particularly attached to the Contents of the Gin-bottle.' (See Appendix I, September 1788.)

POULTON, E.	Woolwich	1874
POYNARD, NOAH	Rochester	1847
PRATT, JOHN	Dover	Free 1793

Recorded in Loomes as gaining his Freedom in 1793

PRATT, THOMAS	Canterbury	c1800-55

Noted in Loomes as a clockmaker and Baillie records a travelling clock in the Virginia Museum. Recorded in 1847 at premises in Best Lane.

PREBBLE, CHARLES	Margate	1823-51

PREBBLE, EDWARD	Folkestone	1858
PREBBLE, JAMES	Margate	1826-74
PRESS, J.	Rochester	1874
PRETTY, JOHN	Deptford	1796-CC1807

Baillie records his apprenticeship in 1796 and 1807 as the date he gained his Freedom from the Clockmakers' Company.

PRICE, GRIFFITH	Ashford	1799

Recorded by Baillie as a clock and watchmaker. but nothing else is known at present.

PRICE, DANIEL	Wye	1832-66
PRICE, JAMES	Ashford	1855-74
PRICE, JOHN	Ashford	1852-74
PRICE, JOSEPH	Dover	apprenticed 1814

Loomes notes this man as beginning his apprenticeship in 1814.

PRICE, JOSEPH	Chatham	1823
PRICE, RICHARD	Hythe	1858-74
PRICE, T.	Wye	1855
PRICE, THOMAS R.	Maidstone	1845-74

Recorded in *Bagshaw's Directory* of 1848 and noted in *Vivish's Directory* of 1872 as trading at 39 Week Street.

PRYOR, E. J.	Dover	1866
PULLEN, H.	Bromley	1874
PUNNETT, Mrs JANE	Lynsted	1866-74
PUNNETT, JOHN	Cranbrook	1669

There is a reference to John Punnett in the Cranbrook churchwardens' accounts in 1669 noting that he kept the church clock, succeeding Thomas Punnett who had moved to Rye. The relationship between them is still not clear, but it is assumed that they were father and son.

PUNNETT, THOMAS	Battle & Cranbrook	pre-1656-c69
	Rye (Sussex)	c1669-died 1713

Although his origins are unclear it is possible that he did come from the Cranbrook area, as the earliest reference to him appears on the Battle Church records where 'Thomas Punnett of Cranbrook' was paid £10 for supplying a new clock to the church in 1656, so we can assume that he was working at a much earlier date. The clock that Punnett supplied to Battle may not have had an outside dial as there is an interesting reference to Robert Cornish of East Grinstead carrying out work on the clock in 1686-87 and receiving £1 10s 0d in payment, presumably for supplying the extra wheelwork and external dial.

The contract between Punnett and the churchwardens is important as it is one of the earliest records of an outside dial being provided. The majority of the early church clocks were installed inside the church and only sounded the hours on a bell. At Mayfield Punnett installed two dials, one inside and one outside. The cost of the actual clock together with the outside dial was £10 and £1 10s 0d for the inside dial.

At Cranbrook he carried out various repairs to the clock between 1659 and 1669, which is probably the year he moved to Rye, as the responsibility for looking after the clock was taken over by John Punnett. Thomas's signature appears in the Rye vestry minutes in 1676, 1694 and 1696, and after his death in 1713 he was buried in Rye.

PURSEY, WILLIAM	**Sheerness**	**1847**
PUTLEY, FRANCIS	**Dover**	**1874**
PYKE, J.	**Chatham**	**nineteenth century**

Noted in Loomes as a clockmaker.

PYSING, JOHN   **Canterbury**   **pre-1756**

A watch signed by this maker is reported as being stolen (see Appendix 14-18 February 1756), but nothing else is known. He does not appear in the Canterbury records.

QUAIFE, THOMAS   **Hawkhurst**   **1866-74**

A very late example of a watch with a verge escapement bearing the assay marks for 1879 has been recorded. He was known as a watch manufacturer and had a small factory on the Hastings Road, near Loose Farm, where he took out a patent for making watch cases in 1853. He subsequently moved to Battle, Sussex, where he looked after the church clock.

Loomes also mention a Thomas Quaife of Mountfield in Sussex also moving to Battle during the middle of the nineteenth century. Is it possible that they are one and the same?

QUAIFE, THOMAS SAMUEL   **Canterbury**   **1874**

QUESTED, THOMAS   **Wye**   **pre-1780-87**

Noted in Baillie as a watchmaker, but nothing else is known other than the following advertisement in the *Kentish Gazette*, 1-4 November 1780, when his business must have failed:

> THE ASSIGNEES of THOMAS QUESTED, of Wye, Watchmaker, do hereby give Notice, that they will attend at the KING's HEAD, in Wye aforesaid, on Tuesday next, the 7th Day of November instant, at Two o'Clock in the Afternoon, in Order to make a Dividend of the Monies arising from the Sale of Mr. QUESTED's Effects; when and where the Creditors are required to attend to receive the same.   Wye, Nov. 1, 1780

However, in March 1787, a lost watch was to be returned to 'Thomas Quested, Watchmaker at Wye', but it is not clear if he was back in business by then or not (see Appendix I).

A finely-engraved arch-dial 8-day five-pillar movement with moon-phase was sold in Canterbury in 1995 (Figs 101-102). The signature is on a cartouche in the dial centre and the movement is in a good walnut case in a typical late-eighteenth century style peculiar to Kent and particularly to the Goulden workshop. Although, presumably dating from the 1780s, the dial has many of the features associated with clocks of some 20 years earlier (Figs 7/101-102).

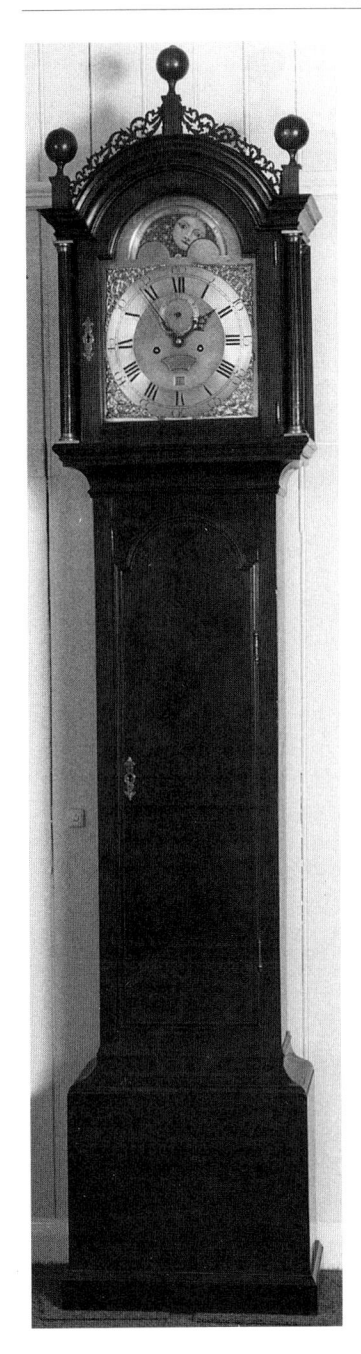

Figs 7/101-102 An 8-day five-pillar clock with moon-phase in the arch, by Thomas Quested of Wye. The original walnut case has a pierced cresting and later finials

RAE, ALEXANDER   **Tunbridge Wells**   c1810

Baillie notes this maker as working in the early part of the nineteenth century and records a bracket clock by him. Nothing else is known.

RAGGETT, JAMES   **Ramsgate**   1823-7

RALPH, GEORGE	Sandwich	1832
RALPH, WILLIAM	Sandwich	1826
RANDALL, A.	Greenwich	1866-74
RANGER, THOMAS	Chipstead	c1740-73

Baillie records a watch by this man, which presumably has the assay marks for 1773. However, he was probably working as early as 1740 judging by two 30-hour posted frame movements seen by the author. Both of these clocks had plain matted 10in dials with mask spandrels and single hands and were housed in oak cases typical of the period. A later 8-day clock with an arched dial in an oak case with parquetry inlay, with the name inscribed in the arch, and dating to no later than 1770, has also been inspected.

This particular clock case obviously came from the same cabinetmaker's workshop as two Baker clocks with an unusual method of fixing the glass to the hood door; the only fixing being a bottom rebate and pins to hold the sides and top of the glass flush against the hood door instead of housing the glass in a rebate with putty, as is the normal practice (se page 52).

RATHBORN, JOHN	Deptford	early nineteenth century

Baillie records a watch by him.

RAWFINGER, JOHN	Deptford	1622

Named as one of the 'alien' clockmakers in the petition made to the King in 1622 by sixteen of the leading London makers asking for restrictions to be placed on their method of working and that the importation of foreign clocks be prohibited. Nothing else is known.

RAWLINGS, EDWARD	Canterbury	died 1750/1

A clocksmith who died in January 1750/1, when his goods and tools included 'Clocks new and old, Watches, an engine for cutting Clock-Wheels, a Barrel Engine, Clock Lades [lathes] and Vices' (see Appendix I). The barrel engine would be for cutting grooved winding barrels. No work by him is known.

READ, EDWIN	Sevenoaks	1874
READER, THOMAS OLIVER	Cranbrook	1805-died 1845

Loomes gives his dates as 1839-45, however it is now known that Reader was taken into partnership by Thomas Ollive of Cranbrook in 1805. His exact date of birth has not definitely been established, but it is believed to be during the 1780s. Owing to the similarity between the two names the two men may have been related, although this has yet to be substantiated.

During his partnership with Ollive, a number of clocks were produced and these all bear the signatures of Ollive & Reader. At least one longcase clock with a musical movement has been noted. After the death of Thomas Ollive in 1829, Reader carried on working alone and he is listed in *Pigot's Directory* of 1832 as a clockmaker and silversmith. He is believed to have died in 1845 in Cranbrook.

REED, JAMES	Rochester	1849
REEVES, BENJAMIN	Lamberhurst	pre-1774-died 1790

These dates are taken from Baillie, but judging by some of the numerous clocks that have survived, he was probably working very much earlier than this. In addition to 30-hour clocks dating from the 1740-60 period, numbers of longcases have been recorded, usually in pagoda-topped cases, both in oak and mahogany finishes. Lacquer examples have also been noted. Reeves also made some fine watches and one particularly good verge movement in a silver pair case is known.

He is also recorded as working on the church clock at Battle, Sussex, where the churchwardens' accounts note that he was paid £11 3s 0d for repairs to the clock in 1774.

Although he is known to have died in 1790 more research into his early background needs to be done, particularly to investigate a possible link to William Reeves of Rye, an early eighteenth-century maker, noted in Baillie as a watchmaker.

REEVE(S), SAMUEL   Chatham   1725
Baillie records his bankruptcy in 1725 and notes him as a clockmaker. No other information at present.

REUBEN, JACOB   Dover   1823-32

REVELL, E.   Eltham   1866

REYNOLDS, RICHARD   Dover   c1710
Recorded in Loomes as a clockmaker.

RICH, THOMAS   Chatham   1792

RICHARDS (RICHARDSON), THOMAS   Westerham   c1730
Recorded in Loomes as a clockmaker, but nothing else is known. A 10in birdcage 30-hour movement with the signature 'Thomas Richardson' on a cartouche in the dial cente has been recorded. Probably the same maker.

RIEDEL, LUDWIG   Chatham   1866-74

ROBATS, ALEXANDER — see ROBERTS, ALEXANDER

ROBBINS, J.   Ramsgate   1866

ROBBINS, THOMAS   Canterbury and Barham   born 1819-78
Son of William Robbins (*qv*).

ROBBINS, THOMAS I   Buckland   born 1795-1866
Loomes gives dates of 1839-66, but he is now known to have been the younger brother of William Robbins and was born in 1795 at the Kingston Forge. In later church records he is noted as a clocksmith. He subsequently moved to Buckland and worked from premises in Chapel Lane with his son, also Thomas.

ROBBINS, THOMAS II   Buckland and Walmer   mid-nineteenth century
Son of Thomas I of Buckland, with whom he worked for a while, then on his own in Walmer, but later moved back to Buckland.

ROBBINS, THOMAS   Chatham   c1740-80
Noted in Loomes as a clockmaker. Nothing else is known.

ROBBINS, WILLIAM Kingstone, Canterbury and Barham
born 1787-1832

Recorded by Loomes as a watchmaker with dates 1826-32. It is now known that he was the son of John Robbins, blacksmith of Kingston (near Barham). In 1781 the family moved there from Buckland, where John's father was the blacksmith. Whether William was apprenticed to his father as a blacksmith is not certain, but he started business as a clockmaker in Kingston, and then moved to Canterbury in 1815, where he worked from premises in St Georges Terrace. His son Thomas, born in 1819, joined him in due course in the business, both becoming Freemen of the City. They subsequently moved to Barham, still trading as clock-

Fig 7/103 Late eighteenth-century and early nineteenth-century watch papers by (clockwise from top left): French of Yalding, Roberts of Charing, R. W. Farmer of Hollingbourne and R. Neale of Hunton

makers, Thomas being noted as such in *Kelly's Directory* of 1878. A watch with a painted dial is in Maidstone Museum (Fig 7/7) and at least two 8-day painted dial longcase clocks with mahogany cases in the Lepine/Goulden style are known.

| ROBERTS, ALEXANDER | Charing | 1826-55 |

Watch paper known (Fig 7/103). The Alexander Robats of Charing listed by Loomes with adate of 1826 is probably the same man.

| ROBERTS, JAMES | Ashford | pre-1803 |

An arched-dial movement with date and calendar work in an oak case has been recorded. The dial is painted in typical Kentish style with floral spandrels and sprays of flowers in the arch, signed 'Jas Roberts'. Nothing else is known, other than his reported bankruptcy as noted in the following advertisement in the February 1803 *Kentish Gazette*:

> [for Sale] A substantial Freehold Messuage in High-street Ashford, with stable and garden; also with the field behind; late in the occupation of Mr. James Roberts, Silversmith, watchmaker and Stationer; with the Stock in Trade ... including several new and old eight-day and thirty-hour clocks, in mahogany and wainscot cases ... and a Library, of more than 1,000 volumes.

Later that month the contents, belonging to 'Mr. Roberts — bankrupt' were offered for sale.

| ROBERTS, JOHN | Folkestone | 1847-55 |

| ROBERTS, ROBERT EDWARD | Canterbury | 1874 |

| ROBINS, E. W. | Gravesend | 1874 |

| ROBINSON, WILLIAM | Canterbury | Free 1761 |

He is listed in the Canterbury records as a watchmaker from London and obtained his Freedom by marriage to Catherine Oakeshott on 24 May 1761.

| ROCE, J. H. | Ramsgate | 1855 |

Chronometers signed by this maker have been recorded.

| ROMNEY, GEORGE | Sandwich | 1891 |

| ROSE, JAMES H. | Ramsgate | 1866-74 |

| ROSS, WILLIAM | Dover | apprenticed 1827 |

This man appears in the Dover apprentice lists as starting his indentures in this year, but nothing else is known.

| ROWLAND, THEOPHILUS | Dover | apprenticed 1829-55 |

Recorded as apprenticed in 1829 and known to have worked until at least 1855

| RUCK, JOHN HENRY | Chatham | 1847 |

| RUFFELL, WILLIAM | Tunbridge Wells | 1874 |

| RUMSEY, JOHN | Woolwich | pre-1723 |

Recorded in Baillie as bankrupt in 1723.

RUSSELL, FREDERICK	Maidstone	1872

Vivish's Directory of 1872 gives 61 King Street as his business address.

RUSSELL, R.	Upper Sydenham	1866-74
SALMON, H. S.	Chatham	1855
SAMUEL, J.	Tenterden	1811

Noted in Loomes as bankrupt in 1811. Nothing further is known of his previous working life.

SARTIN, G.	Lewisham	1866
SAVAGE, HENRY	Canterbury	pre-1766

The entry in Baillie, which presumably refers to the assay date marks on a watch, is the only record of this maker.

SAWYER, JOSEPH POBJOY	Maidstone	1839

In the *Topograph of Maidstone* 1839 he is noted as a clockcase maker.

SCAULTHEISS, J. T.	Chatham	1874
SCHEBBLE, JOSEPH & PAUL	Dartford	1855
SCHELBLE, JOSEPH	Dartford	1866-74

The entry in Loomes almost certainly relates to the Joseph Schebble noted in the previous entry.

SCHULER, M.	Bromley	1851
SCHWERSENSKY, ISAAC	Chatham	1832
SCOTT, JAMES	Rochester	1823
SCOTT, JOHN	Sandgate	1776-95
SCOTT, NICHOLAS	Maidstone	1780-95

Recorded in Loomes as a watchmaker.

SCOTT, STEPHEN	Elham	born 1725-died 1798

The only record of this man is that he was born in 1725 and died in 1798. Nothing else known.

SCOTT & GILES	Maidstone	c1800

Probably a partnership between Nicholas Scott and Nicholas Giles. See entry under Giles.

SEBBER, ANN	Dover	1823
SEDDON, ALFRED	Forest Hill	1874
SEDDON, ALFRED	Sydenham	1851-74

Probably the same man as the previous entry.

SELFE, F.	Chatham	1858
SELFE, FRANCIS WILLIAM	Erith	1874
SELFE, F.	Sheerness	1851
SELFE, WILLIAM	Gravesend	1866-74
SELFE, WILLIAM	Sheerness	1855
SHARP, —	Faversham	c1780

This entry is taken from Baillie. He may be the same man as the following maker. A bracket clock has been recorded.

SHARP, JOHN BUNYEA	Faversham	1823-32
SHARPEY, JOHN	Canterbury	1714

Noted in Loomes as a watchmaker, but nothing else is known.

SHEARER, JAMES	Bromley	1839
SHEATHER, DANIEL	Cranbrook	1713

Noted in the churchwardens' accounts as repairing the church clock in 1713, but it is not known whether he was in fact a clocksmith. No other reference to him has been found.

SHEPHERD, BENJAMIN	Edenbridge	1874
SHERWOOD, JOHN	Faversham	1845-55
SHERWOOD, JOHN	Hythe	1839
SHILLING, E.	Milton	1866-74
SHILLING, E.	Milton	c1800-38

Noted in Loomes as a clockmaker. He is probably the same man as the following entry.

SHILLING, EDWARD	Sittingbourne	1839
SHILLING, JAMES	Boughton	eighteenth century

Watch and clockmaker.

SHILLING, JAMES	Milton	1823-8
SHILLING, JOHN	Ashford	1847
SHILLING, Mrs. HARRIET	Milton	1847-55
SHILLING, THOMAS	Selling	pre-1767

Watchmaker. A watch by him was reported lost in December 1770 (see Appendix I).

SHINDLER, THOMAS I	Canterbury	born c1657-died pre-1705

Fig 7/104 A lantern clock by Thomas Shindler I, converted to anchor escapement from crown-wheel and verge, about 1695

Fig 7/105 The dial of a very fine 8-day, five-pillar clock by Thomas Schindler II of Canterbury, about 1715-20. The 11in square dial has mask spandrels, ringed winding holes and calendar aperture with engraved birds, flowers and basket of fruit

SHINDLER, THOMAS II	Canterbury	Free 1707-died 1751
SHINDLER, THOMAS III	Canterbury	Free 1747-64
SHINDLER, THOMAS	Romney	pre-1746

Probably the same as Thomas III of Canterbury.

THE SHINDLER FAMILY

The only reference that has been found to the father of Thomas I is in the record of Thomas gaining his Freedom by Patrimony on 13 September 1676, where he is described as a watchmaker, the son of Thomas, a proctor. Presumably the father held this position in the cathedral administration, as in addition to its educational and legal connotation the term 'proctor' can also describe a representative of the clergy in the Church of England Convocation. His mother Pheobe died in 1690 and in the inventory taken after her death she is described as a widow. His father therefore died before 1690 although no record of his death has been found. In her will, dated 23 July 1690, there are bequests to various members of her family and to her godson John Fowle, but her son Thomas I is not mentioned. The rents of a property in St Mildred's Parish (which she leased from the Dean and Chapter of the Cathedral) are left to her grandson Thomas II, together with the residue of her estate which she directs to be sold

Colour Plates XII and XIII An early eighteenth-century 8-day longcase clock rack striking by John Wimble of Ashford, about 1710. The caddy-topped case has red and gold lacquer with chinoiserie decoration. The London-style square brass dial has herringbone engraving around its edge and typical doves and a basket of fruit — symbolising 'peace and plenty' — round the square calendar aperture

by her executor, the clockmaker Richard Greenhill. She also left £15 to pay for her grandson's apprenticeship when he was 14 years old. Among the witnesses to the will is another Canterbury clockmaker, James Feild, who is also mentioned in Elizabeth Shindler's will of 1705.

On 26 September 1684 Thomas I, aged 27, married Elizabeth Fowle; although no definite connection has been made, there may be a possible link with the early seventeenth-century maker Thomas Fowle of Canterbury. The inventory taken after Pheobe's death totalling £96 1s 6d describes a comfortably-off household compared with the average seventeenth-century list of effects left by shopkeepers and tradesmen and even includes a quantity of silver and gold items. By 1706 an inventory taken after Elizabeth's death indicates that the family fortune had reduced to £38 10s 6d, although the fact that Thomas II was involved as executor may have something to do with the apparent diminution of the family estate!

Elizabeth Shindler's will, dated 10 December 1705, makes it clear that her husband had died some years previously and she appears to have had a business relationship, and maybe a personal one, with James Feild. 'My will and mind is that my son Thomas Shindler and my friend James Feild who has lived with me several years may continue to dwell and live together and keep on the trade in like manner and upon such Termes and Conditions as the said James Feild and I do and have done', but if they decide to separate then James Feild was to be given £5 immediately and £8 a year for life. After making other bequests to various relatives, the bulk of her lands, tenements and real estate is left to her son Thomas Shindler II. However if he did not carry out her wishes his legacy was to be set aside and everything left to her nephew John Fowle. John Barret, another Canterbury clockmaker, was a witness to the will.

As far as is known only one clock by Thomas I exists, a fine lantern clock dating from the 1690 period, now in the Canterbury Museum (Fig 7/104).

Thomas II does not appear on the apprentice lists, but it must be assumed that he was trained by his father, obtaining his Freedom in 1707 and he also is noted as a watchmaker in the records. Numerous references to the baptism of children of Thomas II and his wife Katherine appear in the registers of St George the Martyr, where he is described as an Alderman, a position he held as early as 1715. However, the record of his marriage has not yet been traced, but his death was reported in the *Kentish Post* for 1-5 June 1751 (see Appendix I). He occupied a prominent position in the affairs of the city for, in addition to his duties as an Alderman, he was a Justice of the Peace and a member of the Court of Guardians responsible for the workhouse that had been set up in the fourteenth-century building in Stour Street, formerly housing the Poor Priests Hospital. This building now houses the fine Heritage Museum.

In 1728 the workhouse minutes record that Thomas was commanded to supply a turret clock at a cost of £20 and that he should receive 10s a year to keep it in good repair. The turret clock movement made and supplied by Shindler has since been replaced by a modern clock, although the old movement still exists, languishing in a museum store room inaccessible to the general public. Hopefully the museum authorities can be persuaded to return it to Stour Street where it rightfully belongs, and place it on public exhibition.

Baillie records a watch by Thomas II in the Stads Museum in Amsterdam, but the only other examples known are a fine 8-day longcase movement with an arch-dial which was stolen in 1993 from premises in Hythe, and a good 8-day 11in dial and movement (Fig 7/105). So far, there is no trace of any watch by Thomas I.

Loomes records a third Thomas, the son of Thomas II, as gaining his Freedom in 1747 and he is probably the same man as the Thomas Shindler listed by Loomes as working in Romney before 1746. In August 1751 he took over his late father's shop opposite Butcher

Lane in Burgate Street, Canterbury, and announced that he 'regularly learnt his Business in one of the principal Shops in London, and makes and repairs all sorts of Clocks and Watches, in the best and newest Method' (see Appendix I). Like a number of other watchmakers and clockmakers, he also sold 'Linnen-Drapery and Haberdashery Goods, and Stockings &c'.

In November 1753 the *Kentish Post* (Appendix I) reported a robbery at his shop:

> Canterbury, Nov. 10.
> On Thursday Morning last, about One o'Clock, the Shop of Mr. Thomas Shindler, Watch-Maker in this City, was broke open and robbed of several Watches, viz. one Gold Watch with a white enamel'd Dial-Plate, green Shagreen Case, a Womans Chain and Hook; one Si[l]ver Glass-hoop'd Watch, Silver Pedestals, Maker's Name Ed. Cockey, Warminster; the others at present cannot be described as to the Makers Names.

The missing names were provided in the *Kentish Post* for 17-21 [and 28] November 1753:

> Whereas it was mentioned in this Paper the 10th Inst. that the Shop of Thomas Shindler ... [was robbed ...] The others are as follows; one Silver Watch, Maker's Name Ward, London, one ditto John Berry, Cant. two ditto Lade, Cant. one ditto Shindler, Cant. with a Silver Chain broke, one very small Watch, Box and Case in one, Moran, London, one ditto Name not known, with four Pair of Cases; all which are supposed to be concealed in or near this City. Whoever can meet with any of the said Watches, and bring them to me, shall receive five Guineas, and for the whole Number ten Guineas Reward.
>
> THOMAS SHINDLER.
> NB. One Thomas Smith having been taken up on Suspicion of the said Robbery: This is to give Notice, that if he has sold or pawn'd any of the said Watches, that if the Persons will bring them to Mr. Shindler, they shall have their Money again, and be satisfied for their Trouble.

Thomas Smith was later found guilty of the felony and condemned to death, but reprieved due to his age and sentanced to seven years' transportation, possibly to Maryland where most Kent criminals seem to have been sent.

It seems inconceivable that only three examples of this family's output survive and it is hoped that more clocks and watches made by them will come to light in the future.

SHORE, A.   Canterbury   1851-5

SHORTMAN, J. S.   Newnham   c1820
Painted dial longcase clocks and bracket clocks have been noted by this maker.

SHORTMAN, JOSIAH WILLIAM   Newnham   1842

SHORTMAN, SAMUEL   Newnham   1830-50
Loomes also notes this man as a cabinet maker.

SHORTMAN, SAMUEL & WILLIAM   Newnham   1856-70

SHORTMAN, WILLIAM   Newnham   1879
The references to the above members of the Shortman family are incorrect as both Baillie and Loomes have confused Newnham in Kent with a village of the same name near Bristol in Gloucestershire. See *Gloucestershire Clockmakers* by Graham Dowler.

SIEDLE, A.   Greenwich   1855

SIEDLE, L.	**Woolwich**	**1851-66**
SILK, JOHN I	**Elmsted**	**c1670-1700**
SILK, JOHN II	**Elmsted**	**c1760-c79**
	Stowting	pre-1779
	Hythe	pre-1795
SILK(E), JOHN & SON	**Cheriton**	**1780**

No references to the members of this family have been found in either the county or cathedral archives. Although Loomes describes John I as a clock and watchmaker it seems unlikely that he made his living solely in that trade, as Elmstead is a tiny village in an isolated valley in the North Downs with very few inhabitants.

It must be assumed that there were at least two, and possibly three makers of the same name. John II is almost certainly the clockmaker from Elmsted who also made clocks signed at Stowting (another nearby isolated downland village) and who moved to Hythe before 1795. Whether he also is the same man as the one recorded in the *Kentish Gazette*, 15-19 April 1780, as moving to Cheriton in 1780 is not clear. However, geographically, the towns and villages mentioned are all within a few miles of each other and it seems reasonable to suppose that they are one and the same.

> JOHN SILKE and SON,
> From ELMSTED, CLOCK and WATCHMAKERS,
> Removed to a Shop at SANDGATE, in CHERITON, Near HYTHE,
> and not far from FOLKSTONE,
> BEG Leave to inform all Persons, who will favour them with their Commands, that they will make and repair all CLOCKS and WATCHES at as reasonable an Expence as possible ....

The following month, and in March 1787, advertisements appeared for lost watches signed by John Silke of Elmsted (see Appendix I), although no surviving watch by him is known.

Very few clocks have been recorded other than an early lantern clock dating from the 1680 period and a 30-hour birdcage movement with a plain matted dial and mask spandrels. Both are signed at Elmsted. Another longcase movement signed at Stowting has also been noted.

Loomes mentions a Robert Silk working in London about 1675 as a clockmaker and gunsmith. Whether there is any connection has yet to be proved.

SIMS, AVERY	**Canterbury**	**1750**

Recorded as a Freeman of Canterbury in 1750. Nothing else is known.

SIMS, GEORGE	**Canterbury**	**c1710-20**

Loomes records this man at this date, but the author has been unable to trace him.

SIMS, GEORGE	**Canterbury**	**Free 1745-after 1760**

George Sims obtained his Freedom by Redemption on 1 October 1745 and his bondsman, probably his uncle Henry Sims, is noted as a linendraper. His marriage, which presumably took place before he came to Canterbury, has not been traced, but in 1751 he took his eldest son George as an apprentice and subsequently his second son Henry in 1758. Numerous clocks signed by him have survived. Examples have been noted in both lacquer and mahogany cases in the pagoda-topped style of the 1750s. A number of clocks have been recorded with plain all-over silver dials dating from much later and they may well be the work of George junior. In August 1750 and December 1756 he was described as a 'Watch-maker, in St. Georges, Canterbury'. For some unaccountable reason, in July 1757 he was acting as the agent for the sale of the building materials from a demolished 'Large old Timber Manshion House, at Upper Wilsley near Cranbrook'.

In January 1772 the *Kentish Gazette* reported that a missing person had ' a Watch in his Pocket, Maker's Name George Sims, No. 656, Canterbury', and one of his watches was reported lost in April 1773 (see Appendix I).

SIMS, GEORGE Jnr   Canterbury   Free 1760-late eighteenth century
Apprenticed to his father in 1751 and obtained his Freedom in 1760.

SIMS, HENRY   Canterbury   apprenticed 1758-died pre-1790
Henry, the younger son of George senior, was apprenticed to his father on 9 May 1758. Like his brother he was known as a watchmaker, although longcase clocks signed by him have also been recorded. The *Kentish Gazette* of 25 February-1 March 1769 reported that 'Part of the Household Furniture and Stock in Trade of Henry Sims, Watch-maker, St. George's Street Canterbury' was to be sold by auction as he was 'leaving off Trade'. Included were: 'a very handsome Twelve-tune Chime Clock, and Common Ditto, and Larums [alarms], Watches, etc. &c. a neat Mahogany Work-board and Drawers, with Tools and all Utensils belonging to the Trade'. The shop was then taken over by Benjamin Smith who had been apprenticed to Sims (*qv*) in 1761. Henry Sims died prior to 1792.

SIMPSON, GEORGE   Maidstone   1839
The *Topograph of Maidstone* published in 1839 records an address in Carey Street.

SLEATH, JOHN   Sandwich   apprenticed 1747
The only record of this maker is that he was apprenticed in 1747 to John Gardner by his father Joseph for the sum of £15. Whether he completed his indentures is not known as John Gardner is believed to have died in 1752 and his son William died in 1758. John Sleath may have completed his apprenticeship under William Gardner, but this is not confirmed. There may be a connection with the John Sleath who gained his Freedom in London in 1758.

SMITH, BENJAMIN   Canterbury   apprenticed 1761-88
Bemjamin Smith was apprenticed to Henry Sims on 5 March 1761 and gained his Freedom of the city of Canterbury in 1768. On 26-29 October 1768 the *Kentish Gazette* records that he had opened his own business in Christ-Church Yard (within the Cathedral Precincts) where he:

> MAKES and repairs all Sorts of Church, Musical, Spring and Common Clocks; all Sorts of Horizontal, Repeating, Skeleton and Stop-Watches. — And, besides his own Knowledge of the Business, which he has acquired by a regular Apprenticeship in this City, has also engaged, as an Assistant, a very eminent Hand from London, where he also, under a very able Master, has much improved in the Knowledge of the Business ....

Six months later the *Kentish Gazette* (26-29 April 1769) recorded his move to St George's Street, taking over his former master's shop:

BENJAMIN SMITH, WATCH AND CLOCK-MAKER,
REMOVED from Christ-church yard,
to St. George's Street, near the King's Arms Printing-office, Canterbury, makes, sells, and repairs, all sorts of church, turret, musical, spring, and common Clocks. Also all sorts of repeating, horizontal, skeleton, seconds, and other Watches. Both Town and Country served on the most reasonable terms … having taken the shop, late Mr. Sims, ….
N.B. The House to be Lett late in the Occupation of Mr. Smith, in Christ-Church Yard.

On 16 September 1773 he was robbed of six of his new watches, and in September 1775 a similar notice to the above also advertised for an apprentice (see Appendix I). In the window tax returns of 1788 he was paying for a house in the parish of St Andrew with fifteen windows.

Numerous clocks and watches have been recorded by him, including an 8-day painted dial movement in a mahogany case, signed 'Benjamin Smith, Sittingbourne'. The rear of the front plate of this clock is engraved '1 ET'. This is quite possibly the monogram of Edward Tiddeman who may have worked as a journeyman for Smith following his apprenticeship to James Warren and prior to setting up in business in Canterbury on his own. The figure '1' probably denotes that this is the first movement made by him for Smith.

SMITH, —	Chatham	late eighteenth century

An 8-day white dial movement in an oak case has been recorded. Nothing else is known.

SMITH, C.	Tonbridge	1866
SMITH, E. S.	Sittingbourne	1851-5
SMITH, JOHN	Wye	1803

Those requiring particulars of the cornmill and house at Elham, for sale in 1803, were to enquire of 'Mr John Smith, watchmaker, at Wye' (see Appendix I). No other reference known.

SMITH, JOHN P.	Sittingbourne	1839-45(47)
SMITH, PETER	Deal	1838-51
SMITH, SAMUEL	Sandwich	1752

Samuel, the son of Henry Smith, was apprenticed to John Gardner for the sum of £6 in 1752. There appear to be no other references to him in the Sandwich archives.

SMITH, WILLIAM	Broadstairs	1838-74
SMITH, WILLIAM	Canterbury	1845-55

48 St Peter's Street in 1847 directory.

SMITH, WILLIAM	Herne Bay	1845-55
SMYTH, CHARLES	Tonbridge	1838-74
SNASHALL, GEORGE	Marden	1847
SNATT, JOHN	Ashford	died 1780

A watch and a 30-hour longcase clock by this maker are noted. Nothing else is known other than a notice in the *Kentish Gazette* on 1-5 July 1780 after his death that the creditors and debtors of the late John Snatt were to send their accounts to Stephen or Catherine Snatt (see Appendix I).

SOLOMON, EDWARD   Margate   1769-1800
Noted in Loomes as a watchmaker.

SOLOMON, EMANUEL   Canterbury   1785-1826
Another of the group of fine Jewish makers centred on Canterbury and East Kent during the late eighteenth century. Although known principally as a watchmaker like his father Edward of Margate, longcase clocks signed by him have been recorded.

A particularly fine example was sold at auction in Canterbury in the late 1980s. The case was of the highest quality mahogany with frets to the side of the hood and the trunk and base were decorated with shell inlays. The arch-dial movement was attached by a false plate to a very pretty dial with bird and flower decoration.

In *Hilden's Triennial Directory* published in 1805 he is described as a clock and watch-maker with an address in St Dunstans Street.

SOLOMON, MIRIAM & BELLA   Canterbury   1838

SOLOMON, NATHANIEL   Margate   pre-1783-95
Watches have been recorded by this man.

SOLOMONS, S. & L.   Chatham   1839

SOMES, J.   Margate   1874

SOMES, WILLIAM   Margate   1838-74

SOUTHEE, H. S. E.   Canterbury   1866

SOUTHEY, —   Rochester   Free c1790
Baillie records this maker as being free of the Clockmakers' Company in London about 1790, but nothing else is known.

SOUTHGATE, WILLIAM   Maidstone   1750
Noted in the Maidstone archives as being apprenticed to Robert Cutbush in 1750, but no other reference has been found. May be the son of Robert Southgate. See Kollsall.

SPASSHATT, JOSEPH   Dartford   1845-7

SPURGE, J.   Faversham   1866

SPURGE, J.   Sheerness   1874

SPURGE, WILLIAM   Woolwich   1826-55

STACE, JOHN   Folkestone   1847-66

STACE, Mrs M. A.   Folkestone   1874

STANBURY, E.   Canterbury   1866-74

STANDVEN, THOMAS	Broadstairs	1839-40
STANDVEN, THOMAS	Ramsgate	1840-51

Possibly the same as the Broadstairs man.

STANDVEN, THOMAS EMERY	Ramsgate	1832-47
STATON, JOHN	Chatham	1845-7
STEBER, DAVID	Dover	1823-33

Watches signed by this maker have been recorded. Son of John Steber.

STEBER, JOHN	Dover	1806-died 1819

Came from Germany to Dover in 1806 and died there in 1819.

STELERT & DOTTER	Bromley	1874
STEPHENS, THOMAS	Forest Hill	1874
STEVENS, ROBERT	Milton	1781-CC1795

In December 1781 Robert Stevens, 'Clock and Watchmaker at Milton next Sittingbourn', advertised for a 'sober, steady Man' as a journeyman clockmaker and also for an apprentice (see Appendix I), but Baillie does not record him as Free of the Clockmakers' Company until 1795.

STEVENSON, WILLIAM	Maidstone	1756-61

William Stevenson is recorded as a watchmaker in East Lane in February 1756, when his shop was broken into and five silver watches and two watch cases were stolen (see Appendix I). He was named as a Freeman of Maidstone in the 1761 Poll Book. Where and with whom he received his training as a clockmaker is not known, although there are records of his family in the Freemen Lists as far back as 1713. Various members are noted as distillers, tallow-chandlers and tinplate workers.

STIERT, L.?	Bromley	1866
STONE, H.	Maidstone	1855
STORKEY, J.	Ramsgate	1866
STRAHAN, JOHN	Canterbury	Free 1790

This maker from London is recorded as obtaining his Freedom by Marriage to Mary Elizabeth Spratt on 30 March 1790. Nothing else is known.

STRELLY, JOHN	Greenwich	pre-1684

Baillie records a table clock by this little-known early maker. Whether there is any connection with Francis Strelley, the London maker working between 1666 and 1691 is not certain, but it is highly likely.

STRICKLAND, WILLIAM	Tenterden	c1790-3

Loomes records this man as a clockmaker actively working about 1802 and an arch-dial 30-

hour birdcage movement with Arabic numerals housed in a mahogany case has been noted. The advertisement placed in the *Kentish Gazette* on 4 January 1803, announcing his takeover of Woolley's business, makes it clear that he had worked for Woolley for 12 years, possibly as an apprentice and journeyman.

WILLIAM STRICKLAND,
Watch, Clock, and Gun-Maker, and Silversmith,
TENTERDEN;

TAKES the Liberty most respectfully to inform the Inhabitants of Tenterden and the neighbouring Parishes, that he has taken the above business of Watch-maker, Silversmith, &c. lately carried on by Mr. WOOLLEY; and having been in the shop himself, now more than twelve years ....
N.B. — A steady LAD, as an Apprentice in the above branches, is wanted immediately, and, if approved of, but a small Premium will be required.
Mr. WOOLLEY ... now begs leave respectfully to recommend Mr. STRICKLAND to them, as a steady and deserving young man.

SULLIVAN, GEORGE	Greenwich	1866-74
SULLIVAN, GEORGE J.	Greenwich	1847
SULLIVAN, RICHARD	Deptford	1829-47
SUTTON, CHARLOTTE	Tonbridge	1823
SUTTON, JOHN WINDRAM	Maidstone	1826-39
SUTTON, THOMAS	Maidstone	apprenticed 1775-1823

Loomes gives a date of 1790 for this maker who was probably the best of the late eighteenth-century Maidstone makers. However the records show that he was apprenticed to Richard Cutbush in 1775 by his father Benjamin, who was a husbandman of Maidstone. In 1790 Thomas Sutton took William Walmesley as apprentice.

Sutton was a very prolific maker, particularly of longcase clocks, usually with finely-painted dials, housed in good quality oak cases (Figs 7/106-107). The movements are invariably of a very high standard, as are those of all of the men who received their training from members of the Cutbush family. He also made a number of small wall-hanging clocks with alarm work both with all-over silver dials and painted dials (Figs 2/10-11 and 7/108). Maidstone Museum has a number of his clocks.

SWABY, JACOB	Romney	1795

Recorded by Baillie as a clockmaker at this date.

SWABY, ISRAEL	Dover	1775-85

Noted by Loomes as a watchmaker and silversmith working in Strand Street. The following advertisement appeared in the *Kentish Gazette* on 8-11 October 1783:

ISRAEL SWABY,

Respectfully informs his Friends and the Public in general, that he has removed from the QUAY at DOVER, and taken a House in KING's HEAD STREET; where he continues to deal very extensively in the WATCH, JEWELLERY, and SILVER BUSINESS ....

There is almost certainly a connection with Jacob Swaby, who is thought to be his son.

Fig 7/106 An 8-day painted dial longcase clock with a centre calendar hand by Thomas Sutton, Maidstone, about 1785. Centre calendars are more usually found on clocks from the northwest of England

Fig 7/107 (above right) An 8-day painted dial longcase clock with a five-pillar movement and a more conventional calendar, by Thomas Sutton of Maidstone, about 1785. The dial is attached to the movement by a false-plate signed by Wilson of Birmingham

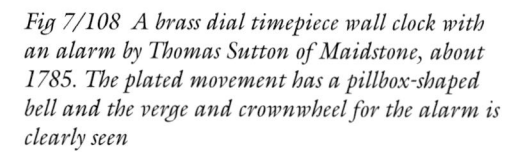

Fig 7/108 A brass dial timepiece wall clock with an alarm by Thomas Sutton of Maidstone, about 1785. The plated movement has a pillbox-shaped bell and the verge and crownwheel for the alarm is clearly seen

SWAN, —	**Margate**	**pre-1754**

A watch by this maker reported lost in July 1754 (see Appendix I).

SWINHOGG, —	**Maidstone**	**early seventeenth century**

Whether he was actually a clocksmith is not certain. The only record that has been found is the following reference to him in the Chamberlain's Accounts for 1621-2: 'Pay'd Mr. Swinhogg for mending the clock and glas in the cundit and for lath silver'. This must refer to the main town clock at the conduit in the High Street. 'Lath' presumably means 'dial' in this context.

SWINYARD, GEORGE	**Maidstone**	**1750**

Noted in the Maidstone records as entering an apprenticeship to Nehimiah (Jeremiah?) Wimble in 1750. Wimble is recorded as a watchmaker and is believed to be the son of John Wimble of Ashford. Swinyard may not have completed his apprenticeship as there is no other reference to him. A Mary Swineyard is noted in the marriage records as marrying Nehimiah Wimble in Maidstone in January 1746. She is possibly the sister of George Swinyard.

TATE, JAMES	**Maidstone**	**1840-7**

Bagshaw's Directory records him as a silversmith at 7 High Street.

TAILLOUR [TAYLO]), JOHES [JOHN]	**Canterbury**	**1428/9**

A memorandum in the City Accounts for 1428/9 records him as 'Clokmaker de Cantuar'. This is the earliest known record of a Kent clockmaker

TAYLOR, JOHN	**Deptford**	**1791**

Noted in Baillie as a Watchmaker at this date. Nothing else is known at present.

TAYLOR, JOHN	**Dover**	**1858**
TAYLOR, JOHN	**Gravesend**	**1838**
TAYLOR, RICHARD	**Foots Cray**	**1866-74**
TAYLOR, T. W.	**Sittingbourne**	**1874**
THATCHER, GEORGE	**Cranbrook**	**pre-1716-died 1773**

An interesting maker from the Weald about whom very little is known, other than that he must have been working at an early date in the eighteenth century as it is known that Obadiah Body was apprenticed to him in 1716

Owing to the striking similarity between one of Thatcher's lantern clocks and another example by John Kingsnorth, who is believed to have moved to Tenterden from London after 1695 when he completed his apprenticeship, it is highly likely that George Thatcher was trained by Kingsnorth. At least two lantern clocks (Colour Plate III and Fig 7/109), two 30-hour longcase clocks with simple matted dials and mask spandrels (7/110) and one 8-day longcase clock have survived, and all exhibit work of a high quality. The 30-hour birdcage movements have substantial brass frames with the upright pillars set at an angle, rather than parallel as is the usual fashion. All of the longcases are housed in very good quality oak cases, almost certainly from the same workshop, with the characteristic low caddy top that is common to many Kentish cases of the 1700-30 period.

His son Thomas was also a clockmaker, presumably trained by his father as the two 30-

hour examples seen signed by him are identical in every way to the father's work. George Thatcher died in June 1773 (see Appendix I) and Thomas probably predeceased him, as in 1774 the family business was taken over by Abraham Body, who was the son of Obadiah and may well have also been trained by Thatcher. There must certainly have been close ties for Body to have taken over the business. The connection with the Body family from Battle in Sussex is given further credence by the existence of a lantern clock signed on the chapter ring 'G. Thetcher, Battle, 1735', sold at the sale of contents of Groombridge Place in 1992.

Fig 7/109 The movement of the rare rack-striking lantern clock by George Thatcher of Cranbrook shown in Colour Plate III. The rack is at the rear, with the snail and starwheel at the front of the movement

Fig 7/110 Single-handed 30-hour clock by George Thatcher of Cranbrook, about 1730. The 11in square dial has a well matted centre and an engraved calendar aperture. The birdcage movement is housed in its original dark oak case

Figs 7/111-112 A two-handed 30-hour oak longcase clock by Thomas Thatcher of Tenterden, about 1730. This style of oak case with a low caddy top is also typical of those used by the Baker family of West Malling

Fig 7/113 Iron hinge on the trunk door of the Thatcher case

THATCHER, THOMAS	Tenterden	died pre-1772

Son of George Thatcher and probably trained by him. He is thought to have died before 1772. At least two 30-hour clocks (Figs 7/111-113), a watch and a lantern clock have been recorded. See previous entry.

THOMSON, MARK GRAYSTONE	Tonbridge	1847-74
THOMSON, WILLIAM JAMES	Ashford	1874
THORPE, F.	Tunbridge Wells	1874
TIDDEMAN, EDWARD	Canterbury	apprenticed 1779-1801

Apprenticed to James Warren on 30 June 1779, obtaining his Freedom in 1787. Traded as a parnership (with whom is not known) until 1789 when Tiddeman (unfortunately no christian name is given, but presumably Edward) 'opened a Shop, Opposite Butchery Lane, on the Parade' (see Appendix I). Numerous watches signed by him have survived, two 30-hour clocks with painted dials in oak cases and an 8-day clock (Fig 7/114) have been noted. See entry for Benjamin Smith.

TIDDEMAN & CO	Canterbury	1787-9

Clock and watchmakers at the Parade (see Appendix I, October 1787). The partnership ceased in December 1789.

TIGHT, GEORGE	Deptford	1822-4

A watch has been recorded.

TILDEN, JOHN	Cranbrook	1713

Recorded by Loomes as making a dial for the church clock. Whether he was a clockmaker is not known as no other reference to him has been found.

TINDALL, GEORGE HENRY	Woolwich	1845-7
TIPPEN, J.	Lenham	1851-5
TIPPEN, JAMES	Ashford	1874
TIPPEN, JAMES	Headcorn	1866-74

These three are very probably the same man.

TIPPEN, WILLIAM	Charing	1866-74
TODE, GEORGE	Canterbury	1858
TOLPUTT, JAMES	Dover Castle	c1714

Loomes notes him as a clockmaker. At present no other information is known.

TRIGGS, R. W.	Blackheath	1866-74
TRIMNELL, WILLIAM HENRY	Canterbury	early nineteenth century-1855

Numbers of watches have been recorded by this man. *Bagshaw's Directory* of 1847 notes an address at 7 Parade and also states that he was a silversmith.

Fig 7/114 8-day painted dial signed 'Tiddeman, Canterbury' with raised gilt decoration, about 1790. The dial has no falseplate

Fig 7/115 (right) Verge watch movement No 32077, by Jacob Vile of Beach Street, Deal

TUCK, JOHN	Margate	1874
TUNNELL, JOHN	Deptford	1847-9
TWICHELL, GEORGE HEARN	Maidstone	1839

Noted in directories as a working jeweller in King Street.

TYLER, —	Ash	c1760

An 8-day 12in square brass dial with chapter ring and four-pillar movement has been noted. The clock is signed 'Tyler, Ash' on a cartouche on the dial, and is housed in an oak case. Nothing else is known of this hitherto unrecorded maker.

USHER, —	Town Malling	c1820
USHER, G.	West Malling	1838

Believed to be the same man as previous entry.

USHER, C.	West Malling	1838
USHER, O.	Wrotham	1851-5
USHER, OLIVER	West Malling	1832-51

| USHER, WILLIAM | West Malling | 1845-74 |

These makers from West Malling and Wrotham are members of the same family. The names Uslar, Usman, Usmar and Usmer in Loomes have been mis-spelt, and should be 'Usher'.

| VAUGHAN, J. | Staplehurst | 1855 |

| VENNALL, CHARLES | Deal | 1866-74 |

| VENNALL, JAMES | Deal | 1845-55 |

| VENNALL, JAMES | Sandwich | 1820-40 |

Loomes records this man as a clock and watchmaker with dates of 1832-40. The Sandwich records note that a son, Thomas, was born to James and Mary Vennall on 12 November 1820, so it is reasonable to suppose he was working by at least 1820. He appears in the Sandwich directories until 1839. *Pigot's Directory* (1826) gives an address in Market Street.

| VENNALL, THOMAS | Deal | 1826-9 |

Recorded by Loomes as a watchmaker, but any relationship to the other members of the Vennall family is not clear.

| VENNALL, THOMAS | Smarden | nineteenth century |

An arch-dial brass wall-clock with alarm work signed 'Thomas Vennall, Smarden' has been reported. The dial is engraved with a sailing-ship and a house. It may be that this is the same man as the previous entry.

| VENTISAN, THOMAS | Canterbury | 1682 |

Noted in the apprentice lists as entering his indentures to Charles Johnson, clockmaker, on 20 March 1682/3. Nothing else known.

| VIDION, JOHN | Faversham | pre-1774-1801 |

Recorded by Baillie as a watchmaker. Several examples are known to exist. He was in business before 1774 as the following advertisement appeared in the *Kentish Gazette* on 13-16 July:

JOHN VIDEON,
WATCH-MAKER, and SILVERSMITH, At FAVERSHAM,
BEGS Leave to inform the Public, that he continues to take in Light English Gold Coin ....

| VIGOR, C. | Tunbridge Wells | 1874 |

| VILE, JACOB E. | Deal | 1809-47 |

Noted by Loomes as a watchmaker. A number of examples are known (Fig 7/115).

| VINCENT, JAMES | Gravesend | 1858 |

| VIRGO, JAMES | Sevenoaks | 1874 |

| WADDLE, — | Marden | 1858 |

| WAGHORNE, C. | New Brompton | 1866-74 |

| WAGHORNE, JAMES | Maidstone | 1826-47 |

WALL, JOHN	Chatham	1812

Recorded by Loomes at this date as a watchmaker. An 8-day clock with an arched silvered dial in a good quality oak longcase has been recorded.

WALMESLEY, WILLIAM	Maidstone	1790-1800

Recorded by Loomes as being apprenticed in 1800. However there appears to be a discrepancy in the records as the Maidstone archives also note Walmesley being apprenticed to Thomas Sutton in 1790. Charles Cadwell is recorded as being apprenticed to him in 1800. Longcase clock known (Fig 7/116).

WALTER, W. J.	Woolwich	early nineteenth century

Baillie notes this man as a watchmaker.

WALTER, WILLIAM JOHN	Woolwich	1847

Noted by Lomes as a watchmaker. Almost certainly the same person as preceding entry.

WARD, EDWARD	Maidstone	1874

WARREN, C.	Canterbury	1770

This entry appears in Loomes, but no reference to him has been found in the records and this probably refers to James Warren.

WARREN, EDWARD DE	Sandwich	born 1808-61

Loomes notes this man as being born in Poole, Dorset in 1808. He appears in the 1851 Census as a watch and clockmaker with three sons and one daughter. The daughter was born in Canterbury. Whether there is any connection with the Warrens is not known. His working address in Sandwich is given as 45 The Butts.

WARREN, G.	Woolwich	1866

WARREN, JAMES	Canterbury	1767-93

James Warren and his son, also James, are probably the most well-known of the late eighteenth-century Canterbury makers and many examples of their work have survived, including watches, bracket clocks and longcase clocks with painted dials (Figs 7/117-119).

James senior was apprenticed to Henry Sims and gained his Freedom in 1767. He subsequently started his own business just outside the city walls by St Georges Gate.

In 1771 he took John Engeham as an apprentice, who is believed to be the maker by this name who worked in Yalding. After finishing his apprenticeship Engeham did not petition for his Freedom, but must have left the city. He subsequently obtained his Freedom by Patrimony some years after on 26 May 1789.

Edward Tiddeman was apprenticed to Warren on 30 June 1779, gaining his Freedom in 1787, and James junior was apprenticed to his father on 1 July 1785, carrying on the family business after his father's death, signing his clocks and watches 'James Warren & Son'.

In March 1777 watch No 4225 was reported stolen, as was watch No 8157 in October 1791 (see Appendix I).

The following entry appeared in the *Kentish Gazette* on 11 January 1793:

> Saturday evening last, between six and seven o'clock, a pane of glass in the shop window of James Warren, without St. George's Gate was suddenly broke, with great violence, and a man's

Fig 7/116 (left) An 8-day longcase clock by William Walmesley of Maidstone, about 1800. The painted dial has seated women represnting the four seasons in the spandrels and moon phase in the arch. The mahogany case is 7ft 8in tall

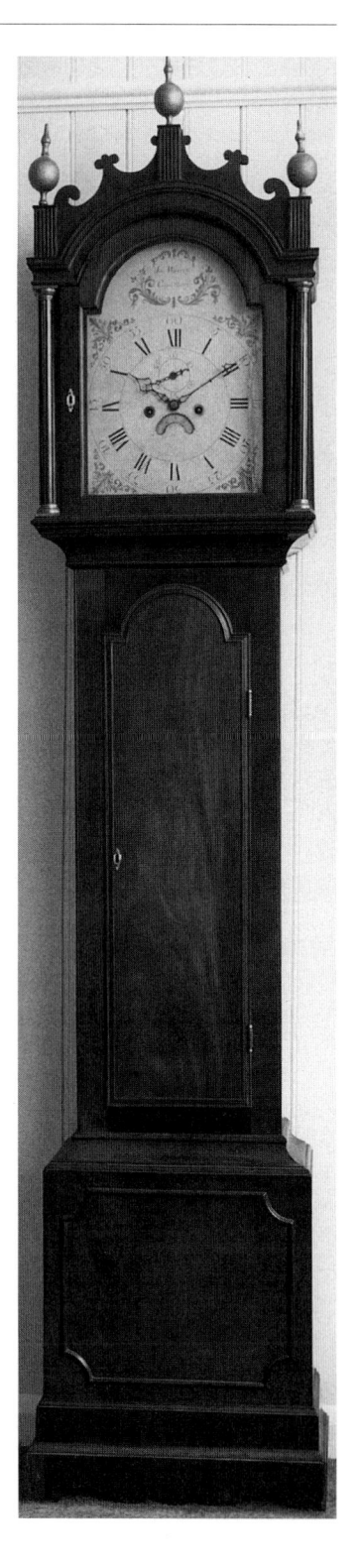

Fig 7/117 (right) A five-pillar 8-day mahogany longcase clock by James Warren of Canterbury, about 1775. The early painted dial has scrolling foliage

hand thrust in, who took therefrom a gold watch and a silver watch with which he made off indiscovered notwithstanding Mrs Warren was in the shop outside the window that was broke and gave immediate alarm.

Two men of suspicious appearance have been confined on the occasion but without any discovery of the property taken.

WARREN, JAMES Jnr	Canterbury	apprenticed 1785-1832

Free in 1800. See previous entry.

WARREN, JOHN	Canterbury	1874
WARREN & SON	Canterbury	1838-45
WARREN, R. J.	Woolwich	1866-74
WARWELL, JOHN	Deal	1787

A watch of his reported lost in March 1787, to be returned to him (see Appendix I), so presumably at work then.

Fig 7/118 An 8-day dial with floral decoration, signed by James Warren, Canterbury, about 1785

Fig 7/119 A Regency bracket clock in mahogany inalid with brass, signed by James Warren, junior, of Canterbury

Fig 7/120 A turret clock from an unknown location in the Canterbury area, signed on the bell by John Cooper, a well known local architect, and on the setting dial by James Warren. Dated 1815. Whether Warren actually made the clock or if it was supplied by one of the specialist turret clock manufacturers is not known

WATTS, INIGO   Canterbury   early eighteenth century

This signature appears on an early eighteenth-century 30-hour single-handed clock. As nothing else is known it is reasonable to suppose that the name is a variant of John Watts.

WATTS, JOHN   Canterbury   Free 1720/1-died 1775

The best of the clock and watchmakers working in Canterbury during the early and middle of the eighteenth century was undoubtedly John Watts, whose origins are still unclear, although there may be a connection with Richard Watts, the late seventeenth-century watchmaker from London.

    The earliest reference known is a note in the Canterbury records that he obtained his Freedom on 16 February 1720/1 by Redemption and his bondsman was named as James Becket[t], clockmaker of Dover, of whom very little is known. Whether he was trained with Beckett is not certain, but his surviving work is of a high quality and includes a lantern clock (Fig 7/121), watches (Figs 7/126-7), longcase clocks in oak and mahogany cases (Fig 7/122), and a fine arch-dial 8-day movement in a good blue laquer case has been noted. He also made what is probably the earliest known Kent bracket clock to have survived (Figs 7/123-124),

Fig 7/121 An eighteenth-century lantern clock by John Watts of Canterbury, about 1730

Fig 7/122 The London-style 12in wide dial of a good quality rack-striking 8-day clock with five ringed pillars by John Watts of Canterbury, about 1725. The hour hand is later. The arch casting, which includes the royal coat of arms (presumably commemorating the accession of George III), has been added using part of an old chapter ring. The clock was probably in a lacquer case originally, but is now in a later mahogany case dating from the period of the added arch

now in Canterbury Museum. Though signed without a placename, the lantern alarm in a very slim walnut case shown in Fig 7/125 is thought to be by this maker. Full-size lantern clocks in cases are known, but cased miniature lantern alarms are very unusual.

On 31 May 1736 he took his son John junior as an apprentice, who in turn obtained his Freedom on 23 April 1744, the same year in which John senior was Mayor of Canterbury.

All of his work that has been seen is of typical London quality, which makes the mystery of who trained him in the craft even more intriguing. There is a strong possibility that he is the same man who was apprenticed to Thomas Feilder in London in April 1698, gaining his Freedom from the Clockmakers' Company in November 1712, and recorded as dying in the same year, 1775, as the Canterbury maker. If they are one and the same, it is interesting to note that both Watts and Beckett began their apprenticeships in the same year, ie 1698.

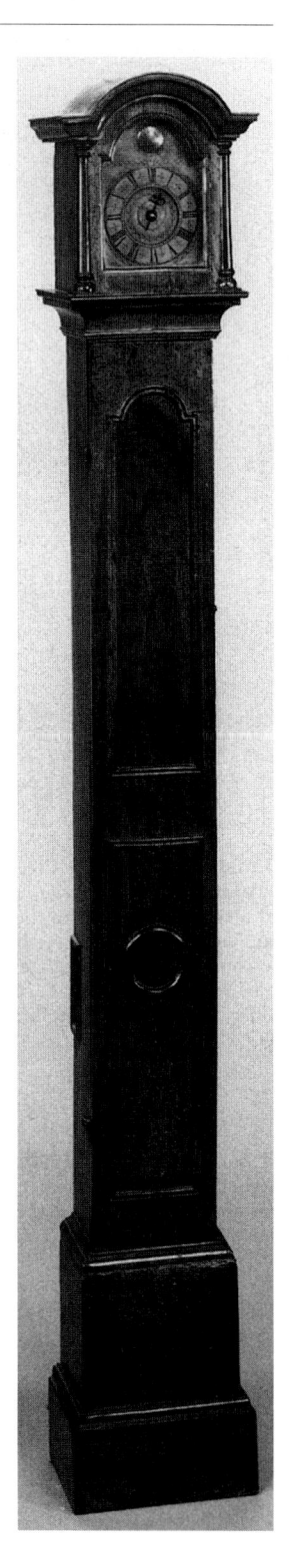

Figs 7/123-124 (left) Bracket clock in an ebonised case, by John Watts of Canterbury, about 1740. It is the earliest-known surviving bracket clock by a Kent clockmaker. Originally it had a crownwheel and verge, now converted to anchor escapement

Fig 7/125 (right) 30-hour miniature lantern clock by John Watts, about 1730. The 5¼in dial has a single hand, central alarm-setting disc, a sun-flower boss in the arch, all surrounded by wheat-ear engraving. The narrow walnut case has a lenticle in the trunk door and slots to allow the pendulum bob to swing freely

Figs 7/126-127 An early verge watch movement by John Watts of Canterbury, about 1725. Note the unusual pillars

WATTS, JOHN Jnr   Canterbury   apprenticed 1736
Apprenticed in 1736 to his father, John Watts, and obtained his Freedom on 23 April 1744. See previous entry.

WATTS, JOHN   Canterbury   early nineteenth century
Noted in Loomes as a watchmaker in 1836.

WATTS, WILLIAM   Gravesend   1845

WEBB, ABRAHAM   Wye   pre-1741
                                       Ashford   after 1741
Took over Arthur Hurt's shop in Ashford following the latter's death and: 'furnished all sorts of Clocks, Jacks, Guns, and Watches, at reasonable Rates; also Watches repaired, and all Business in the Trade of a Clock-Smith and White-Smith carefully done' (see Appendix I, June 1741).

WEBB, RICHARD   Chatham   1874

WEBBER, JOHN   Woolwich   1800-(1832-47?)
Noted in Baillie as working about 1800 and may be the same man as Loomes mentions in 1832-47.

WEBBER, JOHN II?   Woolwich   1866-74
Listed by Loomes and thought to be the son of the Webber in the previous entry.

WEGG, N.   Deptford   1849

WELLBY, JOHN   Canterbury   1839

WELLBY, JOHN   Whitstable   1845-51

WELCH, D.	Queenborough	**1847**
WELLS, J.	Deptford	**1849**
WELLS, WILLIAM	Gravesend	**1847-66**
WEST, DANIEL	Dover	**died 1687**

A recently discovered inventory of this hitherto unrecorded watchmaker, who died in 1687, assesses his estate at a value of £170. Among his effects were two clocks, clock lines, £43-worth of locks and metal worker's sundries, including a dozen bird cages and 'his working tools'.

WEST, RICHARD	Bexleyheath	**1839**
WEST, WILLIAM	Bexley	**1847**
WHALEY, WILLIAM	Maidstone	**apprenticed 1779**

Appears in the Maidstone Corporation List as entering his apprenticeship to Richard Cutbush in 1779. Other than the fact that he came from Goudhurst, nothing else is known.

WHATMAN, ARTHUR	Stalisfield	**died 1691**

Noted in Kent records as a clocksmith. Both his father and grandfather were blacksmiths and a copy of his will is held at Maidstone. Nothing else is known.

WHEELER, WILLIAM	Lee	**1866-74**
WHITEHEAD, EDWARD	Sevenoaks	**1823-38**
WHITEHEAD, EDWARD & SONS	Sevenoaks	**1847-66**
WHITEHEAD, HENRY	Farningham	**1851-74**
WHITEHEAD, J.	St May Cray	**1839**
WHITEHEAD, JAMES	Sevenoaks	**1823-38**
WIGG, NATHANIEL	Deptford	**1847**
WILCOX, JOHN	Woolwich	**1838-55**
WILLBY, JOHN	Canterbury	**1839**
WILMSHURST, —	Sandwich	**c1695?**

Nothing is known of this maker and the only reference known is that a watch by him has been noted. Possibly the same maker as the following entry, but given an earlier date in error.

WILMSHURST, THOMAS	Deal	**apprenticed 1713-died 1777**

Loomes mentions a Thomas Wilmshurst from Mayfield, Sussex, entering an apprenticeship to John Finch's widow Mary in 1713. (John Finch was a noted London watchmaker during the late seventeenth century and served as Master of the Clockmakers' Company in 1707.) The Deal man is believed to be the same person. There is no record of Thomas being freed in

London, but a number of watches and longcase clocks signed Deal are known to exist; a watch is in Canterbury Museum. He retired in September 1764 (see Appendix I), and died 5 August 1777.

WILSDON, T. W.	Canterbury	1865-74
WILSON, JOHN	Chatham	1845-51
WIMBLE, GEORGE	Ashford Faversham	born 1717-37 & 1740-died 1741 1737-40
WIMBLE, JOHN	Ashford	c1680, died 1741
WIMBLE, JEREMIAH (NEHIMIAH?)	Ashford and Maidstone	c1720-1750
WIMBLE, THOMAS	Ashford	born 1694-apprenticed 1716

THE WIMBLE FAMILY

There are numerous references to the members of this family in the parish registers of St Mary's Church, Ashford, but very few clocks made by them are known to have survived.

The first record is of Ann, the daughter of John, being baptized in 1623. Whether he was a clocksmith is not known as his occupation is not stated in the registers and he is unlikely to be John Wimble's father as his date of birth is thought to have been about 1680. Unfortunately there are a number of gaps in the registers, particularly during the Interregnum, and no reference to John's father has been traced.

John was married to Dorothy, the daughter of George Browne, a hatter by trade, and they are known to have had the following children: George, James, Jeremiah, Mary, John, Martha, William, Anne and Hannah. Both George and Jeremiah followed their father's trade as clockmakers and were both presumably trained by him.

In September 1740 George, who had worked in Faversham with John Baldwin for the previous three years, opened a shop opposite the Corn Market in Ashford where 'new Clocks or Watches [were] made or mended' (see Appendix I). An interesting connection with the Faversham clockmaker is the mortgaging of the land his parents owned behind a property of theirs in Ashford High Street (now number 17). This was mortgaged to secure a loan of £15 7s 6d in 1741 to John Baldwin. George's stay in Ashford was short-lived as he died in June 1741. The inventory taken after his death details an estate worth £22 10s 0d. Among his effects were 'One clock and Case' valued at £3 0s 0d, 'one old clock' at 15s 0d, 'Clock Brass' at 15s 0d and his 'Working Tooles' £1 5s (Fig 7/128).

Tragically, John Wimble died soon after, together with two other children, probably as the result of an outbreak of smallpox which had been rife during 1740 and 1741 in the Ashford area. The properties in Ashford are known to have been sold in 1742. There is also a reference in the manor records to another house in the High Street (now number 7) stating that it was 'previously Wimbles', presumably also in John and Dorothy's possession.

No clock or watch has been traced by either George or his brother Jeremiah, who it is believed is the Nehemiah Wimble mentioned as marrying Mary Swineyard in Maidstone in January 1746, and taking George Swinyard, Mary's brother, apprentice in 1750.

Loomes records Thomas Wimble's birth in 1694 and his apprenticeship to his father in 1716, but this is not confirmed as he does not appear in any of the records consulted.

An Inventory of the Goods of George Wimble of Ashford in the County of Kent Clock and Watch Maker lately Deceased taken and Apprized by Us whose Names are under Written this 16th Day of June 1741 As follows

	£	s	d
First For One Clock and Case	3	0	0
Also For One Old Clock	0	15	0
For a Mare	3	0	0
For 8 Cane Chairs	0	16	0
For One Bed and Furniture	1	10	0
For One Vice	0	5	0
For Working Tools	1	5	0
For Clock Brass	0	15	0
For a Small Table	0	2	0
For a Porridge Pot	0	7	6
For a Dozen of Pewter Plates and Two Dishes	0	10	0
For Fire Pan and Tongs, Bellows and Andirons	0	4	0
For a Gridiron	0	1	0
For a Tea Kettle	0	2	6
For his Wearing Apparell	2	0	0
For a Chest	0	5	0
Debts good and bad	4	19	9
Things unseen and forgotten	0	5	0
For Watch Keys and Glasses	0	7	0
For One Silver Watch	2	10	0
	22	19	9

June 16th 1741
Apprized by Us

Fran.s Willis
Allen

To John Wimble of Ashford [Father?] Clockmaker
dec.d 4 Days

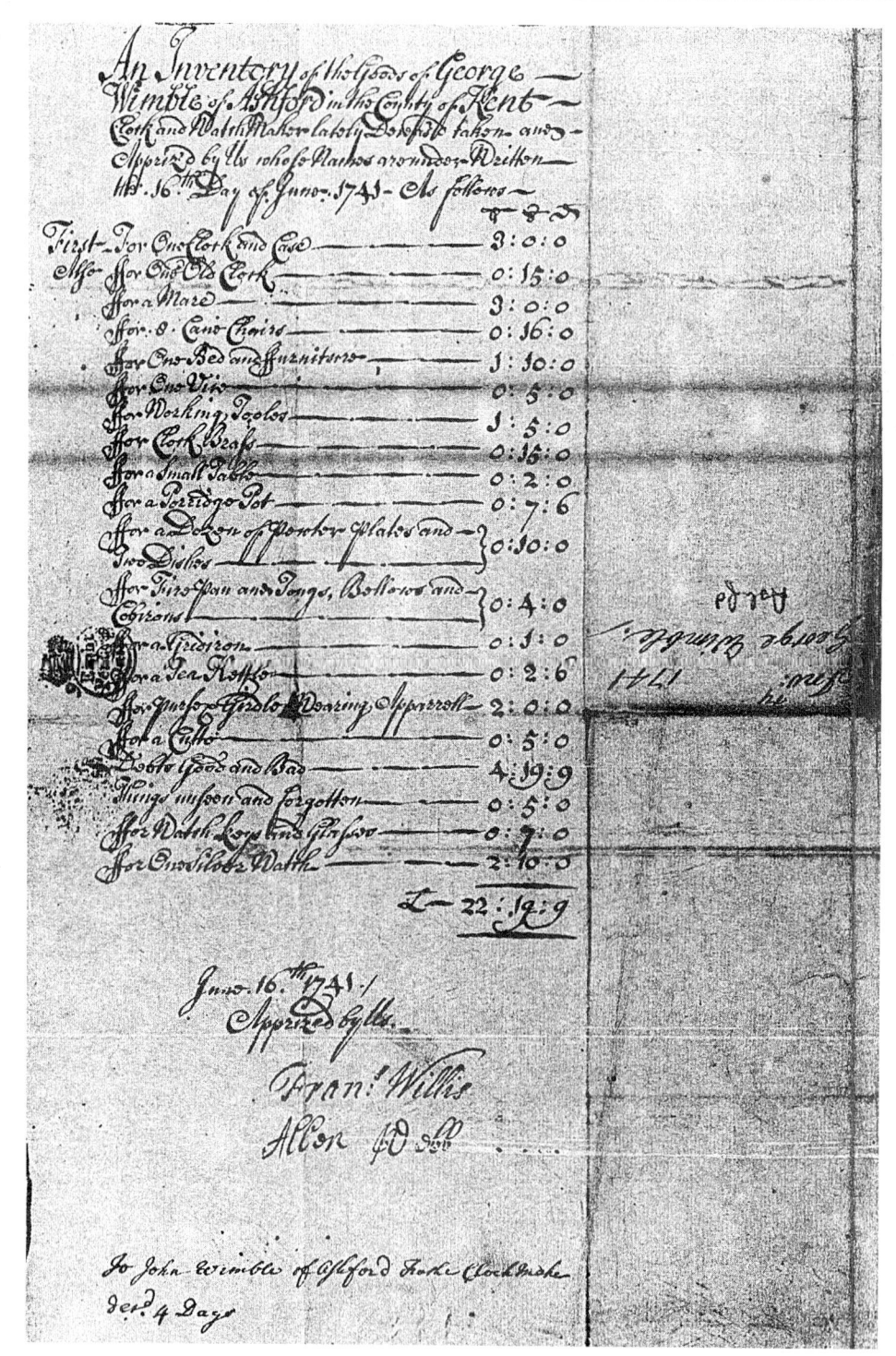

Fig 7/128 *The inventory of George Wimble of Ashford, taken on 16 June 1741*

At least four longcase clocks are known by John Wimble, including an 8-day arch-dial in an oak case and two fine lacquer examples, one red, one black. The red lacquer case, in its original condition, houses a good four-pillar movement with rack striking and although it dates from the 1715 period judging by the dial decoration, the case shows a late use of the convex moulding on the hood which had gone out of fashion in London by 1700-5 (Colour Plates XII and XIII).

In 1730 John supplied All Saints Church, Lydd with a new single-handed turret clock and there are numerous references in the churchwardens' accounts to payments to him in connection with work carried out on the clock, though, they do not indicate the actual cost of the clock. The last reference to him is in 1741, the year of his death. He may also have been responsible for making and installing the new clock in St Margaret's, Bethersden in 1737.

To date, there is no record of from whom John Wimble learnt his trade, although the Greenhills of Ashford would appear to be the most likely. However, the business connection with John Baldwin and the fact that George Wimble worked with Baldwin, suggests that the Faversham maker may have been his tutor.

WINCHESTER, J.   Dover   1813
Recorded in Loomes at this date as a watchmaker.

WINDER, JOHN CHRISTOPHER   Canterbury   1847-74
Recorded in *Bagshaw's Directory* of 1847 in Church Street.

WINTER, J. C.   Canterbury   1851
Probably a mis-spelling of the previous entry.

WINTERHALTER, D. & G.   Maidstone   1865-74

WITHERS, JOHN   Greenwich   1778
Recorded in Baillie as bankrupt in 1778

WOOD, HENRY   Canterbury   1826-8

WOOD, HENRY SAMUEL   Canterbury   1832

WOOD, THOMAS   Tunbridge Wells   1776
Recorded as a 'Goldsmith, and Clock and Watchmaker, the Bottom of the Walks' (see Appendix I, 11-15 May 1776).

WOGDEN, STEPHEN   Greenwich   apprenticed 1713-CC 1724
Noted in Baillie as being apprenticed in 1713 and obtaining his Freedom from the Clockmakers' Company in 1724. A clock has been recorded.

WOODRUFF, CHARLES   Dover   1874

WOODRUFF, CHARLES   Margate   1838-66

WOODRUFF, W. C. & C.   Dover   1855-66

WOODRUFF, WILLIAM   Dover   1858

Fig 7/129 A 11in square painted dial with a 30-hour birdcage movement by Woolley of Tenterden, about 1790

WOODRUFF, WILLIAM	Margate	1874

WOOLLETT, JOHN	Maidstone and London	c1782-90

Recorded in Baillie as being free of the Clockmakers' Company about 1790 and noted in Loomes as a watchmaker. He is thought to be a member of the famous engraver William Woollett's family, possibly a younger brother.

WOODWARD, JOHN	Ashford	1857-67

WOOLLEY, THOMAS	Charing	c1750

An 8-day five-pillar movement in an arch-dial laquered longcase was sold by Phillip's, Sevenoaks in 1996. The good quality movement had a sun-burst in the arch and was signed on a cartouche on the dial centre. The case, and in particular the hood with its characteristic fretted top, were similar to many other Kent clocks of the mid-eighteenth century.

Although previously unrecorded he may be the father of the clockmaker in the following entry. He may also be the maker named Woolley in the Bethersden Church accounts for 1722.

WOOLLEY, W. C.	Tenterden	pre-1789-c1810

Noted in Loomes as marrying in 1789 and soon afterwards formed a partnership with Thomas Wraight of Charing, although there is some confusion as to the exact dates of the partnership. William Strickland is known to have taken over the Woolley business in 1803. (See entry on Strickland.) A number of clocks are known, painted-dial longcases and small wall-hung silvered dial timepiece alarms among them. A particularly fine tavern clock signed 'Wraight and Woolley' is featured in Fig 7/130. This clock dates from the 1790 period and is unique in that it has a date aperture. Watches have also been recorded, in addition to painted dial 30-hour and 8-day longcase movements in well-proportioned oak cases signed solely by Woolley (Fig 7/129).

Fig 7/130 A longcase clock by Wraight & Woolley of Tenterden, about 1795. The mahogany case is 7ft 4½in tall. The dial, painted with roses and strawberries, is by Wilson of Birmingham

Fig 7/131 A tavern clock by Wraight & Woolley of Tenterden, with the unique feature of a date aperture, about 1795. The movement has tapered plates and four pillars

WRAIGHT, GEORGE Eastry 1826-32

Noted in Loomes as a watchmaker. He is thought to be the son of Thomas Wraight of Charing and Tenterden.

WRAIGHT & WOOLLEY Tenterden c1790-1810

A partnership between Thomas Wraight and Woolley, which produced a number of fine longcase clocks (Fig 7/130), timepiece alarms and a notable tavern clock (Fig 7/131).

WRAIGHT, THOMAS Charing c1780-c1790

Before entering into partnership with Woolley at Tenterden he worked on his own in Charing. At least three timepiece/alarm wall clocks are known. A fine example with a 5in dial with floral engraved spandrels in its original hooded oak case with a triangular pediment was sold at Sotheby's in 1986.

WRIGHT, RICHARD Woolwich 1845-51

YOUNG, F. Sittingbourne 1803

Noted in Loomes as a watchmaker at this date.

Appendix I

CLOCKMAKERS & WATCHMAKERS IN EARLY KENT NEWSPAPERS

During the course of research for this book much useful information was gleaned from the pages of the early Kent newspapers, the *Kentish Post* and its successor, the *Kentish Gazette*. Many of the issues between 1726 and the early 1800s feature advertisements for the recovery of lost or stolen watches or notices placed by various clockmakers regarding their businesses, both in relation to location and the services they offer.

This appendix includes all the references to Kent makers, but not to clocks or watches from outside the county unless they are of well-known makers, or of special interest. The entries will prove of great interest both to horologists and students of social history alike.

Between the years 1726 and 1803 there are over ninety reports in the two papers of watches by Kent makers either lost or stolen, which is further evidence of a thriving watchmaking industry in the county, particularly in the middle years of the eighteenth century. In many cases the number of the individual watch is noted. Where a maker, ie Chalklin, used a numbering system starting at 1 it is a guide to the probable number of watches produced by the individual craftsman. Although we cannot be sure just how many were produced it would appear that some makers, particularly from Canterbury, made quite large quantities from what were in most cases small workshops manned by individual craftsmen, possibly with the aid of only one apprentice.

Of general interest is an Act of Parliament in 1754 to prevent the embezzling, selling or pawning of clocks and watches left for repair. The penalty was £20 for the first offence and a fine of £40 for subsequent offences with imprisonment and whipping if the fine was not paid.

22-26 October 1726 *Kentish Post* [p3]

Richard Perkins, near the Market-Place in Dover, maketh and selleth all sorts of large and small Plate, Rings for Funerals of all sorts, good Watches, &c. And buys any of the above mentioned Goods.

12-15 March 1729 *Kentish Post* [p4]

Wereas a Silver Watch and Chain was lost on Wednesday in the Afternoon the 5th instant, between Wingham and Canterbury, with the Name of Watts on the Watch; Whoever hath taken it up and will bring it to the Printing Office in Canterbury, shall receive a Guinea Reward.

27-30 May 1730 Kentish Post [p4]

JOHN DAVIS, late from Stroud, is Removed to ASHFORD, and continues Selling all manner of Small Wares in Silver, such as Pint Cans, Cups, Salts, Spoons, Spurs, Buckles, and many other Smaller Things. Also sold Gold Rings plain or stone, Silver Watches new or second-hand as Cheap as their Goodness will admit. Likewise all manner of Hard Ware, English and Dutch Toys, Walking Canes, fine China Ware. Any Person may also be furnished with Draughts of Pictures, of any Size and Form, for Stair Cases, Rooms and Halls, of either fine or ruff Paintings. Note also, at the same Place, Watches are carefully Clean'd and Mended, and Jewellers Work perform'd by Me, to the Satisfaction of all that has been pleased to make Tryal. If any Person is disposed to Raffle for any one Thing, of what is before mention'd, or what my Shop will afford, it shall be set up as Cheap as if I was to Sell it otherways and have no farther Trouble with it.

18-22 July 1730 Kentish Post [p4]

LOST on Friday the 10th of this instant July, between Fordwich and Wingham, a Silver Watch, with the Day of the Month on it, and the Name of Michael Lade; having a Steel Chain, a straight Key, and one Seal: Whoever brings the said Watch to Tho. Holness, Baker at Wingham, or to Mr. Michael Lade, Gold-smith in Canterbury, shall have 15s. Reward.

3-7 April 1731 Kentish Post [p4]

This is to give Notice, that on Monday in Easter Week, the 19th Instant, two Clocks, one of Twelve Pounds Value, and the other of Ten Guineas Value, will be raffled for at the Dolphin in New Romney; the Persons who raffle for the Clocks to make up the Money at the Time of raffling: The Person who wins either of the Clocks to spend a Guinea, and a Guinea to be spent by Tho. Freebody, who keeps the House.

5-8 May 1731 Kentish Post [p4]

Samuel Kissar, over-against the Mermaid in St. Margaret's Street, Canterbury, designing to leave off Shop-keeping, will sell off his Stock by Wholesale and Retale, at very reasonable Rates: Consisting of Linnen Cloth, and Haberdashery Goods, and the best of new Clocks and Watches; also Counters and Shelves in the Shop. All Persons will be kindly used.

10-13 November 1731 Kentish Post [p4]

Canterbury, November 10.
Mr. Stephen Denston, late of London, and now of Dover, being at the House of James Ellis Watch-maker at Dover the beginning of October last, there gave to the Value of eleven Pounds English Money for French Two-penny Pieces; which when he arriv'd at Bullogn in France were found to be false Coin, he having put off some of the said Money, without the least Thought of its being a counterfeit Coin. Upon which, in the said Stephen Denston's Absence, his Chamber where he lodg'd was broke open, and the rest of the Pieces were seiz'd, with all his other Money, Goods, &c. By good Fortune he was met by a particular Friend, just off the Hills, who told Mr. Denston what had happen'd, advised him to make his Escape, saying that if he was taken he certainly would be broke upon the Wheel. He was struck with surprize upon hearing this melancholy News, and wander'd about the Country for ten Days

in Disguise. After suffering extream Hardships, being in a deplorable Condition, the Drums beating and describing his Person in the neighbouring SeaPort Towns for apprehending him, he by great Providence got to a Vessel, which brought him safe to Dover, tho' at a great Expence. So hopes every Person will be careful how they take any more of that Money. Mr. Denston's Friend, who assisted in his Escape, was afterwards taken up, and 'twas thought would be hang'd. The aforemention'd James Ellis was seiz'd at Chatham, and being carried before a Justice of Peace, his Mittimus was made to be sent to Maidstone Jail.

21-24 June 1732 *Kentish Post* [p4]

LOST on the 2nd of this inst. June, between Brookland and Rye, a Silver Watch, with a Silver Chain to it, made by Reeves at Rye in Sussex, Numb. 552. Whoever brings it to Capt. Nathaniel Pigrams at Rye, or to William Green, Grocer, in Smarden, shall have Half a Guinea Reward. If offer'd to Sale, 'tis desired it may be stopp'd, and Information given of it.

21-24 June 1732 *Kentish Post* [p4]

This is to give Notice, That all sorts of Watches are made, mended and sold by Samuel Kissar, who is lately Removed from St. Margaret's-street to the Crown and Dial in Burgate-street, Canterbury.

N.B. He has a Watch to sell made by Mr. Thomas Tompion, it being one of the best Watches in Kent.

19-23 August 1732 *Kentish Post* [p4]

Samuel Kissar, Watchmaker, at the Crown and Dial in Burgate-street, Canterbury, hereby gives Notice to Persons who have Fleece Wooll to dispose of, that he will buy it, and will give the full Market Price, and pay ready Money for the same.

18-21 October 1732 *Kentish Post* [p4]

ALL sorts of Watches made, mended, and sold by SAMUEL KISSAR, at the Crown and Dial in Burgate-street, Canterbury.

N.B. The Watches that the said Sam. Kissar sells for 5 Guineas, he will warrant to be good, and will keep them in Repair for 7 Years.

29 August-1 September 1733 *Kentish Post* [p4]

A Letter from a Quaker to his Watch-maker Friend,
I Have once more an erroneous Watch, which wants thy speedy Care and Correction, since the last time he was at thy School, I find by Experience, he is not benefited by thy Instructions. Thou demandeth for thy Labour the fifth part of a Pound Sterling, which thou shalt have, let thy Endeavours first Earn it: I will Board him with thee a little longer, and pay for his Table if thou requirest it; but pray thee let thy whole Endeavours and Observations be upon him, for he is mightily Erring from the Principles of Truth: I am afraid he is foul in the inner Man, I mean his Springs; prove and try him with the adjusting Tools of Truth, that if possible, he may be drawn from the Errors of his Ways; by the Index of his Tongue he is a Lyar, and his Motions very varying and unsetled; perhaps his Body may be foul and corrupted; Brush him well with cleansing Instruments from all Polutions, that he may Vibrate with Regularity and

Truth; admonish him Friendly with Patience, and be not too hasty and rash in thy Corrections, least by endeavouring to reduce him out of one Error thou mayest fling him headlong into another, for he is young and of a malliable Temper, and may be wrought to do the Things that are Right; let him visit often the Motions of the Sun, and regulate him by her Table of Equation; and when thou consentest they agree, send him Home with the Bill of Moderation to thy Friend.

19-23 February 1736-7 *Kentish Post* [p4]

LOST on Saturday last, being the 12th Instant, between Boughton Hill and Canterbury, about 5 a-Clock in the Afternoon, a Silver Watch, Mercer, Hithe, on the Dial Plate. — In the inside, John Mercer, Hithe: Whoever has found it, and will restore it to the Owner, Robert Andrews, Carpenter at Milton, or to the Printer of this Paper, shall have a very good Reward. If offer'd to be sold or pawn'd, pray stop it, and give Notice as above.

6-9 April 1737 *Kentish Post* [p4]

LAWRENCE BOYS, Watch-maker, near the Post-Office in Deal, makes and sells all sorts of Watches and Clocks; where any Person may have them good, and as cheap as in any Town in Kent. He mends and cleans the same, and buys second-hand Watches and Clocks.

3-6 August 1737 *Kentish Post* [p4]

LOST on Monday the 25th of July last, between Eastwell and Chilham, a small Silver Watch, with a blue Ribbon, a Key, and a Silver Seal, the Words Michael Lade being on the Plate: Those that have taken it up, are desir'd to bring it to the Printing Office, where they will receive half a Guinea Reward, and no Questions ask'd.

N.B. 'Tis suppos'd to be Conceal'd by a Person at Challock; if it is offer'd to be pawn'd or sold, pray stop it, and Satisfaction will be given for so doing.

31 May-3 June 1738 *Kentish Post* [p3]

LOADER BOURN, Goldsmith, at the Sign of the Three Golden Candlesticks in Burgate-street, Canterbury, in the Shop where Mr. Robert Anslow the Cutler lately liv'd, maketh and selleth all sorts of large and small Plate; Watches of all sorts; Rings for Funerals of all sorts, as Skeletons, Flower'd Rings, Motto Rings, &c. the Work as good, neat, and cheap, as can be performed in London.

Note, He buyes any of the above-mentioned Goods, and will give as good a Price as can be allow'd.

15-19 July 1738 *Kentish Post* [p4]

LOST on the 24th of June last on Eythorn Down, a Silver Watch, with a little Dent on the Side, and the Name of Jenkinson, Sandwich, on it; without a Chain. Whoever brings it to Mr. Wallis, Master of the Red Lyon Tavern in Canterbury, or to Mr. Taylor's at Langden Abbey, or to the Sign of the Victualling Office in Dover, shall receive half a Guinea Reward. If offer'd to be pawn'd or sold, pray stop it, and the same Reward will be given.

6-10 September 1740 *Kentish Post* [p4]

GEORGE WIMBLE, Clock and Watch-maker, who has work'd with Mr. Baldwin of Feversham for these three Years past, is lately removed from Feversham aforesaid to keep a Shop opposite to the Corn-Market in Ashford; where all Persons may have new Clocks or Watches made or mended, at the most reasonable Rates, and best Manner, and quick Dispatch, by their humble Servant,

GEORGE WIMBLE.

27-31 December 1740 *Kentish Post* [p3]

LOST on the 8th of this Instant December, between Nonnington Church and Sir Thomas D'Aeth's at Knowlton, a Silver Watch, with a Steel Chain, a Brass Seal and Key, and a Piece of Iron on the Top of the Key, the Maker's Name Jenkinson, Sandwich: Whoever will bring the said Watch to Mr. Sampson Creake, Shopkeeper, at Nonnington, shall have Half a Guinea Reward, and no Questions ask'd. If offer'd to be pawn'd or sold, pray stopt it, and give Notice as above.

N.B. Any other Person that has lost a Silver Watch may hear where one is in Pawn, by enquiring of the said Sampson Creake, or of Mr. Anthony Ovenden at Petham.

12-15 August 1741 *Kentish Post*

John Peckham, Clock and watchmaker at the Sign of the Spring Clock in the town of Ashford makes, mends and sells all sorts of repeating and Plain Clocks and Watches at the cheapest rates ....

24-27 June 1741 *Kentish Post* [p4]

Abraham Webb, Clock-Smith and White-Smith, from Wye, has taken the Shop of Mr. Arthur Hurt, late of Ashford, Clock-Smith, deceased; where all Gentlemen and others may be furnished with all sorts of Clocks, Jacks, Guns, and Watches, at reasonable Rates; also Watches repaired, and all Business in the Trade of a Clock-Smith and White-Smith carefully done, with quick Dispatch, by   ABRAHAM WEBB.

9-12 March 1742/3 *Kentish Post* [p4]

DOVER,

SAMUEL LEE, from London, Watch-maker, Makes, Mends, Cleans, and Sells all Sorts of WATCHES Plain and Repeating, in the Best and Cheapest Manner; being Entirely Bred to the Finishing Part of that Trade.

23-26 January 1744/5 *Kentish Post* [p4]

LOST Yesterday, in a plow'd Field a small Distance from Bridge Hill House, on the upper Side of Barham Down, near the Distance Post, or between that Place and St. Stephen's, A Gold Watch, with a square Gold Chain, an Enamel'd Dial-Plate, and a Crystal Seal with the Hale's Arms, the Maker's Name Robert Higgs: Whoever has found it and will carry it to Alexander Hales's, Esq: at Charlton-House; to Mr. Wallis at the Red-Lyon in Highstreet, Canterbury, or to Mr. Fuller's at St. Stephens's, shall receive Two Guineas Reward.

8-12 June 1745 *Kentish Post* [p4]

Canterbury, June 12.

On Saturday Night between 10 and 11 of the Clock, at Winding-hill near Sittingbourn, Mr. Lindsey, a Butcher of Milstead, was robb'd of three Moidores, two or three Guineas, and a pretty deal of Silver, with a Silver Watch, on it these Words, Mic. Lade, Canterbury, No. 549. The Rogue is a stout Fellow, and had a large Club in his Hand. If the Watch is discover'd by any Person, and restor'd, he will be rewarded.

16-20 November 1745 *Kentish Post* [p4]

Canterbury, Nov. 20.

George Robinson, Servant to Mr. Tilden of Rodmersham near Sittingbourne, was robb'd on Sunday last, November the 17th, about Seven o'Clock in the Evening, by three Men in Seafaring Habits, between Rainham and Chatham-Hill, who took from him all his Money and a Silver Watch made by William Avenal of Gravesend, as also a red spotted Silk Handkerchief mark'd G. R. The Watch had lost the Minute-Hand.

16-19 July 1746 *Kentish Post* [p1]

To be Sold, at Thomas Shindler's, Watch-maker,
In Burgate-street, Canterbury,
ALL Sorts of Linnen-Drapers and Haberdashers Goods, Gloves and Stockings,
at prime Cost or under, he leaving off that Business.

30 July-2 August 1746 *Kentish Post* [p3]

LOST on Tuesday last the 29th of July on Barham Downs, or on the Road cross the Country to Sandwich, A Silver Watch, the Maker's Name on it, Shindler, Canterbury; Whoever has found the said Watch, and will bring it to the Printing Office in Canterbury, shall receive half a Guinea Reward.

NB. If offer'd to be Pawned or Sold, pray stop it, and the same Reward will be given.

4-7 March 1746/7 *Kentish Post*

LOST between Molash and Charing Heath on Tuesday Night last, a Silver Watch, Crathorne, Maidstone, on it, with a Steel Chain, and two Seals, one Brass and the other Steel; Whoever brings it to the Swan at Charing, or to the Printing Office at Canterbury, shall receive half a Guinea Reward with Thanks.

26-30 September 1747 *Kentish Post*

LOST at Goodneston Fair on Monday the 15th Instant, a Silver Watch, the Maker's Name Lade with the Day of the Month on the Dial Plate; and an old Ribband and a Key ty'd to it: Whoever brings the said Watch to John Moat, at Mr. John Newman's, Collar-maker, at Nonnington near the Halfway House to Dover, shall receive half a Guinea Reward. If offer'd to be pawn'd or sold, pray stop it and give Notice as aforesaid.

7-10 December 1748 *Kentish Post*

We hear from Cranbrook that last Saturday night about seven o'clock Mr Thomas Cackett a watchmaker of that place was, near his own house, knocked down, and robb'd of a Guinea and two watches, by persons unknown; and that he lies dangerously ill of the bruises he recieved.

7-10 February 1749/50 *Kentish Post* [p1]

LOST the second of this instant February, between Godmersham and Lee, a Silver Watch, the Maker's Name Parquait, Canterbury, with a blue Silk String and Brass Seal; Whoever hath found it, and will bring it to the Printer of this Paper, or to William Southee, at Humphry Pudner's, Esq; shall receive Half a Guinea Reward.

    N.B. If the Person that hath found it lives at a Distance, and will bring it to either of the abovesaid Persons, all reasonable Charges shall be defray'd, exclusive of the Reward above-mention'd.

27-30 June 1750 *Kentish Post* [p4]

ALL Sorts of Clock and Mathematical Work; also Brass Plates for Coffins, or Inscriptions for Monuments; and all Manner of Brass Ornaments, curiously Engraved, Varnished, or Silver'd, or Plated with Silver, by

                    MANDEVILE SOMERSALL,
                  Clock Engraver and Varnisher,
            in Fore-Street, near Moorgate, London.

Where Country Chapmen may be furnished with all sorts of Clock Dial-Plates compleately fitted up: As also all Sorts of Tools and Materials for Clock and Watch-making, at the lowest Prices.

    N.B. The Report of my being dead, was occasioned by my long Illness, and the Death of my Father, Mr. George Somersall, Clock Engraver and Varnisher, of Little Moorfields, lately deceas'd.

[Similar advertisements appeared in a number of provincial newspapers, including the *Ipswich Journal* in 1750 and the *Derby Mercury* in 1754.]

8-11 August 1750 *Kentish Post* [p1]

LOST on Sunday July the 29th, as is suppos'd, in the Parish of Bridge, Patridgbourne, or Beaksbourne, a Silver Watch, with an enamel'd Dial; (by some call'd a China Plate or Dial): the Maker's Name, on the upper Plate, in the Inside, Loder Bourn, Canterbury, No. 16. Whoever has found this Watch, and will bring it to George Sims, Watch-maker, in St. George's, Canterbury, shall have Half a Guinea Reward.

9-12 January 1750/1 *Kentish Post* [p1]

JUST arrived from Venice, and now to be seen and heard, till Sold, at the Sign of the FLEECE in St. George's, Canterbury,
A most magnificent and matchless Piece of Mechanism. This curious Piece of Art represents two grand Statues; the one an Indian, the other a Moor; who at the Word of Command, each takes up his Instrument, and can perform twenty different Pieces of Musick, with a thorough

Base, in a very elegant Manner, most of which are entirely new, and compos'd on Purpose for this Machine; all actually perform'd by Clock-Work. This Work is judged by all who have seen it worthy to adorn the Palace of a Prince, as it exceeds whatever has been done of this Kind, it being nine Feet high and fourteen wide. Also at the same Place is to be seen a most excellent Composition of Rock and Grotto Work ....

Also other Curiousities too tedious here to mention. The Whole is to be sold for Two Hundred Pounds, the Proprietor having been bid One Hundred and Fifty. To be seen from Eight in the Morning till Nine at Night, coming four or more in Company.

By particular Desire of several Gentlemen and Ladies, who have not yet had the Opportunity of seeing this curious Piece of Art, it will remain in Town till Thursday next.

N.B. The Curious are desir'd to take Notice, it is not an Imposition, as the Person's was who was here last Summer with his Subterraneous World, Petrefied Caverns, &c.

30 January-2 February 1750/1 *Kentish Post* [p1]

For SALE by Auction, On Saturday the 9th instant February, at the House of Mrs. Sarah Rawling, in the great Street in Dover, THE Shop, Goods and Tools late of Edward Rawling, deceased, in his Trade of a Clock-Smith; consisting of Clocks new and old, Watches, an Engine for cutting Clock-Wheels, a Barrel-Engine, Clock-Lades [lathes], and Vices, and other Tools and Things in the said Trade.

N.B. The Sale to begin at Ten o'Clock in the Forenoon.

20-23 February 1750/1 *Kentish Post* [p4]

Canterbury, February 23.

On Wednesday last about Two of the Clock in the Afternoon the Canterbury Stage-Coach, coming from London, was robb'd by a single Highwayman near the top of the Gravel-Hill coming to Greenhithe. He took from the Passengers about three Pounds in Silver; and a Silver Watch, with a round Ringle in the room of a Chain; on the Plate was ingrav'd, Watts, Canterbury, with a Lion and Unicorn gilt, the Figures pretty much defac'd. He seem'd to be middle ag'd, thin fac'd, a sharp Nose, of a brown Complection, and meanly dress'd, in a Surtout Duffil Cape Coat of a lightish brown Colour. He appear'd by his Hands and Dress to be a Working Man, as a Blacksmith, or some other dirty Employment. He was mounted on a light thin Horse, about 14 Hands and a half high, Hog-buttock'd, and a Flig Tail, of a light Bay (or Chesnut) Colour.

1-5 June 1751 *Kentish Post* [p4]

Canterbury, June 5. Last Sunday dy'd Mr. Shindler, one of the Aldermen of this City.

15-19 June 1751 *Kentish Post* [p1]

Whereas a Silver Watch was lost between Canterbury and Dover Yesterday in the Morning, the 18th Instant, about Three or Four o'Clock, with a black String ty'd to it and a Seal, the Maker's Name, John Watts, Canterbury; Whoever has found the said Watch and will be so kind as to send it to Henry Beal, one of the Servants at the Red Lion in this City, shall receive Half a Guinea Reward.

19-22 June 1751 *Kentish Post* [p1]

    WILLIAM CLEMENT, Junior, Clock-maker, from London,
     who now keeps a Shop in Preston-Street, FEVERSHAM,
MAKES and Repairs all sorts of Clocks and Jacks, after the newest and most approved Method. He also Sells and Repairs all sorts of Guns or Gun-Locks; and mends Plate, or any sort of Jewels. And as to Watch-making and Mending, he has procured William Bird, a Watch-Finisher, who has serv'd some of the greatest Masters in London, to please his Customers, and who, 'tis believ'd will give general Satisfaction; as it will always be the utmost Endeavours for that Purpose of   Their most humble Servant,   WILLIAM CLEMENT, Junior.

10-13 July 1751 *Kentish Post* [p1]

LOST about a Week ago, a Silver Watch, between Hearn-Hill and Davington; Whoever hath taken it up, and will bring it to Mr. Lade's, Silversmith, at Faversham, shall be satisfied for their Trouble.

17-21 August 1751 *Kentish Post* [p1]

Whereas on Monday the 5th of this inst. August was dropt, as is suppos'd, in the Yard of the Ship-Inn at Dover, or coming from the Castle, a Repeating Watch, Dobson the Maker, the inner Case is Pinchbeck-Metal, the outer Case is Shagreen, with a Steel Chain and three Seals; Whoever brings it to Mr. Streeting, Master of the said Inn, shall have Five Guineas Reward, and no Questions asked.

21-24 August 1751 *Kentish Post* [p1]

THOMAS SHINDLER, Watch-Maker, Having taken the Shop of his late Father, opposite Butchery-Lane, in Burgate-Street, Canterbury, begs Leave to acquaint the Publick, that he regularly learnt his Business in one of the principal Shops in London, and makes and repairs all sorts of Clocks and Watches, in the best and newest Method; and will endeavour to give entire Satisfaction to all who will favour him with their Custom.

 N.B. All sorts of Linnen-Drapery and Haberdashery Goods, and Stockings, &c. are sold at the said Shop, at very reasonable Rates.

7-11 September 1751 *Kentish Post* [p1]

LOST the 4th of this instant September, between the Halfway House going to Sturry and Stuppington-Hill, a Silver-Watch, mark'd on the Dial-Plate C. Charlson, having on a black Ribbon, and a green Seal with a Turks Head, the Watch-Key mended with a bit of Steel, and a piece of Scarlet Cloth in the outer Case of the said Watch. If brought to Mrs. Jael Hobbs in Sheepshank-Lane, or to the Printer of this Paper, the Person will receive Half a Guinea Reward.

29 January-1 February 1752 *Kentish Post*

    WILLIAM BIRD, Watch-maker, from LONDON,
  Who, for many Years past, has served some of the Principal Masters there.
    Having NOW taken a Shop in Court-Street,
   opposite to Mr. Cobb's, near the Market-Place in FEVERSHAM,
Makes, Mends, Cleans, and Sells Watches of every kind, either in Gold, Silver, or Shagreen

Cases, at very Cheap and Reasonable Prices. Repeating Clocks, and Watches, and Spring Movements, of every Kind, are likewise carefully Cleaned, and Repaired, as Well, and in as Great Perfection, as any where in London. All Gentlemen, and others may depend on being Honourable, and Faithfully served, by their most obedient Servant.   WILLIAM BIRD.

N.B. He also buys, and gives, the most Money for OLD WATCHES, RINGS, And PLATE of every Kind.

16-20 June 1753 *Kentish Post*

LOST between Boughton and Ospringe, on Saturday last, a middle siz'd Silver Watch, Thomas Jenkinson, Sandwich, on the Dial Plate; a short piece of old Ribbon with a Key and a piece of a Seal to it; Whoever has found this Watch, and will bring it to the Printer of this Paper, shall be satisfied for their Trouble.

7-10 November 1753 *Kentish Post*

Canterbury, Nov. 10.
On Thursday Morning last, about One o'Clock, the Shop of Mr. Thomas Shindler, Watch-Maker in this City, was broke open and robbed of several Watches, viz. one Gold Watch with a white enamel'd Dial-Plate, green Shagreen Case, a Womans Chain and Hook; one Si[l]ver Glass-hoop'd Watch, Silver Pedestals, Maker's Name Ed. Cockey, Warminster; the others at present cannot be described as to the Makers Names.

17-21 [and 28] November 1753 *Kentish Post*

Whereas it was mentioned in this Paper the 10th Inst. that the Shop of Thomas Shindler ... [was robbed ...] The others are as follows; one Silver Watch, Maker's Name Ward, London, one ditto John Berry, Cant. two ditto Lade, Cant. one ditto Shindler, Cant. with a Silver Chain broke, one very small Watch, Box and Case in one, Moran, London, one ditto Name not known, with four Pair of Cases; all which are supposed to be concealed in or near this City.

Whoever can meet with any of the said Watches, and bring them to me, shall receive five Guineas, and for the whole Number ten Guineas Reward.   THOMAS SHINDLER.

NB. One Thomas Smith having been taken up on Suspicion of the said Robbery: This is to give Notice, that if he has sold or pawn'd any of the said Watches, that if the Persons will bring them to Mr. Shindler, they shall have their Money again, and be satisfied for their Trouble. [The stolen watches were found in a vault belonging to the poor house in the borough of Staple-gate (15-19 December 1753), and Thomas Smith was found guilty of the felony and condemed to death (2-5 Jan 1754), but owing to his age reprieved by the Secretary of State and sentanced to seven years' transportation (16-19 Jan 1754).]

20-23 March 1754 *Kentish Post*

LONDON, March 20

We hear that by the new Act of Parliament for preventing Persons employed in the Manufacture of Clocks and Watches from embezzling, selling or pawning, the Clocks and Watches, or such Part thereof as shall be entrusted to them by their Employers, the Penalty is Twenty Pounds for the first Offence, and Imprisonment and Whipping in case of Non Payment; and for a subsequent Offence Forty Pounds, with like corporal Correction for Non Payment, at the Discretion of the Justice of the Peace.

23-27 March 1754 *Kentish Post*

LOST on Monday last, about two or three o'Clock in the Afternoon, between Lydden and Dane Hill, a Silver Watch, Name of Watts, Canterbury, on the Dial Plate, with a Key ty'd to a piece of black Leather; Whoever has found the said Watch and will bring it to Mr. Bromley's, Farmer, at Womenswould, shall receive Half a Guinea Reward. If offer'd to be pawn'd or sold be pleased to stop it, and give Notice as above.

27-30 March 1754 *Kentish Post* [p4]

LOST between Canterbury and Liminge, on Saturday the 23d Instant, between the Hours of One and Five o'Clock in the Afternoon, a Silver Watch, name of Jenkinson, Sandwich, within it; having a Tape String ty'd to it, with a Silver Seal dotted, and a broken piece of Seal: Whoever has found this Watch, and will bring it to Thomas Lilly's at the George in Stonestreet, or to Mr. Ivesson's at the Flying Horse, Canterbury, or to the Printer of this Paper, shall have Half a Guinea Reward. — If offer'd to be sold or pawn'd, please to stop it and give Notice as above.

8-11 May 1754 *Kentish Post*

LOST or Stolen out of a Fob-Pocket near Sittingborne, the 24th of April last, a Silver Watch and Silver Chain, Maker's Name James Harris, Maidstone, with a most remarkable Dial Plate, for instead of the usual Letters round it, are the twelve Following, viz.

I T R A D M E K S H A M

Whoever brings the said Watch to Mr. John Denne at the Post-House at Sittingborne, or to the Printer of this Paper, shall receive One Guinea Reward. If offer'd to be pawn'd or sold, be pleas'd to stop it, and give Notice as above.

11-15 May 1754 *Kentish Post* [p4]

LOST the second Instant, going from Ash to Deal the Country Way, A Silver Watch, Maker's Name Williamhurst [Wilmshurst], Deal: Whoever has found the same, and will bring it to Mr. Saven at the Black Horse in Deal, or the Printer of this Paper, shall be satisfied for their Trouble.

17-20 July 1754 *Kentish Post* [p4]

LOST on Wednesday last, between Gore-Street in the Parish of Monkton in the Isle of Thanet, and Wall-End, a middle-siz'd Silver Watch, with a breaded [=braded] flaxen String, has a small Crack on the Dial-Plate, and the Maker's Name, Swan, Margate: Whoever has found it, and will bring it to William Citchaman, Servant to Mr. Collard of Gore-Street aforesaid, or to John Simmons at Sarr, shall receive Half a Guinea Reward.

27-30 August 1755 *Kentish Post* [p1]

ALL Persons indebted unto Thomas Jenkinson, late of Sandwich, Watch-Maker, deceased, are required to pay the respective Debt's unto Mrs. Sarah Jenkinson, his Widow and Administratrix, at her House in Sandwich, on or before Michaelmas next. And such Persons, who have any Demands on the Estate of the said Thomas Jenkinson, are desired to bring in an Account thereof to the said Mrs. Jenkinson, in order to be paid the same.

7-11 February 1756 *Kentish Post* [p1]

Whereas the Shop of William Stevenson, Watch-maker, in East-Lane in Maidstone, was broken open on Saturday Night last, and five Silver Watches and two Pair of Silver Watch Cases were stolen therefrom; If any of the said Watches or Cases be offer'd to Pawn or Sale, please to stop them and the Person, and give Notice to Mr. Stevenson aforesaid, who will give all reasonable Satisfaction. — NB. The Name Harris, Maidstone, was on one of the Watches; on another, Craythorne; and one was a Box and Case Watch, pretty flat, with a very large middle Piece to the Dial Plate.

14-18 February 1756 *Kentish Post* [p1]

Stolen from out of the Shop of Richard Goodchild, Watchmaker, in St. Dunstan's-Street, Canterbury, on Monday about Noon, a plain Silver Watch, the Name of John Pysing, Canterbury, on it, with a Paper on the Inside, all sorts of Clocks and Watches made and sold by R. Goodchild, Canterbury; the Fellow was a young Man, of a middle Size, dress'd like a Farmer's Servant, with a ragged Frock: Any Person giving Notice to Richard Goodchild aforesaid, so that the Watch may be had again, shall have all reasonable Satisfaction. If offer'd to be pawn'd or sold, please to stop it, and give Notice as above. — N.B. The Rogue pretended that he lived with a Neighbouring Farmer, who had sent him to desire he would come and put his Clock in order.

14-17 April 1756 *Kentish Post* [p1]

To be Sold at Winchilsea in Sussex, A Very good Parcel of Watch-makers Tools, lately belonging to Mr. Thomas Huniset of that Place, deceas'd. Inquire of Thomas Huniset of Stone in the Isle of Oxney in Kent.

22-25 December 1756 *Kentish Post* [p1]

LOST, supposed to be dropt, on Monday the 20th Instant, either in the Choir of the Cathedral Church in Canterbury, or in coming from thence into St. George's Street, an old plain Gold Watch, with a white Dial, the Maker's Name D. Quate [Daniel Quare] London, No 687; had on when dropt a Steel Chain, with a small Cornelian Seal, the Impression a Lion rampart: Whoever has found this Watch, and will bring it to George Sims, Watch-maker, in St. George's-street, Canterbury, shall receive a Guinea Reward, and all reasonable Charges. If offer'd to be pawn'd or sold, please to stop it, and the like Satisfaction will be paid; and, if already pawn'd, your Money again with Thanks. — N.B. It's a very old Watch, and the Cases slight and defective, of no more Value than the Weight of Gold.

13-16 July 1757 *Kentish Post*

To Be SOLD as it Stands,
And to be taken down within such Time as shall be agreed unto
for carrying the Materials off the Premises

A Large old Timber Manshion House, at Upper Wilsley near Cranbrook, some Time since in the Occupation of William Stunt[?].... For further Particulars inquire of Sherlock Thorpe, at Hartley-Street, near Cranbrook; or of George Sims, Watch-maker, at Canterbury.

N.B. The Timber is of a stout Scantling, and fit for modern Building; Girders of about 20

Foot long; a good high-pitched Roof, with stout Rafters of one Length; and a large Parcel of very good Tiles.

20-23 June 1759 *Kentish Post* [p1]

To be SOLD. Part by Auction and Part by Hand,
On Thursday the 28th Day of June instant,
At a House in St. James's-Street, Dover,

Divers Household Goods and Furniture, amongst which are a very fine large Looking-Glass, a large Clock, goes a Month, made by Peter King, one Repeating Table Clock, made by Quire, [Daniel Quare] ...

The Goods to be seen on Tuesday and Wednesday before the Day of Sale, by applying to Mr. Farbrace, Attorney at Dover.

9-12 January 1760 *Kentish Post* [p1]

WANTED immediately, as an Apprentice,

A Youth who is naturally Ingenious, and that has an inclination to learn the Art of Watch and Clock Making in all their Branches. Also all manner of House and Ship painting, Herald painting, Drawing Perspective, &c. &c. Such a Youth that has an inclination to learn the above Arts, either separate or the whole, may apply to Thomas Mercer at Folkestone, as soon as conveniently he can, to treat about the Conditions.

N.B. The said Thomas Mercer sells very good new Watches, at three Guineas each, for the benefit of Country young Men, who are often imposed on by Jews, &c. with bad Goods. Also Mr. Hammond's School is almost at the next Door to the abovesaid, where such a Youth as shall like to be an Apprentice, shall be permitted by the said Thomas Mercer, to learn all sorts of Practical Mathematics at convenient Times.

26-30 July 1760 *Kentish Post* [p1]

Stolen out of the Workshop of Mr. Edward Burgess at Canterbury, on Friday the 25th of July inst. A small Silver Watch with an enamell'd Plate, the Maker's Name Michael Lade, the No. 2241; it had to it when stolen, a Scarlet Silk breded String, with a Silver Seal, the Impression a Ship: Whoever will give any Intelligence to the Owner of the abovesaid Watch, so as he may have it again, and the Offender or Offenders brought to Justice, shall receive a Guinea for their Reward. The Owner may be heard of by applying to the Printer of this Paper.

10-13 September 1760 *Kentish Post* [p1]

LOST on Monday last, the 8th of September, at Feversham, or in the Road to Chilham, a large Silver Watch, the Maker's Name Reves, at Rye, with a Purple String mixt with Gold Thread, and a Key; Whosoever brings the same to the Printer of this Paper shall receive One Guinea Reward.

29 April-2 May 1761 *Kentish Post* [p4]

LOST between Sittingbourn and Boughton Hill, on Thursday last, the 30th of April, A Silver Watch, the Makers Name John Brice, and W. H. on the out side of the in-side Case, being the two letters of the Owners Name; a Leather String, with a common ordinary Seal, and Key: Whoever has found the said Watch, and will deliver it to Mr. Tiddeman, at the Blue Anchor at Ospringe, shall receive Half a Guinea Reward: And if offered to be pawn'd or sold, pray stop it, and send to William Harason, at the George Inn, Sittingbourn.

14-17 October 1761 *Kentish Post* [p1]

JOHN CHALKLEN, Watch-Maker, in Burgate-street, Canterbury,
WILL Attend at Mr. Richard Elgar's, Cabinet-Maker, in Folkstone, on Monday the 26th Inst. and every Monday Fortnight, or oftener if requir'd: Where Persons may be supply'd with Clocks or Watches; Also Old ones cleaned and repaired in the best Manner, on the same Terms as at his House in Canterbury. Orders taken in at any Time by Mr. Richard Elgar; which will be punctually observed, and the Favours gratefully acknowledged.

By their humble Servant,

JOHN CHALKLEN.

14-17 October 1761 *Kentish Post* [p1]

WILLIAM MERCER, Clock and Watch-maker,
TAKES this Opportunity to inform his Friends, that he is coming to settle at Folkstone, to carry on the several Branches of that Business; Those Gentlemen, Ladies, and others, who are so kind as to oblige him with their Commands, may depend on being served on the best Terms, and their Favours gratefully acknowledged by

Their humble Servant,

WILLIAM MERCER.

23-26 December 1761 *Kentish Post* [p1]

LOST on Saturday the 19th of December, between Canterbury and Chilham, a Silver Watch, the Maker's Name John Chalklen, Canterbury, No. 68. Whoever hath found this Watch, and will bring it to John Chalklen, Watch-maker, in Burgate-street, Canterbury, shall have half a Guinea for their Reward.

13-17 March 1762 *Kentish Post* [p4]

To Miss ——. On a W A T C H.
By Miss Elizabeth Carter.
WHILE this gay Toy attracts thy sight,
    Thy Reason let it warn;
And seize, my Dear, that rapid Time,
    That never must return.
If idly lost, nor Art or Care,
    The Blessing can restore:
And Heav'n exacts a strict Account
    For ev'ry mis-spent Hour.

Short is our longest Day of Life,
    And soon its Prospects end,
Yet on that Day's uncertain Date,

Eternal Years depend.
Yet equal to our Being's aim,
The space to Virtue giv'n:
And ev'ry Minute well improv'd,
Secures an Age in Heav'n.

22-25 September 1762 *Kentish Post* [p1]

WILLIAM NASH, Watch-Maker at Bridge, will attend at Mr. Bushel's, Cornfactor, opposite the Cattle-Market, Canterbury, every Saturday, where Persons may supply themselves with Watches and Clocks, and Old ones carefully Repair'd, at the same reasonable Terms, as at his House at Bridge: Orders taken in at Mr. Bushel's, every Day in the Week; and which will be gratefully acknowledg'd, by Their humble Servant,

WILLIAM NASH.

22-25 September 1762 *Kentish Post* [p1]

On the 11th of October next, will be open'd at BRIDGE,
near CANTERBURY:
A Variety of new and second-hand Watches, to be sold at the lowest Prices; a Man will be constantly kept there to repair Clocks and Watches, by

JOHN CHALKLEN, Watch-Maker, at Canterbury.

16-19 February 1763 *Kentish Post* [p1]

WHEREAS William Nash, of the Parish of Bridge, Clock-smith, did, on Wednesday the 12th of [*sic*] Day of January last, send his Apprentice William Knight, to Littlebourne, to put up two Clocks, which the said Apprentice accordingly went and did, and afterwards absconded, and took with him a little Bay Pony, about twelve Hands high, with a Star in his Forehead: And the said William Nash hath not since seen or heard anything of the said Apprentice, or Horse. — If any Person can inform me, where the said Apprentice or Horse are to be found, they shall be satisfied for their Trouble, and the Favour will be gratefully acknowledg'd, by me, THOMAS KNIGHT. Dover, the 16th of February. 1763.

 N.B. The said Apprentice is a short Lad, of a brown Complection; about thirteen Years of Age, had on when he went away a black Wig, and a blue great Coat.

23-26 February 1763 *Kentish Post* [p1]

LOST last Saturday, the 19th of this inst. February, about Four o'Clock in the Afternoon, between Sturry and Canterbury, a Silver Watch, with a China Face, Maker's Name Nash, Bridge: Whoever has found the same and will bring it to the White Swan in Sturry, shall receive Half a Guinea Reward.

24-27 August 1763 *Kentish Post* [p1]

STOLEN on Tuesday the 16th Instant, a Dumb repeating Watch, in a Silver Case, and Silver Cap over the inside Work, the Maker's Name (engraved) Daniel Gregnion, 661, a Steel Chain, two Steel Seals, on one is engraved three Oaken Leaves between a Cheveron, the other an indifferent Head, a Steel Watch Key, with a Hook, and a Brass Watch Key: If offered to Pawn or Sale, pray stop the same with the Person offering it, and give immediate Notice to Mr. Thomas Tomlyn, Attorney, at Chatham; and they shall receive Two Guineas Reward.

27-31 August 1763 *Kentish Post*

Brown-Dun Gelding was found between Sturry and Hearne ... by John Banks, Clocksmith of Hearne ... if the owner don't soon apply, he will be sold to pay charges etc.

7-10 September 1763 *Kentish Post* [p4]

Canterbury, Sept. 10.

On Thursday last, in the Evening, about Eight o'Clock, Robert Spratt, Carpenter, was stopt in Dover Road coming to Canterbury, about 30 Rods on this Side of the Lane that goes to Bishopsborne, by two Footpads; one of the two, a tall Man, took hold of his Horse's Bridle, and bid him to deliver his Watch and Money; and while he was giving him some Money, he snatch'd his Watch out of his Fob, and struck him on the Temples with a Hedge-stake, and bid him get along. The Watch is of a middling Size, with a China Dial-plate, the Maker's Name on the inner Plate, John Watts, Canterbury.

12-16 November 1763 *Kentish Post* [p1]

LOST on Monday the 24th of October, between Tenterden and Maidstone, A Silver Watch, Maker's Name Cackett, Cranbrooke: Whoever has found the said Watch, and will deliver it to Mr. Drury, at the White Lion at Tenterden, shall receive Half a Guinea Reward.

28 January-1 February 1764 *Kentish Post* [p4]

LONDON, Jan. 30.

The article inserted in the Papers, concerning the curious repeating watch, in a ring, making for his Majesty, is not making by Mr. Reynolds, but by Mr. Arnold, in Devereux-Court, Temple-Bar. This curious piece of mechanism is less in circumference than a silver twopence; it repeats the hours, quarters, and half quarters, and runs upon diamonds; the horizontal wheel and pinions are equal in weight only to the 16th part of a grain of gold, and the pendulum and spring equal in weight only to the 16th part of a grain.

7-11 April 1764 *Kentish Post* [p1]

LOST on Tuesday the 3rd Instant, between Eythorne and Elminstead, A Silver old-fashioned Watch, with a Red Ribbon; Maker's Name, Shindler, on it. Whoever has found it, and will bring it to the Printer of this Paper, shall have Half a Guinea Reward.

8-11 August 1764 *Kentish Post* [p4]

Canterbury, Aug. 11.

At the General Session of the Peace and Goal Delivery held for the City of Canterbury, on Thursday last, ... one Fells Garner was convicted of grand larceny, in stealing a pair of box cases of a watch, the property of Thomas Shindler, a Watchmaker in the said city ... the said Fells Garner was burnt in the hand.

22-26 September 1764 *Kentish Post* [p1]

THOMAS WILMSHURST,
Watch-Maker at Deal, leaving off Trade,

HATH got a Parcel of Clocks and Watches, Silver Buckles and Buttons, Gold Rings new or second-hand, to be sold at prime Cost, By me,

THOMAS WILMSHURST.

———

27 February-2 March 1765 *Kentish Post*

LOST between Barham Brick-Kiln, Mr. Page's, and Wingmore Court, the 22d of February, 1765, A Silver Watch, Maker's Name John Silke, of Elmsted; whoever hath taken it up, and will deliver it to Mr. Thomas Fox, at Wingmore Court, in Elham, shall receive Half a Guinea, and if not Damaged One Guinea for their Trouble.

———

21-25 September 1765 *Kentish Post*

The watches to shew the three-hundreth part of a minute, were not invented by Mr. Arnold, but by Mr. Chalklin of Canterbury, who hath compleated two of that sort; the first in the year 1754, and the second in the year 1763, which Mr. Chalklin shew'd to Mr. Arnold, soon after it was finished.

———

22-25 January 1766 *Kentish Post*

Whereas Mr. John Jackson, Gun-maker, late of Cranbrook, being deceased, his Son, Owen Jackson, begs leave to inform Gentlemen and others, that he carries on the Gun-business in all its Branches, opposite the Crane in the said Town; where they may be supplied with any Sort of Guns or Pistols mounted in Silver, Steel, or Brass, according to the best Improvements and newest Fashions.

N.B. Watches, Rings, and Plate, as usual, at the lowest Prices.

———

1-5 March 1766 *Kentish Post*

Whereas it hath been industriously reported, that the Business at the old Shop, late John Jackson, Gun-maker at Cranbrook, was stopt, and the Trade carried to another Shop; therefore, as such Report tends greatly to my Prejudice, I Thomas Jackson, Son and Administrator of the said John Jackson, deceased, beg leave to inform all Gentlemen and others, that I carry on the Business which my Father did, in all its Branches, at the said Shop, known by the Sign of the Golden-Gun, next Door to the George and Bull Inns in Cranbrook aforesaid, where, as well as at my Shop in Maidstone, all Persons may depend on being served with the best Goods, on the cheapest Terms, and that their Favours will be gratefully acknowledged,

Maidstone,   By their much obliged,
Feb. 20, 1766.   And most obedient Servant,
                                                                            THO. JACKSON.

N.B. Wanted as an Apprentice, an active ingenious Boy, not exceeding fourteen Years of Age. For Particulars, enquire as above.

———

19-22 March 1766 *Kentish Post*

JUST come to the City of Canterbury, to a Commodious Room, at the GEORGE, in High-street, a GRAND PIECE of MACHINERY; being a most curious Astronomical and Musical SPRING-CLOCK, otherwise call'd, "The Theatre of the Muses", made by the famous Mr. Pinchbeck. This wonderful Machine gives a general Satisfaction, and equally surprizes and delights all that ever see and hear it. It is most beautifully composed of Musick, Architecture, Painting and Sculpture, with such diverting Variety of moving Figures in the Front, that renders it a very entertaining Piece of Art.

ALSO, The ROYAL WAX-WORK, from Fleet-Street ... [details given]
from Nine in the Morning till Nine at Night, Price Six pence, each.

21-24 May 1766 *Kentish Post*

LOST, A Pinchbeck-case Watch, wrought, with a Chiane [*sic*] Dial-plate, with a single black Line for each Hour, no Minutes marked, Maker NASH; and two Seals, one of Silver P. S. the other a Bloodstone and Head in Silver, and a Steel Chain: Whoever brings them to Peter Scalder, of Bossington, in Adisham, shall be paid Ten Shillings for their Trouble, with Thanks.

30 July - 2 August 1766 *Kentish Post*

LOST, last Wednesday, (supposed on Barham Downs) A SILVER WATCH, with an enamel'd Dial Plate, the Maker's Name John Chalklen, No. 17. — Whoever has found the said Watch, and will bring it to the Printer of this Paper, shall receive Half a Guinea Reward.

1-4 October 1766 *Kentish Post*

MARGATE,
JOHN POPE, Watch-Maker,
TAKES this Opportunity of returning his sincere Thanks to his Friends, and the Publick, for their past Favours when in Business with Mrs. Swan: He also begs Leave to acquaint them, that he hath taken a Shop next Door to Mr. Covell, Hoyman, in Highstreet, where he hopes for a continuance of their Favours. He intends to Make and Repair all sorts of Clocks and Watches, and warrant them to perform very Correct, on moderate Terms: Those Persons who please to favour him with their Commands, may depend on having the neatest and best of Goods, and their Orders punctually executed, by

Their very humble Servant,   JOHN POPE.
N.B. A proper Allowance for Ready Money.

22-26 August 1767 *Kentish Post*

WAS found, in the Road between Ramsgate Pier and Saint Lawrence Church, on the 8th Day of this inst. August, a Gold WATCH; the Owner thereof may have the same again by applying to Richard Goodchild, Watch-maker, at Harbledown, near Canterbury, and by giving the account of the Name and Number, and some Reward to the Person that found it.

23-26 December 1767 *Kentish Post*

HENRY GODDARD,
WATCH-MAKER and SILVERSMITH,
BEGS Leave to inform the Public, That he is remov'd from Tenterden to Dovor, and hath taken and furnish'd a Shop in Snargate Street, leading from the Town to the Pier; with an

entire new Assortment of the most fashionable Plate: All such Gentlemen, Ladies and others, who shall please to favour him with their Commands, may depend on being dealt with on the most reasonable Terms; and that their Favours will be gratefully acknowledged, by

Their oblig'd humble Servant, HENRY GODDARD.

N.B. He gives the most Money for old Plate, Watches and Rings.

---

30 December 1767-2 January 1768 *Kentish Post* [p4]

OWEN JACKSON,

Of Cranbrook in Kent; Watch and Gun-maker, and Silver-smith,

BEGS Leave to inform Gentlemen, Ladies, and others, that he has taken the House and Shop late Mr. Goddards at Tenterden, in the said County, where he intends to carry on the above Trades in all their various Branches. Those Gentlemen that please to honour him with their Commands, in any of the above Particulars, may depend on having them duly and faithfully executed, and the Favour gratefully acknowledged, by their most humble and obedient Servant, Owen Jackson.

N.B. As I cannot remove immediately to Tenterden, I purpose opening Shop on New-Year's Day, and attending weekly on Fridays, for the Conveniency of Gentlemen and others, who may have Occasion to deal in any of the abovementioned Articles, and may depend on their being used well.

Cranbrook in Kent, December the 9th 1767.

10-13 August 1768 *Kentish Gazette*

LOST, on Tuesday the 9th of August, between Northfleet and Rochester, A SILVER WATCH, with a Steel Chain and Key; Maker's Name Jos. Harris, Maidstone, No 3109, with a Silver Cock. Whoever will bring it to Mr. Jos. Harris, Maidstone, or to William Read, at the King's Head, Rochester, shall have One Guinea Reward.

---

17 August 1768 *Kentish Gazette*

RICHARD CRAMP,
WATCH and CLOCKMAKER,
In Burgate Street, CANTERBURY,

MAKES and repairs Repeating, Horizontal and Common CLOCKS and WATCHES. He hopes that the Knowledge of the business which he acquired from his late Master Mr. JOHN CHALKLEN, and the Improvements which he has since made in the Service of an eminent workman in London, will recommend him to the Favor of the Public, which will ever be gratefully acknowledged by their very humble Servant, RICHARD CRAMP.

26-29 October 1768 *Kentish Gazette*

BENJAMIN SMITH,
WATCH AND CLOCK-MAKER,
In Christ-church Yard, Canterbury,

MAKES and repairs all Sorts of Church, Musical, Spring and Common Clocks; all Sorts of Horizontal, Repeating, Skeleton and Stop-Watches. — And, besides his own Knowledge of the Business, which he has acquired by a regular Apprenticeship in this City, has also engaged, as an Assistant, a very eminent Hand from London, where he also, under a very able Master, has much improved in the Knowledge of the Business. — And flatters himself, that by charging on the most reasonable Terms, and by executing his Work with the utmost Accuracy and Dispatch, he shall give Satisfaction to all his Friends, whose Favours will be kindly acknowledged with Gratitude and Respect.

25 February-1 March 1769 *Kentish Gazette* [p1]

To be Sold By AUCTION
by WILLIAM MATSON, on Tuesday the 7th of March, 1769,
PART of the Household Furniture and Stock in Trade of HENRY SIMS, Watch-maker, in St. George's Street, CANTERBURY, leaving off Trade: consisting of Bedding, Glasses, Mahogany and other Tables, Chairs, Window-curtains, with Kitchen and other Furniture, with a very handsome Twelve-tune Chime Clock, and Common Ditto, and Larrums [alarms], Watches, &c. a neat Mahogany Work-board and Drawers, with Tools and all Utensils belonging to the Trade. Also a Sign and Sign Iron, with a large Glass Sash to the Shop. — The whole to be viewed on Monday the 6th of March, and to the Time of Sale, and continue till all is sold. The Sale to begin each day at One o'Clock.
Catalogues to be had at the Place of Sale, and at W. Matson's, Burgate-street, Canterbury.

26-29 April 1769 *Kentish Gazette*

BENJAMIN SMITH,
CLOCK and WATCH-MAKER,
REMOVED from Christ-church yard, to St. George's Street, near the King's Arms Printing-office, Canterbury, makes, sells, and repairs, all sorts of church, turret, musical, spring, and common Clocks. Also all sorts of repeating, horizontal, skeleton, seconds, and other Watches. Both Town and Country served on the most reasonable terms. Watches and Orders, sent by the carriers of this Paper are taken in; and great care will be paid to their being carefully repaired and returned by the newsmen with all expedition, carriage free, with thanks. Mr. Smith takes this opportunity of thanking those ladies and gentlemen who have already employed him; and having now taken the shop, late Mr. Sims's, hopes the customers of the shop will continue their favors as usual.
   Most money for old Gold and Silver.
   N.B. The House to be Lett late in the Occupation of Mr. Smith, in Christ-Church Yard.

26-30 August 1769 *Kentish Gazette* [p1]

WILLIAM NASH,
Clock and Watch Maker, at Bridge,
BEGS Leave to inform the Publick, that the Servitude of his Brother, JOHN NASH, is now dissolved. Therefore what Watches he should be instructed with for the future, I will not be accountable for, nor for the Performance. I continue working in the Country as usual, and shall be ever studious in meriting the Approbation of those who will continue their Favours to Their obedient humble Servant,   WILLIAM NASH.
   N.B. Clocks and Watches of all Sorts made in the best and completest Manner.
   WANTED IMMEDIATELY, a Journeyman, or a Young Person, to be instructed in the Clock and Watch Making Business.

2-6 September 1769 *Kentish Gazette* [p1]

JOHN NASH,

TAKES this Opportunity to acquaint the Public, that there is no Occasion for his Brother, Mr. WILLIAM NASH, to be answerable for any Watches intrusted in his Care; for he has not wronged him of any thing, as was imagined by his Advertisement.

New Clocks and Watches to be had a reasonable Terms, by sending to JOHN NASH at Beaksbourne, near Canterbury, who will go round the Country as usual. — All Persons, that please to favour him with their Custom, may depend on their Work being done well, and reasonable,

By their humble Servant,   JOHN NASH.

21-25 October 1769 *Kentish Gazette* [p4]

Canterbury, Oct. 25.
On Thursday last was admitted to the freedom of the ancient town of Margate, Mr. John Pope, an eminent watch-maker of that place, and it's thought a court will soon be called to swear him accordingly.

9-12 December 1769 *Kentish Gazette* [p1]

JOHN NASH, Clock-maker,
at Beaksbourn.

WHEREAS some malicious Reports have been propagated by my Brother, Wm. Nash, Clock-maker at Bridge, representing me as imposing on my kind Employers.— in Particular, that I had overcharged Mr. Drayson of Upstreet, for repairing his Clock, and as such Reports have a manifest tendency to prejudice me in my Business, I have taken the Opportunity to lay the said Charge before some reputable Clock makers of Canterbury; who have confirmed the Equity of the same, and will readily attest, if called upon, the Injustice done to

JOHN NASH.

17-21 July 1770 *Kentish Gazette* [p1]

LOST or STOLEN, A SILVER WATCH,
with a Silver Chain and Seal,
engraved J. G. the Maker's Name J. BALDWIN of Sittingbourne.

A Guinea Reward is offered for the Recovery of the above Watch, which may be had by applying to the Printers of this Paper. If offered to be sold or pawned, pray stop the Person so offering it.

6-9 October 1770 *Kentish Gazette* [p1]

SUPPOSED TO BE LOST,
At Canterbury, or near the River at Sturry, on Friday last, the 5th Instant,
from a Lady's Side, by the Hook breaking,
AN Old Gold WATCH — Maker's Name ROBART; with a White Dial-plate.
Whoever will bring it to Mr. ADAMS, at the Red Lion in Canterbury, or to Mr. BEALE, at Margate, shall receive Two Guineas Reward.

4-8 December 1770 *Kentish Gazette* [p1]

SUPPOSED TO BE LOST,
Between Chilham and Canterbury,
on Saturday the first of December Instant,
A New SILVER WATCH, Maker's Name, T. SHILLING, Selling, and had, instead of a Chain, a Black leather Slip, to which a Brass Key, a Seal set in Silver, and a Hook to hang up the Watch by, were fastened. Whoever hath found the said Watch, and will bring it to FRANCIS GRIGGS, Butcher, at Chilham, aforesaid, shall receive Half a Guinea Reward.

2-5 February 1771 *Kentish Gazette* [p1]

TO BE SOLD
A very good small modern built BRICK MESSUAGE, in thorough Repair, with the proper Appurtenances, advantageously situated in the High-street, in Rochester, near the Guildhall, in the Possession of James Booth, Watchmaker.

The above Purchase is supposed to be particularly worth the Notice of a Watchmaker — the said James Booth being the only one of the Trade in that Place, and has always had a very good Succession of Business. Some very advantageous Preospect in London is the only Occasion of the present Occupier's quitting the Premisses.

For further Particulars enquire of the proprietor, or of Mr. William Stacy, No. 186, in Fleet-street, London.

2-6 April 1771 *Kentish Gazette* [p1]

LOST, A SILVER WATCH,
last Friday Night, between Canterbury and Sturry;
Name R. BRADSTREET; No. 450, with an enameled Dial-plate, and the Emblems of Masonry instead of Hour-figures; and a Steel Chain with a Cornelian Seal. Any Person that will bring it to Mrs. CHALKLEN, Watchmaker, High-street, Canterbury, shall receive Two Guineas Reward. And if any Person will give Information, who has got the said Watch, they shall have a Reward of Five Shillings.

3-6 August 1771 *Kentish Gazette* [p1]

LOST.
Between Woodchurch and Bethersden,
A WATCH with a short steel Chain, china Face, and the Maker's Name Owen Jackson, at Tenterden. Who ever will bring it to WILLIAM HISTED, at the Bull at Bethersden, or to JOHN HAWKINS, at the Bonny Cravat, at Woodchurch, shall receive Half a Guinea Reward.

6-10 August 1771 *Kentish Gazette* [p1]

LOST, On Tuesday the 6th of this Instant, August, 1771. Between the Bridge at Sandwich and Old Mead Farm, in the Parish of Wingham, in the County of Kent.
A SILVER WATCH, with a short dirty Ribbon on it, Makers Name William Nash, Bridge. Whoever will bring the said Watch to William Hawks, at the aforesaid Farm, shall receive One Guinea Reward, with Thanks.

2 November 1771 *Kentish Gazette* [p4]

CANTERBURY, Nov. 2.

We hear there are now travelling in this country a set of sharpers, who pretend some to be foreigners, and cannot speak English; some that they are servants out of place; and others who pretend to be lately come from sea; their chief trade is in the sale of watches, imposing upon the unwary by feigning great distress. They are chiefly lurking in villages and alehouses for farmers servants, and others, who are unacquainted with the value of what they buy; this is therefore inserted as a caution, to prevent the public being imposed on by such rascally impostors.

14-17 December 1771 *Kentish Gazette* [p1]

OWEN JACKSON,
Watch-Maker, Silver-Smith, and Gun-Maker,
at Tenterden, in Kent,

BEGS leave to inform Gentlemen and others that he carries on the above Trades in their several Branches, and purposes, for the Conveniency of his Customers in and about Rye and Cranbrook, to attend each Market once a Fortnight regularly, in the Afternoon, at the Bull Inn in Cranbrook, and the George Inn, Rye; where all Work and Orders will be taken in and returned, and great Care will be used in doing them in a complete Manner and Quick Dispatch. — All Favours will be gratefully acknowledged by

Their most obedient Servant,
OWEN JACKSON.

P.S. I shall be at Cranbrook on Saturday the 21st Instant; at Rye, on Saturday the 28th; and so continue in Turns. — Likewise shall bring with me a neat Assortment of New and Second-hand Watches.

Tenterden, Dec. 14, 1771.

18-21 January 1772 *Kentish Gazette*

WHERAS on Monday Evening the 30th of December last, Mr. GIDION GAMBIER of St. Mildred's in the City of Canterbury, went from his House and has not since been heard of. And whereas on the Wednesday following, a Hat was found in the Water between DEAN's MILL and SHALL OAK, which there is Reason to believe is his, it is to be feared, that by some unhappy Accident he has fallen into the River.

This is therefore to give Notice that any Person who discovers the Body, that upon Proof thereof, THREE GUINEAS will be paid, by applying to the Printers of this Paper.

He had on, when missing, a brown Surtout Coat, a blue-grey Coat and Waistcoat with Metal Buttons, and a Watch in his Pocket, Maker's Name George Sims, /No. 656, Canterbury.

27-31 October 1772 *Kentish Gazette* [p1]

STOLEN, On Wednesday the 28th Instant, about Nine o'Clock at Night,
from THOMAS SPAIN, Waggoner's Mate, at Mr. Harrison's, at Preston Court,
AN almost-new SILVER WATCH, with a China Face, Maker's Name John Godden.
If offered to be sold or pawned, please to stop it; And any Person discovering the Offender, so that he may be brought to Justice, shall receive Two Guineas Reward from Mr. Harrison, at Preston Court.

As the Watch was stolen in the Dark, no other Description can be given of the Thief than that he seemed to be a tall Man, with a round white Frock on.

30 January-3 February 1773 *Kentish Gazette* [p1]

LOST,
Between Ospringe Street and Blacketts, in the Parish of Tonge,
On Friday last, the 29th of January,
A SILVER WATCH, mark'd on the inner Case T. A.
Maker's Name EDWARD ACURS [Acres], Sittingbourn.

Whoever finds the same, and will bring it to Mr. Wm. TURNER, at Bapchild; or to THOMAS AVERY, at Mr. Lipyeatt's, near Ospringe, shall receive Half-a-Guinea Reward, and all reasonable Charges.

14-17 April 1773 *Kentish Gazette* [p1]

LOST,
On Wednesday, April 14, in the High Road from Sandwich to Canterbury,
AN old GOLD WATCH, with a White Dial-plate, the Maker G. Sims, Canterbury, with a Steel Chain, and three Seals; a large White Cornelian, with a Coat of Arms; a small red triangular Cornelian, and a very small black Stone cut on both Sides.

If the Person that has found this Watch will bring it to Mr. SIMS, Haberdasher, by the Church Gate, Canterbury, he shall receive Five Guineas Reward.

8-12 May 1773 *Kentish Gazette* [p1]

THOMAS ANDREWS,
CLOCK and WATCHMAKER,

TAKES this Opportunity of informing the Public, that he has opened a Shop for carrying on the above Business in all its Branches, Near the Market-place, Dovor. He has also laid in an entire new Stock of the SILVER BUCKLES, and many other Articles of the best and newest Fashions. The Favours of the Inhabitants of that Town and the neighbouring Country will be thankfully received, and a punctual Attention paid to all their Orders.

CLOCKS and WATCHES repaired in the most accurate Manner.

12-16 June 1773 *Kentish Gazette* [p1]

To WATCH MAKERS, TO BE LET,
And Entered upon Immediately,
At CRANBROOKE, in Kent,

THE Messuage or Tenement, together with the well-known Shop, of GEO. THATCHER, late of CRANBROOKE aforesaid, Watchmaker, deceased.

N.B. CRANBROOKE is a large Market Town, situate in a good Neighbourhood, and the Trade belonging to the Shop very extensive and large. The Goods and Furniture of the House, and also the Stock and Implements of Trade, are to be disposed of.

For further Particulars enquire of Mr. JOHN THATCHER, of CRANBROOKE, Brazier; or of GEORGE THATCHER, of SEVENOAKS, Brazier.

18-22 September 1773 *Kentish Gazette* [p1]

   ROBBERY, EARLY on Thursday Morning last, the 16th Instant, the House of MR. BENJAMIN SMITH, Watchmaker, in St. George's Street, Canterbury, was robbed of Six New Silver Watches, Maker's Name, BENJAMIN SMITH, Canterbury, …

27-30 October 1773 *Kentish Gazette* [p1]

        WANTED,
   A YOUTH for an APPRENTICE to a
    CLOCK and WATCH MAKER.
For further Particulars please to apply to Mr. SAMUEL OLLIVE, Clock and Watchmaker at TUNBRIDGE TOWN in this Country [sic].

 A Premium will be expected. This will not be advertised any more.

9-12 February 1774 *Kentish Gazette*

      To be SOLD by AUCTION,
 On WEDNESDAY the 16th Instant at Ten o'Clock in the Forenoon,
    and to continue the two following Days,
   By MR. POUT, Upholsterer and Auctioneer,
 ALL the entire HOUSEHOLD FURNITURE and valuable STOCK in TRADE
   of the late Mrs. CHALKLEN, Clock and Watch Maker,
      in High-street, Canterbury;
Consisting of Great Variety of exceeding good Gold and Silver Repeaters; likewise Plain and Chase Gold Watches, Horizontal, Seconds, and plain Silver and metal Watches; several very good Eight Day, Table, Spring, and Thirty-Hour Clocks; also a Variety of Silver Plate, Buckles, Rings, Seals, Trinkets, &c.&c.

 The Goods to be viewed on MONDAY and TUESDAY next.

 All Persons having any Demands on the Estate of the late Mrs. CHALKLEN, are desired forthwith to send their Accounts to Mr. Barham, jun. Attorney at Law, Canterbury. And all Persons who stand indebted to the said Mrs. CHALKLEN, are desired forthwith to pay their respective Debts to the said Mr. BARHAM, who is authorised by the Executors to receive the same.

 Catalogues may be had this present Saturday, at the Rose, Sittingbourn; at the Ship, Faversham; at the Ship, at Boughton, at the Woolpack, Chilham, at the Red Lion, Bridge; at the Antwerp, Dover; at the Old Bell, Sandwich: the Red Lion, Wingham; at the three principal Taverns, the Coffee-houses, the King's arms Printing-office, at Mr. POUT's Upholsterer, and at the Place of Sale, in Canterbury.

23-26 February 1774 *Kentish Gazette*

    To be SOLD to the BEST BIDDER,
 Some Time in March next, at the Crown in High-street,
       CANTERBURY,
ALL the good and substantial Buildings, situated in the High-street of the City of Canterbury.
LOT 1. A House, in the Occupation of Mr. Mark Giles, Butcher.
LOT 2. The House adjoining, late Mrs. Chalklen's, deceased.
LOT 3. The Crown Alehouse.
[and 4 more Lots.]

All which Premises will be sold to the best Bidder lotted out, or all together. Articles to be seen at the Time of Sale.

In the mean Time please to enquire of JOHN HORN, at the Hoath Farm near Canterbury.

2-5 March 1774 *Kentish Gazette*

WILLIAM CHALKLEN,
CLOCK and WATCHMAKER,
On KING's-BRIDGE, CANTERBURY

BEGS Leave to return his sincere Thanks to the Public in general for the great Encouragement they have been pleased to give him in carrying on the above Business; and assures them, that he will ever make it his particular Study to merit the Continuance of their future Favours. — As he never presumed to [solicit] the Custom of one Individual who dealt with his late Sister, Mrs. CHALKLEN, in High-street, during her life Time, and whose Business is now given up; he hopes those Ladies and Gentlemen, who favoured her with their Orders, will now lay their future Commands on her Brother; where they may depend on having their Work well executed, with the greatest Expedition, and on the most reasonable Terms.

He cleans and repairs all sorts of Clocks, and Watches in the best Manner, and at moderate Rates.

13-16 July 1774 *Kentish Gazette*

JOHN VIDEON,
WATCH-MAKER, and SILVERSMITH, At FAVERSHAM,
BEGS Leave to inform the Public, that he continues to take in
Light English Gold Coin, at the Rate of £ 3. 16s. per Ounce.
And Portugal Gold Coin, viz.
Ports at £ 3.16s. per Ounce.
Moldores at £ 3 15s.

And as by Royal Proclamation all Guineas coined before the first of George III. not more deficient in Weight than ONE SHILLING are to be called in and exchanged at the Bank, between the 15th of July and 31st of August next, and as it may be inconvenient to many people to remit their Light Gold to London, he will take in and exchange such Guineas as are not more deficient in Weight than One Shiling, at TWOPENCE per Guinea Discount.

25-28 January 1775 *Kentish Gazette* [p1]

RICHARD CRAMP,
Silversmith, Watch-maker, and Jeweller,
Is removed from his Shop on the PARADE
To the GOLDEN CUP, opposite the Red-Lion in the High-Street,
CANTERBURY:

WHERE he has laid in a large and valuable Assortment of the most fashionable Goods in each Branch of his Profession; and has taken particular Care to select and engage the most capital Workmen in the Jewellery and Watch Trade, which he undertakes to make and repair on the best Principles and most reasonable Terms.

He hopes, therefore, the Public, and his Friends in particular, will accept his sincere Thanks for the many Favours he has already received; and as he is determined to give the utmost Satisfaction, he flatters himself he shall merit their further Commands, which will greatly oblige Their most humble Servant.

Mourning Rings neatly made; also Arms, Crests, or Cyphers, engraved on Plate, &c. &c. on the shortest Notice.

8-12 July 1775 *Kentish Gazette* [p1]

LOST,

On Tuesday, the 4th of July, 1775, between Ospringe Street and Sittingbourn.
A small Silver Watch, Maker's Name Jos. Hand, London, No. 2516, with an old plated Chain, two little small Keys, one Metal Watch Key, and a square Steel Key; the Paper in the Watch-case, Wm. CLEMENTS, Watch maker, Faversham.

If any Person has found the said Watch, and will bring it to Lees-Court, or the Printers of this Paper, shall receive HALF A GUINEA Reward.

15-19 July 1775 *Kentish Gazette* [p1]

THOMAS ANDREWS,

WATCH MAKER and SILVERSMITH,

IS removed from his Shop in the Market-Place, to the QUAY at the PIER in DOVER, where he has laid in a large and valuable Assortment of the best and most fashionable Goods, in each Branch of his Profession, and undertakes to make and Repair Watches and Clocks, on the best Principles and most reasonable Terms.

He hopes, therefore, the Public in general, and his Friends in particular, will accept his sincere Thanks for the many Favours he has already received, and as he is determined to use his utmost Endeavours to give Satisfaction, he flatters himself he shall Merit their further Commands.

19-23 August 1775 *Kentish Gazette* [p1]

LOST, A Silver Watch,

On Tuesday August 22, 1775,

Between Ospringe and Sittingbourn,

The Maker's Name BRICE, Sandwich, A single steel Chain, and straight steel Key. Whoever will bring it to the Printers of this Paper, or to Mr. Perey, at the Blue Anchor, Ospringe, shall receive ONE GUINEA Reward.

23-27 September 1775 *Kentish Gazette* [p1]

BENJAMIN SMITH,

WATCH and CLOCK MAKER,

Next Door to the King's Arms, Printing Office, St. George's Street, Canterbury, MAKES, SELLS, and REPAIRS all Sorts of CLOCKS; and all Sorts of Repeating, Horizontal, Skeleton, Seconds, and other WATCHES, on the most reasonable Terms.

WATCHES and CLOCKS carefully repaired and returned with all Possible Dispatch to any Part of the Country.

N.B. An APPRENTICE is wanted to the above Business.

21-25 October 1775 *Kentish Gazette* [p1]

LOST,
On Saturday last, between Faversham and Hothfield,
A SILVER WATCH,
With the Owner's Name on the Face, William Crouch.
And the Maker's Name, within Side, Richard Bailey, Ashford.

Whoever has found the said Watch, and will bring it to William Crouch, at Westwell, shall receive One Guinea Reward.

15-18 November 1775 *Kentish Gazette* [p1]

WILLIAM CHALKLEN,
CLOCK and WATCH MAKER,
(Late of King's Bridge, Canterbury)

TAKES this Opportunity to acquaint his Friends and the Public in general, that he is removed near to the SHAKESPEAR's HEAD, in BUTCHERY LANE, CANTERBURY; where he Makes and Sells all Sorts of CLOCKS and WATCHES; likewise Repairs and Cleans them in the best Manner. Those who will favour him with their Commands, may depend on their Work being executed on the most reasonable Terms. All Orders will be strictly obeyed, and their Favours gratefully acknowledged by their most humble Servant.

Gentlemen, Ladies, and Others, may have their Clocks or Watches strictly examined, and know their neat Expence, if required, by applying to the above Wm. CHALKLEN.

16-20 March 1776 *Kentish Gazette* [p1]

TO BE LET,
And Entered upon immediately,
At CRANBROOK,

A House and Shop, well situated for Trade, in the Watch and Clock Business, with the Household Furniture and Tools, suitable for every Convenience in that Branch of Business, and now or late in the Occupation of Mr. William Chittenden.

For further Particulars enquire of Mr. George Thatcher, Brazier, at Cranbrook.

10-13 April 1776 *Kentish Gazette* [p1] [*cf* 28 January 1775]

JEWELLERY, HARDWARE, &c.
Now SELLING under PRIME COST,
For the Benefit of the Creditors, by Order of the Assignees,
ALL the neat genuine Stock in Trade of Mr. RICHARD CRAMP; Silversmith Jeweller and Hardwareman (a Bankrupt) at his Shop in High-street, Canterbury.

Consisting of a neat and fashionable Assortment of Silver and Silver Plated Candlesticks, Mugs, Salvers, Pierced and Plain Salts, Pepper-castors, Milk Ewers, Butter-boats Spoons, Cruets mounted with Silver, sundry Buckles and Spurs of the newest Patterns; a genteel Assortment of Jewellery Goods; likewise Table knives and Forks, with Silver, Ivory, Ebony, and Camwood Handles; Pen-knives, Razors, Scissars, Cork-screws, Pontipool Tea-trays and Waiters, Japan Tea Urns, Candlesticks, Plate-warmers and Coal-skuttles; with a Variety of other Articles, all in good Condition.

All Persons, indebted to the Estate of Mr. Cramp, are desired immediately to pay their respective Debts to Mr. Loftie, Linen-draper, in Mercery-lane, Canterbury, one of the Assignees of the said Estate.

24-27 April 1776 *Kentish Gazette*

LIGHT GOLD,

WHEREAS it may be inconvenient to many People to exchange their LIGHT GUINEAS, &c. at Places appointed by Government for that Purpose,

RICHARD BAYLEY,

WATCH MAKER and SILVERSMITH, Ashford,

Begs Leave to inform the Public, he will exchange such GUINEAS, HALFS, and QUARTERS, as specified in the Royal Proclamation, at so small a Premium as ONE PENNY per Guinea Discount.

GUINEAS, &c. under the Weight, and PORTUGAL Coin, bought at 3£ 16s. per Ounce, as usual.

4 May 1776 *Kentish Gazette*

To be SOLD by AUCTION,

By WILLIAM CALDER,

On the Premises on Thursday the 9th Day of May instant, and the following Days, by Order of the ASSIGNEES of Mr. RICHARD CRAMP, Silversmith, Watch-maker and Jewiler, at his House in High-street, Canterbury,

ALL his remaining STOCK in TRADE, Tools, Fixtures, Household Furniture, &c. consisting of a Variety, of Silver, and Silver Plated Goods, Ebony Castors, with cut Glass Cruets mounted with Silver; Watch Movements, Men and Ladies Watch Chains, Watch Keys, Seals, Trinkets, Earrings, Sleeve Buttons, Rings, Buckles, Pontipool Tea Trays, Waiters, Candlesticks, Table Knives and Forks, Cork Screws, with a Number of other Articles in the Hard Ware, Cutlery, and above Branches.

Two large Shew Glasses, a Counter with Drawers and Mahogany Top, a neat Bow Window with inside Shutters and other Fixtures, Linen, China, with good Parlour and Kitchen Furniture.

The Stock is put into small Lots for the conveniency of private Families, and may be viewed on Wednesday next, till the Time of Sale (which will begin each Day at Eleven o'Clock in the Forenoon) where Catalogues may be had.

N.B. In the first Day's Sale will be sold, a very handsome Eight-day Chiming Clock (by Mr. Chalklen) in a Mahogany Case, inlaid with Brass and Brass Ornaments.

11-15 May 1776 *Kentish Gazette* [p1]

TUNBRIDGE WELLS.

May 1, 1776

HIS MAJESTY having been pleased to appoint me to receive the deficient GOLD COIN of this Realm ... I will take in and exchange the deficient Gold Coin as expressed in the said Proclamation, every Tuesday, Wednesday, Thursday, and Friday, at my Shop on the Walks, during the following Hours:

The Nobility and Gentry who frequent the Place from Nine to Ten in the Morning, and Two to Three in the Afternoon.

The Inhabitants of the Place and Market People, from Seven to Nine in the Morning. And all Persons from the Country, from Four to Seven in the Evening ... [Also:] for all Persons, in the intervening Hours of the aforesaid Days, at the Shop of Mr. THOMAS WOOD, Goldsmith, and Clock and Watchmaker, the Bottom of the Walks ....

JASPER SPRANGE.

———

22-25 May 1776 *Kentish Gazette* [p1]

WANTED IMMEDIATELY,

AS an Apprentice to a CLOCK and WATCH-MAKER, GOLDSMITH and CUTLER, in an eligible Situation in the County of Kent, A Lad of good Morals, and genteel Education.

For Particulars enquire of Mr. SPRANGE, Bookseller, Tunbridge-wells.

A Premium will be expected.

———

5-8 June 1776 *Kentish Gazette* [p1]

RICHARD BAYLEY,
ASHFORD,

BEGS Leave to inform his Friends and the Public in general, that he is appointed to exchange the deficient Coin of the Realm, pursuant to the Proclamation of the 12th of April last, according to the following Weights, Gratis:

	Dwts.	Grs.
Guineas coined before 1772, weighing less than	5	8
Not more deficient than	5	6
Half Guin. coined before 1772, weighing less than	2	16
Not more deficient than	2	14
Quarter Guineas, weighing less than	1	7

———

19-22 June 1776 *Kentish Gazette*

LOST,
Supposed to be dropped between Hythe and Folkestone,
A Plain GOLD WATCH, Maker's Name HENRY GODDARD, Dover, within Side, and GODDARD, Dover, on the Face, No. 1395, with a Steel Chain and Key.

Whoever has taken up the Same, and will bring it to the Printers of this Paper, shall receive TWO GUINEAS Reward.

If offered for Sale, beg it may be detained, and Information sent to the Printers, who will give a reasonable Satisfaction for the Trouble.

———

22-26 June 1776 *Kentish Gazette*

LOST,
Last Sunday, the 23d of June instant,
Between LITTLEBOURN and the EIGHT BELLS, at WINGHAM WELL,
A SILVER WATCH,
Maker BRICE, Sandwich, No. 1736, marked on the Case T. B.
on the inside Case T. Burrows.

Whoever has, or may find it, by sending or bringing it to Thomas Burrows, at the Red Lion Inn, Canterbury, shall receive HALF A GUINEA Reward.

A Steel Chain, with a [?Steel] Seal and Key.
* If offered to be sold, pray stop it.

22-26 March 1777 *Kentish Gazette* [p1]

RUN AWAY,
From Mr. T. W. COLLARD of Little Barton Farm,
near Canterbury, on Saturday Evening last,
JOHN LUCK, Bailiff ... and took with him a Watch, the Maker's Name, James WARREN, Canterbury, No. 4225 ....

Two Guineas on his being brought to Justice. Likewise run away from the above Farm,
JOHN NEWIN.

13-16 August 1777 *Kentish Gazette* [p1]

JOHN MANLEY
CLOCK and WATCH-MAKER from London,
BEGS Leave to acquaint the Ladies and Gentlemen, and Public in General, that he has lately opened a Shop, opposite the GEORGE in Globe Lane, Chatham; where he sells all Kind of CLOCKS and WATCHES, in the most neat and fashionable Manner, as cheap as in London; he also sells all Sorts of JEWELLERY GOODS; likewise a great Variety of SILVER and PLATED BUCKLES at the lowest Rates.

Those Ladies and Gentlemen who will please to favour him with their Commands, may be assured of the best [—], and will confer an Obligation on their most obedient humble Servant,
JOHN MANLEY.

Chatham, Aug. 14, 1777.
An APPRENTICE immediately wanted.
A Premium will be expected.

27-30 August 1777 *Kentish Gazette* [p1]

Deal, August 27, 1777.
STOLEN,
Between the Hours of One and Two o'Clock in the Afternoon,
out of the House of Mr. Thomas Mourrilyan, jun.
A PINCHBECK WATCH,
China-faced, Marked ALDRIDGE, Deal, thereon.
If any Person or Persons will discover the Offender or Offenders, so that they are convicted thereof, shall receive Five Guineas Reward [paid by]   THOMAS MOURRILYAN.

If offered for Sale or Pawn, the Person is desired to stop it, and give Notice to me thereof, who shall receive reasonable Satisfaction.

4-8 July 1778 *Kentish Gazette* [p1]

FRANCIS CROW
BEGS Leave to inform his Friends and the Public in general, that he has opened a SHOP near the SHIP in West-street, Faversham; where he intends to carry on the Business of WATCH and CLOCK MAKING.

Horizontal, Repeating, Skeleton, Seconds, and common Watches, Church and Turret Clocks, carefully cleaned and repaired.

Arms, Crest and Cyphers, engraved on all Kind of Seals and Plate.

Those, who please to favour him with their Commands, may depend on their Orders being punctually obeyed, and their Favours gratefully acknowledged,

By their humble Servant.

26-29 August 1778 *Kentish Gazette* [p1]

SHOOTING
SAMUEL OLLIVE,
Gun-maker at Tonbridge

HATH invented a Charger to deposit the Gunpowder into the Chamber of the Barrel, whereby the Inconvenience of the Powder's hanging to the Sides of the Gun is totally avoided.

This Invention is simple, not easily put out of Repair, and sold as low as Two Shillings each, with good Allowance to those who sell again.

Wanted, at the same Place, an APPRENTICE to the BUSINESS of a CLOCK and WATCH-MAKER.

19-23 June 1779 *Kentish Gazette* [p1]

WANTED IMMEDIATELY,
A Journeyman Clock-maker,
Who is a good Hand at new Work.

He may have constant Work and good Wages of JAMES DANELL, Watch and Clock Maker, at New Romney, Kent.

---

29 December 1779-1 January 1780 *Kentish Gazette* [p1]

ONE GUINEA Reward.
LOST.
On Christmas Day, about Eleven o'Clock in the Forenoon,
between Canterbury and the Halfway-House to Faversham,
A SILVER WATCH,

Maker's Name, JENKINS, Rochester; with a Pompadoro Watch String, intermixed with Gold, with a Black Wedgewood Seal, the Impression. — Andromache weeping over the Arms of Hector — and a Steel Key.

Whoever will bring the said Watch to the General Printing Office, Canterbury; or to Mr. Perkins, Glazier, West Street, Faversham; shall receive One Guinea Reward.

9-12 February 1780 *Kentish Gazette* [p1]

LOST,
On Monday, the 7th Instant, near Margate,
A Plain Silver Watch, Maker's Name, COLLINS, Margate; No. 590

Whoever has found the same, and will bring it to Mr. Collins on the Parade, shall receive One Guinea Reward.

26 February-1 March 1780 *Kentish Gazette*

WANTED IMMEDIATELY,

A JOURNEYMAN,

In the CLOCK-MAKING BUSINESS

A good Workman may have constant Employ by applying to THOMAS ANDREWS, Clock and Watch Maker, and Silversmith, at Dover.

Also wanted,

An APPRENTICE,

A Premium will be expected.

15-19 April 1780 *Kentish Gazette*

JOHN SILKE and SON,

From ELMSTED,

CLOCK and WATCHMAKERS,

Removed to a Shop at SANDGATE, in CHERITON, near HYTHE,

and not far from FOLKSTONE,

BEG Leave to inform all Persons, who will favour them with their Commands, that they will make and repair all CLOCKS and WATCHES at as reasonable an Expence as possible; and hope for the Support and Encouragement of the Public, which will be gratefully acknowledged.

Will wait on any Gentleman or Lady at their own Houses, if required.

10-13 May 1780 *Kentish Gazette*

LOST,

On Monday the 8th Instant, between the George at Boughton

and the Sign of Dover without Canterbury,

A SILVER WATCH, with a leather Thong and a Brass Key to it;

Maker's Name JOHN SILKE, of Elmsted.

Whoever has taken it up, and will bring it to Mr.Doorne, of Faversham, or the Printers of this Paper, shall receive TEN SHILLINGS and SIXPENCE Reward.

The Number of the Watch is not known.

1-5 July 1780 *Kentish Gazette*

NOTICE.

THE Creditors of JOHN SNATT,

late of Ashford, Clock and Watch Maker, deceased,

are desired to send an Account of their respective Demands to STEPHEN or CATHERINE SNATT, of the same Place, in Order that they may be paid. — And such Persons, as were indebted to the said JOHN SNATT at the Time of his Death, are desired to pay their respective Debts to the said STEPHEN or CATHERINE SNATT, within fourteen Days.

22-26 July 1780 *Kentish Gazette* [p1]

LOST,

On Wednesday, July 19, between Chilham and Bridge,

A SILVER WATCH,

With a Bit broke out of the Face.
The Maker's Name, Thomas Oldridge.
Whoever has found the same, and will bring it to the Printers of this Paper, or to Mr. Kennett, Wye Court, shall receive Half-a-Guinea Reward.

———

13-16 September 1780 *Kentish Gazette* [p1]

SAMUEL OLLIVE,
TUNBRIDGE, in KENT,

TAKES the Liberty of acquainting the Public, that he makes CLOCKS on an entire new Construction, from an Invention of his own, which for Utility, Convenience, and Cheapness, exceed any Thing of the Kind ever before offered to the Public. — They go eight Days, shew the Hour and Minutes, and by Means of pulling a Line, which may be conveyed from the Clock to any Room in the House, strike the Hours the same as an Eight-day Clock.
Price 2£. 5s. — with a Case, 3£. 6s.

N.B. As the Beating of Clocks is disagreeable to some People, they may (if required) be made to go silent.

———

18-21 October 1780 *Kentish Gazette* [p1]

F. CROW,
Watch and Clock Maker,
West Street, Faversham,

TAKES this Method to inform the Public, that he makes for Sale TIME-KEEPERS on the most simple and best-approved Principles; they go eight Days — shew Seconds, Minutes, and Hours. — They have a compound Pendulum on a new Construction, whereby the Inequalities of Expansion and Contraction are greatly removed.

And likewise every other Article in the WATCH and CLOCK BUSINESS.
ARMS, CRESTS, and CYPHERS, engraved on SEALS and PLATE.

———

1-4 November 1780 *Kentish Gazette*

THE ASSIGNEES of THOMAS QUESTED, of Wye, Watchmaker, do hereby give Notice, that they will attend at the KING's HEAD, in Wye aforesaid, on Tuesday next, the 7th Day of November instant, at Two o'Clock in the Afternoon, in Order to make a Dividend of the Monies arising from the Sale of Mr. QUESTED's Effects; when and where the Creditors are required to attend to receive the same.
Wye, Nov. 1, 1780

———

11-15 November 1780 *Kentish Gazette* [p1]

RICHARD BAYLY, jun.
Watchmaker and Silversmith, From Ashford,

BEGS Leave to inform the Inhabitants of Folkstone and the Public in general, that he has taken a Shop, on the Court leading to the Beach, Folkstone; where he purposes carrying on the Business of a Watchmaker and Silversmith in all its Branches. All Orders will be expeditiously executed, and gratefully acknowledged, by their most obedient Servant,
RICHARD BAYLY.

———

18-21 April 1781 *Kentish Gazette* [p1]

LOST,
Between the HALF-WAY-HOUSE and CLIFF's END,
On Sunday the 15th instant.
A SILVER WATCH,
Maker's Name, Bing, Ramsgate,

Whoever has found it, and will bring it to Thomas Huggett, at the Star, at St. Peter's, shall receive Half-a-Guinea Reward.

———✻———

26-29 September 1781 *Kentish Gazette* [p1]

LOST,
On Tuesday last, at Goodnestone Fair,
A SILVER WATCH,
Marked within-side, LEVI EMANUEL, Canterbury,

with a Silver Chain, a flat Seal with a Cypher A. S. on one Side, and a [Boat] on the other. Whoever has found the said Watch, and will bring it to William Neame, at the Royal Exchange, at Deal; or to Levi Emanuel, St. Dunstan's, Canterbury; shall receive Two Guineas Reward.

———✻———

15-19 December 1781 *Kentish Gazette* [p1]

WANTED IMMEDIATELY
A Journeyman Clock-maker,

A sober steady Man, who can be recommended from his last Place, may have Employ by applying to Robert Stevens, Clock and Watch-maker at Milton next Sittingbourn.

Wanted, An APPRENTICE to the above Business.

Particulars may be known by applying as above.

———✻———

27-31 July 1782 *Kentish Gazette*

JOHN BUCKLEY,
Begs Leave to inform his Friends and the Public in general,
THAT he has taken the late Shop of Mr. Thomas Shindler, Behind the Butter-market: Where the Haberdashery and Millinery Business is carried on, as usual, on the very lowest Terms.
Toys Wholesale and Retale.

J. BUCKLEY likewise returns his sincere Acknowledgments to the Public for the Encouragement he has already met with in the Watch and Clock-making Business, and humbly solicits a Continuance of the same, at his Shop opposite Butchery-lane, Burgate-street, where every Branch of that Business will be executed in the best Manner, on the most reasonable Terms.

———✻———

21-25 December 1782 *Kentish Gazette* [p1]

BAYLY,
Watchmaker and Silversmith,
Opposite the Corn Market, in the High Street,
MAIDSTONE,

Respectfully acquaints the Nobility, Gentry, and the Public in general, that he makes, sells, and carefully repairs, all Sorts of WATCHES, on the most approved Principles; likewise, on very reasonable Terms, every Article in the SILVER and JEWELLERY BRANCHES, neatly and elegantly executed; with a general Assortment of PLATED and other GOODS, in the above Business.

All Orders executed with the utmost Dispatch.

23-26 April 1783 *Kentish Gazette* [p1]

STOLEN, On Friday, the 18th Instant,
A SILVER WATCH,
Out of the Shop of Mr. John Rete, Collar-maker, Margate,
Maker's Name, COLLINS, Margate, No. 810.

One Guinea will be given for the Recovery of the said Watch; and, on the Conviction of the Offender, One Guinea more.   By me, W. DAVIS, Margate.

26-30 April 1783 *Kentish Gazette* [p1]

LOST, On Monday, the 21st Instant,
Between Standford Street and Dymchurch, in Romney Marsh,
A SILVER WATCH,
With a Steel Chain and Silver Seal, with a Turk's Head for Impression,
Maker's Name, Henry Goddard, Dover.

[Bring to Mr. Hawkins Goddard, Westenhanger; … One Guinea Reward.]

18-21 June 1783 *Kentish Gazette*

Tenterden, June 18, 1783.

WHEREAS a Report has been industriously spread in and about the [Environs] of Tenterden, that I have declined the WATCH TRADE, I respectfully take this Method of informing my Customers in particular, and the Public in general, that such Report is void of the least Foundation in Truth; and that I carry on my several Branches of Trade, as usual, at the old Shop, opposite the Prison in Tenterden; where all Orders will be thankfully received and carefully executed, with the utmost Dispatch,   By their most obedient Servant,
OWEN JACKSON.

Neat NEW SILVER WATCHES, at Three Guineas each, equal to any at the Price sold in this Kingdom.

Likewise all Sorts of NEW CLOCKs, and OLD WORK, repaired on easy Terms.

13-16 August 1783 *Kentish Gazette* [p1]

WANTED,
At Cottington near Deal,

AN able Man, who has been used to Barley-mowing, in the Room of one WILLIAM PETERS, who has absented himself, and left a poor young Creature in the Parish of Deal big with Child, whom he promised to marry, and had bought a Ring, as he pretended for that Purpose.

The said PETERS either pawned, sold, or carried away with him, a Silver Watch belonging to his Master, Mr. Thomas Gurney — the Maker's Name, EDWARD ALDRIDGE, Deal….

15-18 October 1783 *Kentish Gazette*

ISRAEL SWABY,

Respectfully informs his Friends and the Public in general, that he has removed from the QUAY at DOVER, and taken a House in KING's HEAD STREET; where he continues to deal very extensively in the WATCH, JEWELLERY, and SILVER BUSINESS. — He returns his grateful Acknowledgements for Favours already received; and flatters himself, that his strict Attention to his Business, and reasonable Charges, will recommend him to future Favour.
    Dover, October 9, 1783.

1-5 May 1784 *Kentish Gazette* [p1]

    WATCH LOST. LOST in Faversham, or between Boughton Hill and Faversham, a WATCH in a Tortoise Shell Case, Makers Name John Buckley, Canterbury, No. 102, a Yellow Chain with a Cornelian Seal, &c.
Whoever has found the same and will bring or send it to the Printers of this Paper, or to Mr. Pain, at the Dolphin Inn, Faversham, shall receive One Guinea Reward.
    If offered to be pawned or sold pray stop it and the same Reward will be given.

10-14 July 1784 *Kentish Gazette* [p1]

    LOST at MARGATE, On Saturday last, the 10th instant July,
A Silver Cap and Day of the Month Watch, Maker's Name Thomas Parnell, Canterbury, Number 15; on the Cap engraved David Price; likewise the Name of Parnell, Canterbury, on the Dial-plate.
Whoever hath found the above WATCH, and will bring it to David Price, of Whitstable, or Mr. Charles Cricket, at the Crown, Margate, shall receive One Guinea Reward.
    Had on the WATCH a Steel Chain, a Silver Seal, Cypher D. P.

18-22 September 1784 *Kentish Gazette* [p4]

BURGLARY.

WHEREAS the House of Mr. THOMAS ANDREWS, Watch-Maker and Silver-Smith, on the Quay, Dover, was on Saturday night last, or early on Sunday morning, broke open, and the following Goods stolen, viz.
A Metal Watch, with an enamel Case, Maker's Name, Mercer, jun. Hythe.
A Silver Watch, Maker's Name, Nash.
A Silver Cap Watch, Maker's Name, Andrews, Dover, with J. G. on the outside Case.
A Gold Watch, Maker's Name, Goddard, Dover, Cypher W. L. (in the Cock).
A Silver Watch, William Chalk, round the Dial Plate.
A Silver Watch, Name, Barr, Dover, with B. B. engraved on the inside Case.
A Gold Watch.
A Silver Watch, in a Tortoise-shell Case, Maker's Name, Woodgrave, No. 5524.
A Silver Watch, Maker's Name, Dolphin, London.
A Silver Watch, Maker's Name, Bull, with only one Case. [and many items of Plate and
    Jewellery, details listed].
    Notice given to Mr. Andrews, who Promises a Reward of Twenty Guineas on Conviction, over and above the Reward allowed by Act of Parliament.   Dover, Sept. 20, 1784.

Canterbury, Sept. 29
We are extremely happy to have so early an opportunity of informing the Public, that the villains who committed the burglary at Dover are discovered, and most of the goods found. From the best account we can get, the circumstances seem to be as follow:— A person on Monday night came to the Ship Tavern at Dover, to take place to go to London in the coach the next morning, and seemed particularly attentive to a box and other package which he had with him, but was told that his baggage must be committed to the care of the Book keeper, in order to be packed up ready for the morning; to this he so reluctantly complied, that some suspicion arose respecting him, and an opportunity was taken, with proper authority, to open the box, when there was found therein a great many of the goods of which Mr. Andrews had been robbed, except the watches. The man who was walking on the Quay, was immediately secured, and being carried before a Justice, pretended to be totally ignorant of what the box contained, and said that he was employed by some Jews in London to fetch up a box from Dover, and that he was told he should find it in a room at the Green Dragon, which he did, and intended to deliver it to his employers. The man is committed, and some persons are gone up to town to make a strict search after his villainous [companions].

[London] Friday, Oct. 1
Dixon, the accomplice of Morgan, in the robbery and murder of the unfortunate Mr. Linton, a month or two since, in St. Martin's-lane, was on Tuesday apprehended at the house of his father, in Lisle-street, Leicester-fields, and brought before Justice Addington, at his office, in Bow-street, where on his examination, a charge was made against him of a burglary, committed in the house of Mr. Andrews, watchmaker, at Dover, on or about the 18th inst. The father was concerned with Dixon, in receiving the property taken at Dover, and it was on that account, that Sir Sampson Wright's officers visited his house in Lisle-street; while they were searching for the stolen goods, in a dark closet, they felt the face of a man, and upon their forcing him from his hiding place, it appeared to be Dixon, after whom they had long been in pursuit. The father is in Dover gaol ... and the sister of Dixon, a decent well-looking girl, was committed to New Prison for further examination. The mother made her escape; but Dixon will himself be re-examined on Friday. He on Wednesday asserted that Morgan and he were both innocent of the murder of Mr. Linton.   (Lond. Pack.)

Canterbury, Nov. 3
Monday night Dixon, who was acquitted for the murder of the unfortunate Mr. Linton, at the last sessions at the Old Bailey, passed through this city, with a strong guard, on his way to Dover, where he will take his trial at the Quarter Sessions for that town, for the burglary at Mr. Andrews's, watch-maker, at Dover. The father of Dixon is now in that gaol, and will appear as principal evidence against the son. There are likewise three other prisoners in the above gaol, for capital offences.

Dover, Nov. 5
[Yesterday Alexander Dixon charged ....] The evidence went to prove, that after the escape of Dixon from prison, where he was confined for the murder of Mr. Linton, he was furnished with money by his friends with an intention to go to France .... The Jury withdrew for a few minutes, and returned with a verdict — Guilty. Death. [Thomas Dixon ... was discharged.]

Dover, Nov. 15.
Dixon and Osbourn are to be executed on Saturday if no reprieve comes.

———

CANTERBURY, Nov. 24.
from Dover, Nov. 20
Saturday morning last exhibited one of the most awful sights that has been seen in this town for a number of years, being the day appointed for the execution of the two unhappy criminals, James Husband and Alexander Dixon, who were convicted of different burglaries committed in Dover. After they had received the Sacrament, the procession began from the gaol, about half past ten o'clock, and moved with great solemnity, amidst a prodigious concourse of spectators; they arrived at the place of execution about eleven o'clock, and spent near an hour in prayer with the clergyman; after which Dixon made the following confession:— That neither he nor Morgan, (the unfortunate man who suffered some time since at Newgate) were guilty of the murder of Mr. Linton; but that he was in company with those who committed the murder early the same evening, though not at the time the murder was committed. He confessed the crime for which he was going to suffer, and that he was the only person concerned in breaking open Mr. Andrews' shop;— his sins were numerous, and the death he was going to suffer but a small atonement for them, but he hoped for mercy at the tribunal he was now going to appear before …. About twelve o'clock they were turned off amidst the prayers of the commiserating spectators. The whole of this awful business was conducted with great decency and decorum; and yesterday an excellent sermon, suitable to the occasion, was preached at St. Mary's Church. Saturday morning a boat from Deal, with [some] persons, mostly tradesmen, coming to see the exection, was overset, and three of them unfortunately drowned; two others were taken up with little hope of recovery.

———

11-14 May 1785 *Kentish Gazette* [p1]

LOST, On Thursday last at Wingham,
A SILVER WATCH,
With a Silver Cap, Steel Chain, a Seal and Key. Maker's Name, John Pope, Margate. Whoever has found it, and will carry it to Mr. Powel, at the Dog at Wingham, shall receive a Guinea and a Half Reward. — If offered to be sold or pawned, please to stop it and give Notice as above. No greater Reward will be offered.

———

3-7 June 1785 *Kentish Gazette* [p1]

BAYLY, Watch-maker, Silversmith, and Hardwareman, High-street, Ashford, Respectfully informs his Friends and the Public in general, that he continues to carry on the above Business in all its Branches, and, gratefully acknowledging all past Favours, humbly solicits their future Indulgence. — He also carries on the WEAVING BUSINESS of the late Mr. JOHN COLLAR, deceased. — Where may be had, on the shortest Notice, all Sorts of Flax, Hemp, Hop-bagging and Pocketing, Corn-sacks, and Cheese cloths, &c.

Likewise to be sold at the same House by his Daughters, E. and M. BAYLY, Milleners, a Variety of the best and most fashionable Articles in the MILLENERY, LINEN-DRAPERY, HOSIERY, and HABERDASHERY BUSINESS.

All Orders will be strictly attended to, expeditiously executed, and thankfully acknowledged,
By their most obedient humble Servant.

———

9-12 August 1785 *Kentish Gazette* [p1]

BUCKLEY,
Watch and Clock Maker,

RETURNS his sincere Thanks to the Ladies and Gentlemen of Canterbury and its Environs, and the Public in general, for the great Encouragement he has met with in the above Business, during his Residence in Burgate Street; and takes this Method to inform them, that he has removed from thence to a more commodious Shop, late Mrs. PARKER's, at the Corner of Mercery Lane, on the Parade; where he has laid in a fashionable Assortment of Silver, Jewellery, Plated, Ivory, Japan, and Cutlery Goods, which he is determined to sell on the very lowest Terms, and all Favours will be gratefully acknowledged.

Elegant Pedestal Image and other Time Pieces. — Repairing done in the best Manner, on reasonable Terms. — Money for old Gold and Silver, or Lace, burnt or unburnt.

2-6 December 1785 *Kentish Gazette*

NOTICE.

ALL Persons, who have any Demands on the Effects of Mr. RICHARD BAYLY, Watchmaker and Silversmith, of Ashford, in the County of Kent, lately deceased, are forthwith desired to transmit an Account of the same to Mrs. ELIZABETH BAYLY, of Ashford, that the same may be discharged. — And all Persons, who stand indebted to the said RICHARD BAYLY, are hereby requested to pay the same to ELIZABETH BAYLY aforesaid, or either of his Sons, namely, Mr. JOHN BAYLY, of Ashford, or Mr RICHARD BAYLY, of Maidstone, who are authorised to receive and discharge the same.

Mrs. BAYLY respectfully returns her sincere Thanks to the Friends of her late Husband, and the Public in general; and begs Leave to acquaint them, that she purposes, through the Assistance of her Son, to continue the Business of Watch-maker and Silversmith, in all its various Branches; likewise, that she will continue the Millenery, Linen-drapery, Haberdashery, and Hosiery Business, as usual. — The Continuance of their Favours will be most thankfully received and gratefully acknowledged,
By their obedient Servant,   ELIZABETH BAYLY.

2-6 December 1785 *Kentish Gazette*

BAYLY.
Watch-Maker, Silversmith, and Hardwareman,
MAIDSTONE.

Respectfully returns sincere Thanks to his Friends and the Public in general, for the very liberal Encouragement he has hitherto met with, and begs Leave to inform them, that he is desirous of giving his Mother every Support in his Power towards carrying on the Business of his late Father; as it must necessarily take him frequently from Home, he has engaged with an experienced Workman from London, in Order that every Thing, which is entrusted to his Care, may be executed with the utmost Dispatch. — The Continuance of their Favours will be most thankfully received and gratefully acknowledged.

16-20 June 1786 *Kentish Gazette* [p1]

LOST, On Saturday, the 17th of June, 1786,

A SILVER WATCH, at Deal, or between Deal and Betshanger, Maker's Name, John Marsh,

Eastry, with a Land-skip on the Face, Black Ribbon, and one Stone Seal set in Silver.

A Reward of One Guinea will be given by bringing it to the Printers of this Paper.

9-12 January 1787 *Kentish Gazette* [p1]

WATCH LOST.

LOST,

On Saturday Night last, about Four o'Clock,
Between Canterbury and Buckwell Farm, in the Parish of Sturry,

A NEW SILVER-CASED WATCH,

No. 186; Maker's Name BUCKLEY, with a Steel Chain, two Keys, and a Steel Seal. Whoever has found the same, and will bring it to Mr. John Holbrook, Walter Lane, Sturry, shall receive One Guinea Reward.

23-27 February 1787 *Kentish Gazette* [p1]

LOST,

On Sunday Evening, the 11th of this instant February,
Between Shandford Street and Fordwich,

A SILVER WATCH;

Maker's Name Wm. NASH, Bridge, and a Cypher on the Outside Case W K, and within Side of the outer Case is engraved Wm. Keen, Sturry, 1784.

Whoever has found the same, and will bring it to the Little Rose, in the Borough of Staplegate, or to the Swan in Sturry, shall receive Half-a-Guinea Reward.

Sturry, Feb. 26, 1787.

13-16 March 1787 *Kentish Gazette* [p1]

LOST,

On Thursday, the 8th of March instant,
Between Wye and Faversham,

A SILVER-WATCH; Maker's Name, John Silke, Elmsted.

Whoever has found the same, and will bring it to THOMAS QUESTED, Watchmaker at Wye, shall receive Half-a Guinea Reward.

30 March-3 April 1787 *Kentish Gazette* [p1]

LOST,

This Day, between Ash and Canterbury,

A SILVER WATCH, with a Steel Chain and straight Key; Mark on the Out-side Case with J. W. — In the Inside Case, JOHN WARWELL, Deal, in full.

Whoever has found the same, and will bring it to the Printers, or to Mr. Warwell's, at Deal. shall receive a Guinea Reward.   Deal, March 29, 1787.

7-10 August 1787 *Kentish Gazette* [p1]

SAMUEL ABRAHAMS,
SILVERSMITH, JEWELLER, and WATCH-MAKER,
Near the Flower-de Luce, Sandwich,

RETURNS his sincere Thanks to those Ladies And Gentlemen who have favoured him with their Custom, and begs Leave to inform them and the Public in general, that he has laid in a large, elegant, and fashionable Assortment of Plate, Jewellery, Watches, Hardware, &c. which he intends selling at very moderate Prices. He has also now on Sale, on the lowest Terms, a great Variety of Articles in the several Branches of Cloathing and Hosiery.

 N.B. Hats and Shoes of all Sorts and Prices.

 The most Money given for old Gold and Silver, and light Gold taken without any Loss to the Customers.

2-5 October 1787 *Kentish Gazette* [p1]

TIDDEMAN and Co,
Watch and Clockmakers, On the PARADE, CANTERBURY,

MANUFACTURE and repair, on the best Principles, all Sorts of Musical, Spring, and Common Clocks, also Repeating, Horizontal, and other Watches. They have likewise laid in a large and elegant Assortment of Silver and Plated Goods, a great Variety of Silver and Plated Buckles, which they intend to sell on the most reasonable Terms.

Mourning Rings and Devices on the shortest Notice.
All Sorts of Silver Plate repaired in the best Manner.
Arms, Crests, and Cyphers, neatly engraved.
The utmost Value given for OLD GOLD and SILVER.
Light Guineas bought at Eight-pence Loss.
CLOCKS repaired at any Part of the Country.

 Mr. TIDDEMAN hopes by his Assiduity and strict Attention to Business, to merit the Favor of his Friends and the Public, as he is determined Nothing shall be wanted on his Part.

2-5 October 1787 *Kentish Gazette* [p1]

BUCKLEY,
Watchmaker and Silversmith,
PARADE, CORNER of MERCERY-LANE, CANTERBURY,

RETURNS sincere Thanks to the Public for past Favors, and hopes to merit a Continuance of the same, as he ever has, and will, make it his Study to oblige his Customers. He has for their Inspection, laid in a very large and fashionable Assortment of Plate, Plated, and Japan Goods, in great Variety, consisting of Tea Trays, Waiters, Caddies, Bread Baskets, Bread and Card Racks, Cruet Frames and Glasses of the newest Pattern, elegant Candlesticks, Tea Pots, Taper Stands, &c. &c. with the greatest Variety of Silver and Plated Buckles of the newest Patterns, and all Kinds of Cutlery, Jewellery, Tortoiseshell, and Ivory Goods in general, which he offers to the Public on the very lowest Terms.

 The utmost Value given for Watches, Gold, Silver, and Jewels, Light Gold, &c.

 Variety of Clocks and Watches, on the best approved Principles, on reasonable Terms. — Likewise Watches, Plate, and Engravings, repaired and done in the best Manner.

 N.B. Umbrellas of all Sorts, as cheap as at the London Warehouses.

13-16 November 1787 *Kentish Gazette* [p1]

WHEREAS a Person, in the Habit of a Sailor,

left a METAL WATCH, single Case, at Mr. THOMAS JENKIN's, Watchmaker, in Rochester,

to be repaired, on which he borrowed Seven Shillings and Six-pence in the Month of July last ... he having not called, according to his Promise, gives the said Thomas Jenkin Room to suspect it being dishonestly come by. Any Person being the Owner, and describing the same, may have it, paying all Costs and Charges.
Rochester, Nov. 13, 1787.

29 April - 6 May 1788 *Canterbury Journal* 1788

BUCKLEY,
Watch and Clock Maker, Silversmith and Jeweller,
PARADE, Corner of MERCERY-LANE, CANTERBURY.

BEGS Leave to acquaint his Friends and the Public, that he is just returned from London with a large and fashionable Assortment of every Article in the above Branches, and has taken great Pains to select (from the different Manufactories of London and Birmingham) the most choice Patterns, and numerous Assortment of Silver and Plated Buckles, together with a great Variety of elegant Japan Tea Trays and Waiters, Bread Baskets, Dressing Boxes, Inlaid and Japan Caddies, Bottle Stands, &c. Ivory, Tortoiseshell, Silver, Filligree, and Glass Smelling Bottles; Toothpick Cases and Etwees [sic]; fine Knives and Scissars, &c. &c. with many other Articles too numerous to insert, which he will sell on the Lowest Terms, and all Favors gratefully acknowledged

By the Public's very humble Servant.
Cheap Watches for Ready Money.

Mr. BUCKLEY, during his Stay in London, having attended the Sale of a Watch-maker's Stock in Trade, has made a Purchase of a large Selection of Watches, considerably cheaper than they were manufactured for, consisting of Gold, Silver, Gilt, and Cover-cased Watches of the newest Fashion; some Second-hand Watches; a Repeating Watch; an exceeding good Spring Clock, in a handsome ornamental Case; all of which he is enabled to sell considerably under the Wholesale Prices.

Whoever may be Purchasers, Mr. B. pledges himself for the Performance of the above, as they are no Auction Job, but sound Watches, the real Property of a Watch-Maker in the Decline of Life, leaving off Trade. Plate and Plated Goods in great Variety. Watches and Clocks repaired as usual. Greatest Choice of Umbrellas, of all Sorts, at the London Prices.

26-30 September 1788 *Kentish Gazette* [p1]

A CAUTION to CLOCK-MAKERS.

WHEREAS on MONDAY last WILLIAM POTTS went away from his Master's (JOHN MARSH, Clock and Watch Maker, Eastry) and left his Work unfinished, taking with him several of his said Master's Tools. He is about five Feet eight Inches high, dark Complexion, his own dark Hair tied short. Had on a dark-coloured Coat, striped Linen Waistcoat and Nankeen Breeches, and is particularly attached to the Contents of the Gin-bottle.

26-30 September 1788 *Kentish Gazette* [p1]

LOST,
On Thursday, the 18th Instant, near Folkstone,
in the Road leading from thence to Canterbury and Dover.
A SILVER WATCH, engraved Case, Steel Chain, and two Stone Seals;

Maker's Name THOMAS PARNELL; Number 25, as supposed.
Whoever has found the same, and will bring it to the Printers of this Paper, or to Mr. Thomas Parnell, Watchmaker, Canterbury, or to Mr.Stace, Miller, at Folkstone, shall receive One Guinea Reward.   If offered for Sale, please to stop it.

27-31 March 1789 *Kentish Gazette* [p1]

WATCH LOST,
LOST,
About the 23d of March, within two Miles of Canterbury,
A SILVER WATCH, Maker's Name, MICHAEL LADE, Canterbury;
has a Silver Cock and Slide.
[bring to Mr. B LEVY; ... a Guinea Reward.]
  N.B. The Reward offered is more than the Value of the Watch, but being an old Family Watch, the Owner has a particular Wish to retrieve it again ....

14-17 April 1789 *Kentish Gazette* [p1]

W. CLARIS,
Cabinet and Clock-case-maker, No. 15, St. Dunstan's street, Canterbury,
Begs leave respectfully to return his Thanks to his Friends and the Public for the Encouragement he has received in the above Businesses, and informs them that he continues to make and sell all Kinds of Watch and Spring Cases, and every other Article in the Cabinet Branch, in the most elegant and fashionable Taste, and on the most reasonable Terms.

21-24 April 1789 *Kentish Gazette* [p1]

STOLEN,
Out of the House of Stephen Ketchley, of Lydd,
Between the Hours of Five and Six o'Clock in the Afternoon of the 19th of March last, A SILVER WATCH, made by BAYLEY, of Ashford; the Maker's Name on the Face of the Watch, and S. K. in a Cypher on the Outside of the outward Case, and S. K. in Capitals on the Outside of the inward Case. There was a Silver Chain and Seal hanging to it.
  Whoever will bring it to the Owner, or inform him where it is, so that he may recover it, shall receive One Guinea for his Trouble.
  A Smuggler from the Country was seen to go into the Kitchen where the Watch was; but finding Nobody there, came out immediately. Ketchley's Wife, who had just before gone into the next Room, returned in a few Minutes after the Man left it, and found the Watch gone.

28 April-1 May 1789 *Kentish Gazette* [p1]

To be peremptorily SOLD by AUCTION,
by JOHN LEITH,
On Monday, the 4th of May, 1789, and following Day,
ALL the neat and genuine HOUSEHOLD FURNITURE, PLATE and CHINA, of
Mr. THOMAS JENKINS,
Watch and Clock Maker (deceased) in the High Street of Rochester; comprising Four-post Bedsteads, with Crimson Check Furniture; a Mahogany Bureau Ditto; Feather beds, Blan-

kets, Quilts, and Counterpanes; a new Dial and Ebony Compass; Chairs, nailed over the Rail, with Satin Hair Seats, Elbow Ditto to match; Mahogany Double Chest of Drawers, Ditto Dining, Pembroke, and Tea-tables; Pier and other Glasses, in carved Mahogany and Gilt Frames; a bright Kitchen Range, Bath and other Stoves; Walnut tree Desk and Book-case; a Pair of Globes, Celestial and Terrestrial; Kitchen Requisites, &c. &c.

GOLD and PLATE.

A Gold Watch and eleven others; about one hundred and ninety Ounces of Plate; Silver Buckles; ninety Gold Rings; Plated and Mourning Buckles, Correls, and a Variety of other Articles too tedious to mention.

The Sale to begin each Day at Two o'Clock precisely.

N.B. The Plate to be sold the last Day.

Catalogues to be had at the Crown Inn, Rochester; at the Rose, Sittingbourne; at the Bull, Maidstone; at the White Hart, Gravesend; and of John Leith, Auctioneer and Appraiser, on St. Margaret's Bank, Rochester.

1-4 September 1789 *Kentish Gazette* [p1]

NOAH LEVI,
Watch and Clock Maker, Working Goldsmith and Jeweller,
South Street, near the Market Place, Ramsgate,

RETURNS Thanks to the Ladies and Gentlemen, and the Public in general, of Ramsgate, Margate, Broadstairs, St. Peter's, &c. for the many Favours conferred on him, and hopes for a Continuance of the same; and acquaints them, that he having removed from his late Shop to No. 12, on the other Side of the Way; where he continues, as usual, to manufacture all Sorts of Musical, Spring, and Plain Clocks; Horizontal, Plain, and Patent Stop Watches; Mourning and Device Rings; Lockets, &c. on the shortest Notice. A large Quantity of Wedding Rings, ready-made; Plate and Jewellery neatly mended. Also, a good Assortment of Cutlery, Hardware, &c. and a large Assortment of Foreign and English Toys, Canes, Sticks, Whips, &c. — The utmost Value given for any Quantity of Diamonds, Pearls, Watches, Gold, Silver, and Gold and Silver Lace; Arms, Crests, Cyphers, &c. neatly engraved; Clocks and Watches repaired on the best-approved Principles, and on the most reasonable Terms. — Ladies and Gentlemen may be served with a Side-board of Plate on the shortest Notice; a regular Plate-book is kept of the newest Fashion and Patterns.

N. LEVI hopes by his strict Attention to Business, and the Reasonableness of his Charges, to merit the Favours of a discerning Public.

To be sold cheap, at the above Shop, a large Cabinet, containing a great Variety of curious and valuable Shells, Petrifications, Skeletons, Insects, a great Number of curious Fish, and many other Things too tedious to mention; the whole forming a complete Museum. Some of the above came from Port Jackson.

2-6 October 1789 *Kentish Gazette* [p1]

BUCKLEY,
Clock and Watch Maker, Silversmith and Jeweller,
Corner of Mercery-lane, Parade,
CANTERBURY,

Respectfully begs Leave to acquaint his Friends and the Public, that he has just returned from

London with a large and fashionable Assortment of every Article in the above Branches; consisting of all Kinds of Plate and Plated Goods of the newest Fashion; the greatest Variety of Silver and Plated Buckles; new and Second-hand Clocks and Watches; Patent and Pocket Umbrellas, with all other Kind of Silk and Lawn Umbrellas, at the very lowest London Prices.

Repeating, Horizontal and Plain Watches, repaired in the best Manner on the most reasonable Terms.

N.B. Clocks cleaned and repaired, at any Part of the Country, at two Days Notice.

Most Money for old Watches, Gold, Silver, &c. Umbrellas repaired, and Plate and Jewellery neatly mended.

8-11 December 1789 *Kentish Gazette* [p1]

WATCH and CLOCKMANUFACTURER,
And SILVERSMITH.
HAMMOND NICHOLLS,
(From LONDON, late apprentice to Mr. SMITH)

Respectfully informs the Public, that he has taken Mr. Smith's Shop, on the Parade, Canterbury; humbly solicits the Continuance of the Customers, his Friends, and the Public in general. — He flatters himself by his Assiduity, and having had five Years Experience in finishing and repairing Horizontal, Repeating, and Seconds Watches, for the first Shops in London, he shall be able to give Satisfaction, as he is determined to pay the utmost Attention to the Performance of those Watches and Clocks entrusted to his Care; and as the Manufactory is carried on at this Shop, hopes to merit the Support and Encouragement of a discerning and generous Public.

He likewise has laid in a new and fashionable Assortment of Silver and Plated Goods, with a great Variety of Plated Buckles, Jewellery, &c. &c. which will be sold on the most reasonable Terms.

Gold, Silver, and Jewellery repaired. — Engraving neatly executed. — Mourning Rings, with Devices, in two Days Notice. — The utmost Value given for Gold and Silver. — Clocks repaired and sent to any Part of the Country.

A Clock-Maker may have constant Employ.

11-15 December 1789 *Kentish Gazette* [p1]

TIDDEMAN,
WATCH and CLOCK-MAKER,
and Silversmith,

RETURNS his sincere thanks to his Friends and the Public, for the many favours he has received during his late partnership, and respectfully informs them, he has entered into business on his own account, and has opened a Shop, Opposite Butchery Lane, on the Parade, CANTERBURY;

where he humbly solicits a continuance of their future Favours — and flatters himself, by his assiduity and strict attention to business, to merit the encouragement and support of a discerning and generous Public. He assures his Friends and the Public, that his whole STOCK in TRADE being entirely new, and having the best connection in London for his goods, he will be able to sell them on the most reasonable terms, as he is determined to pay the most strict attention to make reasonable charges.

TIDDEMAN hopes to attract the notice of his Friends and Customers by just and honest dealings, and not by fictitious names or artful insinuations.

A strict attention will be paid to repairing all sorts of Horizontal, Repeating, Seconds, and other Watches; likewise all sorts of Musical, Spring, and other Clocks. — All Watches sent by newsmen and carriers will be carefully attended to, and returned with quickness.

Mourning and Device Rings on the shortest notice.
Arms, Crests, and Cyphers neatly engraved.
Gold, Silver, and Jewellery, repaired in the neatest manner.
The utmost value given for Old Gold and Silver.

5-8 January 1790 *Kentish Gazette* [p1]

DANIEL BARNARD,

Next the Little Rose, in the Borough of Staplegate, Canterbury.
Respectfully informs the Public, he has taken out a Licence for Pawnbroking. — Takes in all Kinds of Linen and Clothes — also Gold and Silver Plate, Watches, &c. and all Kinds of Metal.

N.B. Buys and sells all the above Articles.

7 October 1791 *Kentish Gazette* [p1]

WATCH LOST,

On Saturday night last, the 1st instant,
In the road from Canterbury to the Chequers, in Stone Street,
A SILVERWATCH; maker's name, Warren. No. 8157.
Whoever has found the same, and will bring it to Mr. James Warren, without St. George's Gate, Canterbury, shall receive One Guinea reward.

14 October 1791 *Kentish Gazette* [p1]

BUCKLEY,

Watch-maker, Silversmith and Jeweller,
Corner of Mercery Lane, Parade, Canterbury,
BEGS leave to inform his friends and the public, that he has laid in a fresh assortment of every article in the above branches, with cutlery, plated and japan goods of all kinds, which he will sell on such terms as will insure the continuance of their favours. — Likewise umbrellas from 6s. each and upwards; there being a considerable drop in that article, he is happy to acquaint the public he has it in his power to sell them at very reduced prices.

N.B. A large assortment of japan tea-trays and waiters, dressing cases, tea urns, &c. &c. on the very lowest terms.

30 December 1791 *Kentish Gazette* [p1]

LOST.

Last Monday, the 16th instant, about two o'clock in the afternoon, in a field close by the footpath, leading from Shoaloak to Broadoak,
A SILVER WATCH, maker's name, WATTS, Canterbury;
was bought of him twenty-four years [ago]; has a pinchbeck chain, seal and trinket to it, but no key. The letters A S in a cypher are engraved on the outside case.

Any person who brings the said Watch to John Spratt, Castle Street, Canterbury, shall receive a reward of One Guinea.

Wanted Immediately,
THREE or FOUR JOURNEYMEN CARPENTERS.
Apply to the said John Spratt.

---

28 February 1794 *Kentish Gazette* [p.4]

[to be sold at auction —

Household furniture of Mr. Hammond Nichols,
WATCH & CLOCK-MAKER, and SILVERSMITH, High-street, Canterbury.
Note: there is no reference to his working stock or tools.]

January 1803 *Kentish Gazette*

JOHN ELLIOTT.

Watchmaker, Silversmith, Stationer and Bookseller opposite the White Hart Inn, High-street, Ashford.

---

4 January 1803 *Kentish Gazette*

WILLIAM STRICKLAND,
Watch, Clock, and Gun-maker, and Silversmith,
TENTERDEN;

TAKES the Liberty most respectfully to inform the Inhabitants of Tenterden and the neighbouring Parishes, that he has taken the above business of Watch-maker, Silversmith, &c. lately carried on by Mr. WOOLLEY; and having been in the shop himself, now more than twelve years, and perceived the custom during that time continually to increase, he begs leave to assure them that his utmost endeavours shall be exerted to execute all orders with that punctuality and dispatch, and on such reasonable terms, as that nothing shall be wanting on his part to merit the continuance of their favours.

N.B. — A steady LAD, as an Apprentice in the above branches, is wanted immediately, and, if approved of, but a small Premium will be required.

Mr. WOOLLEY with great pleasure embraces this opportunity to return his most sincere thanks to his numerous friends in Tenterden and its vicinity, for all favours conferred on him while he conducted the above business, and now begs leave respectfully to recommend Mr. STRICKLAND to them, as a steady and deserving young man.

---

4 January 1803 *Kentish Gazette*

D. BOWEN,
Clock and Watch-maker, Silversmith, Jeweller, &c. (from London)
Nearly opposite the Union Bank, Parade, Canterbury,

MOST respectfully returns thanks to his friends for the very liberal encouragement he has received during the short period he has resided in Canterbury, from which he has been induced to take the above eligible situation, for the convenience of his friends and for the purpose of extending his business. He also acknowledges with gratitude the numerous favours of those who have recommended him as a Watch and Clock Maker, in which department he trusts that he has given sufficient proof that he has a complete knowledge of his business, which has been gained by a constant application to it for many years, and also from receiving every possible instruction from men of the first experience in the trade; he therefore flatters himself that he has it in his power to obviate that very general complaint which is alleged against many of the trade, that Clocks and Watches are oftner spoiled by unskilful workmanship, than by the wearer.

All sorts of musical and church clocks, repeating, horizontal, and plain watches made and repaired on the best principles.

He also takes this opportunity of informing the public that he is just returned from London with a large assortment of Jewellery, Plate, Hardware, Cutlery, &c.&c. consisting of silver, plated and gilt tea and coffee urns of various sizes; silver table and tea spoons; candlesticks, gold chains and seals, wedding and fancy rings; a variety of fashionable beads; elegant tea trays and waiters; pocket pen knives and scissars; trinkets, and a choice collection of every article in the above respective businesses.

In offering this assortment to the notice of the public, he leaves the fairness of his pretensions to their future decision, not doubting, from a long experience, that the public will not only find his collection of the newest and most fashionable patterns, but of a superior manufacture, and such as he will venture to warrant, and from the reasonable terms that he has affixed to every article, with a constant attention to their commands, he hopes to merit a share of their patronage and support.

WANTED,

A respectable Youth, as an apprentice to the above Business, where he will have an opportunity of learning every branch, which is seldom the case in the country, and it is much to be lamented that an apprentice, after serving seven years, should not be capable of following his profession for want of necessary information being given to him during his apprenticeship.

He assures the friends of any respectable youth, that the utmost care will be taken to render whatever may be entrusted to his care, a proficient in every art of his business.

For particulars, apply as above.

February 1803 *Kentish Gazette*

[for Sale] A substantial Freehold Messuage ... in High-street Ashford, with stable and garden; also with the field behind; late in the occupation of Mr. James Roberts, Silversmith, watchmaker and Stationer; with the Stock in Trade, including several new and old eight-day and thirty-hour clocks, in mahogany and wainscot cases ... and a Library, of more than 1,000 volumes.

[later in February 1803: sale of contents of the above, Mr. Roberts — bankrupt.]

18 February 1803 *Kentish Gazette*

Tuesday last died, age 50, Mr. Jas. Levi, senior partner of Mr. Benjamin Levi, watchmaker and goldsmith, Dover. — sympathy offered to grief-stricken widow and family, he was held in high esteem in society.

19 July 1803 *Kentish Gazette* [p1]

TO MILLERS and MEALMEN,

TO be sold by Auction, by ROBERT FINIS, at the Rose and Crown at Elham, on Friday the 22nd Day of July 1803, between the hours of three and five in the afternoon (unless disposed of by private contract) —

All the FREEHOLD CORN-MILL, in complete repair, with a DWELLING-HOUSE, stable, lodge, garden, and small piece of pasture land adjoining; situated in the town of Elham, in the occupation of the proprietor, Mr. John Smith, who will give immediate possesion.

At the same time will be sold, a light cart, with harness; a quantity of sacks, scales, weights, and other articles in the millering business, &c.

For further particulars enquire on the premises; of Mr. John Smith, watchmaker, at Wye; or of the Auctioneers, at Hythe.

19 July 1803 *Kentish Gazette*

On Wednesday last died at Lenham (after long illness) Mr. William Payne, watchmaker, leaving widow and three children.

19 August 1803 *Kentish Gazette*

TO WATCH AND CLOCKMAKERS.

TO BE SOLD, THE HOUSE and premises of OWEN JACKSON, watchmaker and silversmith, situated near the centre of the town of Tenterden, Kent, with or without the stock in trade, tools, &c. with every necessary convenience for carrying on an extensive trade in the above branches, being an old original sh shop for thirty-five years past.

N.B. A desirable situation for a young man and good workman. — Mr. JACKSON is going to retire from business, and will quit at Michaelmas next.

19 August 1803 *Kentish Gazette*

NOTICE is hereby given, that the partnership between Benjamin Levi and Emanuel Levi, of Dover, silversmiths, watchmakers, jewellers, &c. was this day dissolved by mutual consent. — All persons having any claims on the said firm, or on the late firm of Joseph and Benjamin Levi, are requested to send in an account thereof to the said Benjamin Levi, or to Mr. Pain, of Dover aforesaid, attorney — And all persons, indebted to the said firms, or either of them, are earnestly desired to pay their respective debts immediately to the said Benjamin Levi, or to Mr. Pain. — Dated August 15th, 1803.

17 October 1803 *Kentish Gazette*

Lost near Chilham ... A Metal Skeleton Watch, the outer case covered with tortoiseshell; maker's name W. Nash, Bridge.

5 January 1808 *Kentish Gazette* [p4]

Improved Mathematical Spectacles
CHARLES DOWSETT,
Watch-Maker, of MARGATE, ...

8 January 1808 *Kentish Gazette* [p1]

[by this date — if not earlier — D. Bowen, Parade, Canterbury, has no mention of clockmaking or watchmaking ; he merely stocks: 'Elegant Silver and Plated Goods; Jewellery, Gold and Silver Watches &c. &c.']

Appendix II
Kent Clockmakers & Watchmakers Listed by Town

The following lists the clockmakers in Chapter 7 by their place of work, with their known dates and also any other places where they worked. After the placename is the key to the map on the endpapers.

Ash M4
Mepstead, John	1788	
Pain[e], Mark	1791-1862	Sandwich
Pierce, Thomas	1866-74	
Tyler, —	1760	

Ashford J6
Barrett, William	1614-62	
Bayl[e]y, Richard Snr	1752-85	Biddenden
Brushfield, George	1814	
Bull, John	1845-55	
Deale, Thomas	1680-7	
Elliott, George	1866	
Elliott, John & Son	1802-41	
Flint, William Snr	1757-1813	
Flint, William Jnr	1787-1871	
Foster, Henry	1858-74	
Greenhill, John	1644-1706	
Greenhill, Richard I	1616-87	
Greenland, John	1675	
Hayward, Edward	1847-74	Folkestone
Jackson, —	1790	
Jefferys, William R.	1874	
Kew, William	1874	
Peckham, John	1741	
Price, Griffith	1799	
Price, James	1855-74	
Price, John	1852-74	
Roberts, James	1803	
Snatt, John	1780	
Thomson, William James	1874	
Tippen, James	1874	
Webb, Abraham	1741	Wye
Wimble, George	1717-41	Faversham
Wimble, John	1680-1741	
Wimble, Jeremiah (Nehemiah?)	1720-50	Maidstone

Barham L5
Robbins, Thomas	1819-78	Canterbury
Robbins, William	1787-1832	Kingston & Canterbury

Bekesbourne L4
Nash, John	1769	Bridge
Wimble, Thomas	1694-1716	
Woodward, John	1857-67	

Bellingham A2
Graham, Thomas	1785-1827
Graham, William	1785

Belvedere C2
Newman, John	1866-74	Bexley Heath

Bethersden H6
Padgham, John	1855

Bexley C2
Mason, William	1762
West, William	1847

Bexley Heath C2
Bennett, George	1855	
Bromley, E. H.	1851-5	
Bromley, Henry	1847-74	
Bromley, William	1832-9	
Newman, John	1866-74	Belvedere
Oclee, John	1832	

Blackheath B2
Bennett, G.W.	1851
Bennett, William Cox	1866
Triggs, R. W.	1866-74
West, Richard	1839

Boughton J4
Baker, Alfred	1855-66
Baker, Henry	1826-57

303

Baker & Son, M.	1836	
Fenn, John	1874	

BRENCHLEY E6

Brattle, Thomas	18thC	
Moon, —	1845	

BRIDGE L4

Hardeman, Samuel	1794-1839	Canterbury
Hardeman, William Henry	1848-74	
Nash, John	1769	
Nash, William	1762-94	Canterbury

BROADSTAIRS O3

Smith, William	1838-74	
Standven, Thomas	1839-40	Ramsgate

BROMLEY A3

Adams, Samuel	1775-92	
Bailey, W.	1830	
Dow, Alexander	1845	
How, James	1802-74	
How, Martha, Thomas & John	1847-66	
Kirby, John	1730	
Pullen, H	1874	
Schuler, M.	1851	
Shearer, James	1839	
Stelert & Dotter	1874	
Stiert, L.	1866	

BROMPTON F3

Ballard, William	1847-74	
Brady, Daniel Henry	1847-88	
Jackson, John	1826-8	
Kearly, R.	1800	
McKenzie, Alexander	1847	
Waghorne, C.	1866-74	

BUCKLAND N6

Robbins, Thomas I	1795-1866	
Robbins, Thomas II	mid-19thC	Walmer

CANTERBURY L4

Abrahams, Abraham	1838-74	
Allanson, Joseph	18thC	
Aiano, C.	c1840	
Annott, Charles	c1659-82	
Ardgrave, George P.	1846-74	
Barnard, Benjamin	1832	
Barrett, Thomas	1636-died 1662/3	
Barrett, Thomas Snr	1659-1662	
Barrett, Thomas Jnr	1726	
Barrett, John	1700-1733	
Bates, —	19thC	
Beard, William	1667-74	London
Bekelys, Nicholas	1485-95	
Berry, John	pre-1753	
Bing, Mrs	1865	
Binoy (Bing?), David	late 18th/early 19thC	
Blarkley, J.W.	1805	
Bliss, Ambrose	1647	London
Bourn, Loder	1738-88	
Bowen, David	late 18thC-1815	
Bradshaw, Thomas	c1790-1804	
Breckley, John	1784	
Brinkley, William	1756-68	London
Brodie, Hugh	1781	
Buckley, George	1823	
Buckley, John	1782-1805	
Buckley, Thomas	1784	
Cackett, Thomas	1748-74	Cranbrook
Cannon, William	1676	
Chalklin, John	1757-66	London
Chalklin, Mrs Margaret	1766-74	
Chalklin, William	1763-87	London
Chambers, Edward	1656-78	London
Cheese, Thomas	1779	Milton
Chen, Edward	1680	
Claris, W.	1789	
Corringham, R.	1858 66	
Cramp, Richard	1759-99	
Crampton, Edward	1796	Kennington
Dendy, John	1753-60	
Dennis, Thomas	1672-84/5	Maidstone and London
Dobell, Jesse (& E.)	1851-74	Hastings
Ellis, James	1658-80	London
Ellis, James	1764	
Emanuel, Levi	1781	
Engeham, John	1771-89	Yalding
Etcher, William	1845	
Feild, James	1690-1710/11	
Fielding, Augustin	1838-59	
Fowle, Thomas	1609	
Francke, Richard	1700	
Gambier, George	1847-51	
Gate, Arthur	1674-81	
Gee, John	1680-81	
Goatley, Daniel	1823-8	
Godden, Thomas	1832	
Goodchild, Richard	1756	
Goodlad, Richard	1677-1716	London
Gorrum, William	pre-1788	
Goulden, William	1782-1822	
Goulden, William	pre1797-early 19thC	
Gray, Stephen	1682	
Greenhill, Richard II	1648-1705/6	Ashford
Greenhill, Samuel	1684-1723	

Greenwood, John	1832-40		Pysing, Joan	1756	
Hardeman, Edwin Samuel	1838-55		Quaife, Thomas Samuel	1874	
Hardeman, Samuel	1794-1839	Bridge	Rawlings, Edward	1750/1	
Harding, E. H.	1865		Robbins, William	1787-1832	Kingston & Barham
Harris, Walter Snr	1693-1704	Maidstone			
Harris, Walter Jnr	1700		Roberts, Robert Edward	1874	
Hart, James	1752-61		Robinson, William	1761	London
Hayman [Heyman] Robert	1685-1722		Savage, Henry	1766	
Heitzman, Charles	1832-47		Sharpey, John	1714	
Heitzman, John	1840		Shindler, Thomas I	1657-1705	
Heitzman, M. G. & B.	1838-40		Shindler, Thomas II	1707-51	
Heitzman, M. & J.	1847-51		Shindler, Thomas III	1747-64	Romney
Heyman, John	1714		Shore, A.	1851-5	
Higgins, Charles	1865-74		Sims, Avery	1750	
Higgins, John Simms	1859-74		Sims, George	1710-20	
Hobbs, Samuel	1742/3		Sims, George	1745-60	
Homersham, John	1838		Sims, George Jnr	1751-late 18thC	
Horne, J.	1784-1828		Sims, Henry	1758-92	
Hughes, Richard	1784		Smith, Benjamin	1761-88	Sittingbourne
Jagger, Hannah	1823		Smith, William	1845-55	
Jefferys, James	1823-65	Wingham	Solomon, Emanuel	1785-1826	
Johnson, Charles	1667-82		Solomon, Miriam & Bella	1838	
Judge, T.	1866		Southee, H. S. E.	1866	
Kendall, Richard	1838-40		Stanbury, E.	1866-74	
Kissar, Samuel	1700-32	London	Strahan, John	1790	London
Knight, George	1683-96	Faversham	Taillour (Taylor), John	1428/9	
Kuner, S.	1865	Sheerness	Tiddeman, Edward	1779/1801	
Lade, Michael	1723-37		Tiddeman & Co	1787/9	
Lemaitre, Jules	1865	Paris	Tode, George	1858	
Lepine, Charles	late 18thC-1822		Trimnell, William Henry	1855	
Lepine, Henry	1823-38		Ventisan, Thomas	1682	
Levi, Benjamin	1783-91	Dover	Warren, C.	1770	
Levi, Solomon	1788-93		Warren, James	1767-93	
Martin, A.	1846		Warren, James Jnr	1785-1832	
Mason, Charles	1838-51		Warren, John	1874	
Mason & Son	1874		Warren & Son	1838-45	
Masoth, Charles	1847		Watts, John	1720-75	London
Merito, William	1847		Watts, John Jnr	1736-44	
Milles, Isaac	1680/1		Watts, John	early 19thC	
Monti, Anthony	1845-68		Wellby, John	1839	
Monti, John	1847		Willby, John	1839	
Monti, Joseph	1838	Jersey	Willsden, T. W.	1865-74	
Nash, William	1762-94	Bridge	Winder, John Christopher	1847-74	
Nicholls, Hammond	1789-1810		Wood, Henry	1826-8	
Papprill, Joseph	1789	London	Wood, Henry Samuel	1832	
Parnell, Thomas	1773-1805				
Parquot [Parquait], —	1750				
Philcox, George	1823				
Phillips, Solomon	1832				
Pollard, J.	1855				
Pollard, Samuel	1840-65				
Pope, John	1766				
Pratt, Thomas	1800-55				

CHALLOCK J5
Austen, Robert	1730-93
Austen, Thomas	1733-1805

CHARING H5
Flint, William	1733-93
Tippen, William	1866-74
Woollley, Thomas	c1750

Wraight, Thomas　　　c1780-90　　　Tenterden

CHARLTON B2
Newman, John　　　1821

CHATHAM F3
Aldersley, John　　　1832
Ansell, Henry　　　1839-55
Barrett, Daniel　　　1792-1800　　　London
Baurle, L. & F.　　　1874　　　Rochester
Bell, J.　　　1838
Bromfield, —　　　1707
Butler, W.　　　1851-5
Casper, Lewis　　　1832-51
Chapman, Daniel　　　late 18thC-1845
Chapman, Edwin　　　1832-55
Cigill, H. W.　　　1855-66
Crittenden, George　　　1838-47
Delacour, George　　　1823-55
Emanuel, Emanuel　　　1832
Hansell, Henry　　　1839
Hayler, Benjamin　　　pre-1769-1800
Hayler, William　　　1765-1851
Higginson, Samuel　　　1725
Hyman, Philip　　　1866-74
Isaacs, John　　　1839-47
Jefferys, George　　　1748-83
Jordan, James　　　1705-40
Levirn, Louis W.　　　1845
Longhurst, —　　　1790　　　London
Lyon, L. & S.　　　1845　　　Rochester
Mallery, John　　　1832
Manley, John　　　pre-1777-1828　London
Muddle, Edward　　　1707-72
Newman, J.　　　1855
Nye, Thomas John　　　1847-66　　　Rochester
Perse, Frederick　　　1874
Phillox, George　　　1826-8　　　Canterbury
Price, Joseph　　　1823
Pyke, J.　　　19thC
Reeve[s], Samuel　　　1725
Rich, Thomas　　　1792
Riedel, Ludwig　　　1866-74
Robbins, Thomas　　　1740-80
Rock, John Henry　　　1847
Salmon, H. S.　　　1855
Scaultheiss, J. T.　　　1874
Schwersensky, Isaac　　　1832
Smith, —　　　late 18thC
Solomons, S. & L.　　　1839
Staton, John　　　1845-7
Wall, John　　　1812
Webb, Richard　　　1874
Wilson, John　　　1845-51

CHERITON L7
Silke, John & Son　　　1780　　　Elmsted & Hythe

CHILHAM K5
Knowler, Samuel　　　c1770

CHIPSTEAD C4
Ranger, Thomas　　　c1740-73

CRANBROOK F7
Ballard & Co　　　1845
Ballard, Henry　　　1847-58
Ballard, William
　(or Frederick W.)　　　1826-66
Body, Abraham　　　1763-99
Body, Henry　　　1859
Body, Obadiah　　　1702-67　　　Battle
Burton, John　　　1795
Cackett, Thomas　　　1762-74　　　Canterbury
Chittenden, William　　　1773-6
Ferrall, Edward　　　1686-1706
Jackson, Owen　　　1760-1803　　　Tenterden
Marlow, John　　　1732-47
Ollive, Thomas　　　1752-1829　　　Tenterden
Punnett, John　　　1669
Punnett, Thomas　　　1656-1713　　　Rye
Reader, Thomas Oliver　　1805-45
Sheather, Daniel　　　1713
Thatcher, George　　　1716-73
Thatcher, Thomas　　　pre1772
Tilden, John　　　1713

CRAYFORD C2
Franks, E.　　　c1790

DARTFORD C2
Beckwith, William　　　1832-9
Braund, William　　　1823-55
Butterley, Stephen　　　1728-59
Butterley, William　　　1791
Cornish, Robert　　　1720-62
Ealand, B. C.　　　19thC
Edgcumbe, Edwin　　　1866-74
Fowkes, Gabriel　　　1759-1808
Hendrick, Charles　　　1851-5
Herman, Joseph　　　1874
Lowe, Thomas　　　1770-1836　　　London
Page, James　　　1839-47
Page, Thomas　　　1838-47
Porter, Charles　　　1823
Schebble, Joseph-Paul　　　1855
Schelble, Joseph　　　1866-74
Spasshat, Joseph　　　1845-7

DEAL N5
Adams, Robert　　　19thC

Adkins, G. B.	19thC		Wells, J.	1849	
Aldridge, Edward	1704	London	Wigg, Nathaniel	1847	
Aldridge, Thomas	1753-91	London	**DOVER N6**		
Blitz, A.	19thC		Andrews, Richard	1797	
Boys, Laurence	1720-37		Andrews, Thomas	1773-1802	
Cairns, James	1866-74		Barr, —	1784	
Cave, John R.	1847-51		Becket, James	1720/1	London
Cave, William Richard	1874		Carpenter, James	1866-74	
Cockings, Richard	1874		Catchpool, William	1801	
Cranbrook, William	1823-8		Cave, Richard	1826-51	
Day, John	1832-45		Cave, William John & Son	1855-89	
Emanuel, Emanuel	1797-1826		Cave, W. J.	1855	
Fulwell, Thomas	1838-55		Chidwick, William	1832	
Hayward, E.	1866		Collyer, Allen	1845-7	
Humphfreys, William	1832-40		Coveney, Zebulon	1815	
Lepine, Charles	1839-45		Cranbrook, Stephen	1791-5	
Le Plastrier, —	19thC	Ramsgate & Dover	Crosbey, William	1778	
			Crouch, Daniel	1847-55	
Long, Richard	1784		Dorrer, Eagen & Co	1845	
Payne, William	1796	Lenham	Ellis, James	1731	
Smith, Peter	1838-51		Fehrenbach, Emilian	1839-51	
Vennall, Charles	1866-74		Fehrenbach, F.	1851	
Vennall, James	1845-55		Fehrenbach, F. & O.	1855	
Vennall, Thomas	1826-9	Smarden	Goddard, Henry	1730-67	Tenterden
Vile, Jacob E.	1809-47		Greenwood, John	1823-8	
Warwell, John	1787		Grunwald, J.	19thC	
Wilmshurst, Thomas	1713-77	London	Hall, Kennett	1838-66	
DEPTFORD A2			Hopley, George Henry	1851	
Allen, Thomas	pre-1789		Hopley, W. F.	1855	
Bishop, John	1748		Hopley, William	1818-55	
Cockle, Thomas	1866-74		Igglesden, John	1818	Chatham
Dalgety, Alexander	1847		Igglesden, G. R.	1866-74	
Harris, John	1791		Kelvey, Robert	late 18thC	
Hayden, John	early 18thC		Lee, Samuel	1742-65	London
Heeley, Benjamin	pre-1747		Le Plastrier, Robert	1810-40	
Heeley, Joseph, Son	c1715		Le Plastrier, William	1800-2	
Heeley, Joseph	1747		Levey [Levi], Emanuel	1818-51	
Hopkins, Henry	pre-1780		Levi, James	1753-1803	
Hopkins, Henry	1802-24		Levi, J. & B.	1791	
Johnson, Jeremiah	1770-1808		Marsh, John Jnr	1825-47	
Killman, John	1847		Marsh, Charles Hollands	1834-66	
Lyne, —	1747		May, William Alfred	1874	
McClennan, —	1839		Miller, William Frederick	1823	
May, Robert	1828-39		Morris, John	1761	Gravesend
Pretty, John	1796-1807		Moses, —	1801-28	
Rathborn, John	early 19thC		Mummery, Thomas Snr	1808-55	
Rawfinger, John	1622		Mummery, Thomas Jnr	1814-47	
Sullivan, Richard	1829-47		Murden & Son	1874	
Taylor, John	1791		Norwood, George	1826-33	
Tight, George	1822-4		Overall & Bradshaw	1802	
Tunnell, John	1847-9		Penny, Alexander	1845	
Wegg, N.	1849		Perkins, Richard	1726	

Pratt, John	1793	
Price, Joseph	1814	
Pryor, E. J.	1866	
Putley, Francis	1874	
Reuben, Jacob	1823-32	
Reynolds, Richard	c1710	
Rowland, Theophilus	1829-55	
Sebber, Ann	1823	
Steber, David	1823-33	
Steber, John	1806-19	Germany
Swaby, Israel	1775-85	
Taylor, John	1858	
Tolputt, James	1744	
West, Daniel	1687	
Winchester, J.	1813	
Woodruff, Charles	1874	
Woodruff, W. C. & C.	1855-66	
Woodruff, William	1858	

EASTRY N4
Burton, —	1820
Cave, William	1832-51
Marsh, John Snr	1771-88
Pollard, James	1874
Potts, William	1788
Wraight, George	1826-32

EDENBRIDGE B6
Crundwell, William	1838
Hayward, William	1858-66

ELHAM L6
Dunn, Samuel Edward	1819-74
Dunn, Esther, Mrs	1874
Scott, Stephen	1725-98

ELMSTED K6
Silk, John I	1670-1700	
Silk, John II	1760-79	Stowting & Hythe

ELTHAM B2
Arnold, Charles	1824	London
How, Peter	1832-9	
Kimpton, Benjamin	1874	
Mesure, Lionel	1845	
Pike, James	1805-42	
Pike, John	1820-42	
Pike, Lawrence J.	1839	
Pike, Ruth	1839-55	
Pike, Sarah Mrs	1845	
Revell, E.	1866	

ERITH C2
Barnard, James	1851-66
Bell, Aaron	1769-76
Bromley, Thomas P.	1866-74
Brown, Aaron	1769-76
Selfe, Francis William	1874

EYTHORNE M5
Pemble, Gilbert	1760

FARNBOROUGH B3
Heydon, Thomas	1780-1800
Baker, George	1874

FARNINGHAM C3
Hankins, Jacob	1826-45
Whitehead, Henry	1851-74

FAVERSHAM J4
Bird, William	1752	
Carter, Jonathan	1824	Herne Bay
Clements, William	1751	London
Crow, Francis	1770-1803	
Dodd, John	1693-1700	
Farley, Thomas	1778-1802	
Foot, R.	1820-55	
Gates, Thomas	Late 17thC	
Heitzman, —	1858	
Knight, George	1683-96	Canterbury
Knight, John	late 17th-early 18thC	
Martin, James	1847-55	
Martin, John	1818-23	
Morris, James W.	1866-74	
Sharp, —	1780	
Sharp, John Bunyea	1823-32	
Sherwood, John	1845-55	
Spurge, J.	1866	
Vidion, John	1774-1801	
Wimble, George	1737-40	Ashford

FOLKESTONE M7
Bayl[e]y, Richard Jnr	1780-5	Maidstone
Boxer, John Snr	1779-1804	London
Boxer, John Jnr	1779-1852	
Boxer, Michael	1780-1810	
Cattanio, Joseph	1858-74	
Darley, F. H.	1866	
Fagg, Jacob	1823	Ramsgate
Hayward, Edward	1866-74	Ashford
Levi, A.	1799	
Lyons, James	1845	
Mercer, Thomas	1751-1805	Hythe
Mercer, William	1730-91	Hythe
Mummery, Charles	1839	
Prebble, Edward	1858	
Stace, John	1847-66	
Stace, M. A. Mrs	1874	

Foots Cray B2
Taylor, Richard — 1866-74

Forest Hill A2
Francis, W. — 1874
Seddon, Alfred — 1874 — Sydenham
Stephens, Thomas — 1874

Gillingham F3
Harris, Samuel — 1830
Mesure, Henry — 1848

Gravesend D2
Anderson, Frederick B. — 1823-51
Anderson, Hugh — 1839-66
Aunale, — — pre 1743
Avenall, William — 1720-70
Baker, Edmond or Edward — 1858-74
Ball-Baker, — — 1851
Boyce, Thomas — 1830
Chapman, Thomas — 1823-32
Copsey, Samuel — 1839
Couch, George — 1847
Dawes, William M. — 1823-7
Dunstall, James — 1874
Eveleigh, James — 1839
Fitch, George — 1839-45
Forster, William — 1845-7
Forster, William Henry — 1838-66 — Sheerness
Jackson, John — 1832
Morris, John — 1761 — Dover
Parker, Augustus — early 18thC — Milton
Paul, James — 1847
Pirkis, George — 1841 — Stroud
Pixley, John — early 19thC
Pollard, T. S. — 1855
Robins, E. W. — 1874
Selfe, William — 1866-74
Taylor, John — 1838
Vincent, James — 1858
Watts, William — 1845
Wells, William — 1847-66

Greenwich A2
Abraham, Joseph — 1812-24
Beamont, Charles John — 1847-58
Bennett, George — 1802-11
Bennett, John — 1817-24
Blundell, — — 1678-1730 — London
Buregar, S. — 1851
Cannon, John — 1874
Coleman, William — 1805-8
Dann, R. B. — 1874
Dorer, L. & P. — 1866
Dorer, P. — 1874
Dunkley, John — 1847
Ekins, F. G. — 1855
Elliott, Alfred — 1847-55
Elliott, James — 1720
Elliott, Thomas — 1720
Fenn, J. & Son — 1851-74
Fielding, A. — 1855
Fielding, George — 1839-51
Illman, Charles & Son — 1839-74
Johnson, B. — 1851
Ketterer, Charles — 1847
Ketterer, Crispin — 1847
Kibble, Richard — 1847-74
Kirby, John — 1866-74
Leoffler & Co — 1849
Leoffler, Peter — 1839
McGhan, R. — 1851-66
Martin, Alfred — 1866-74
Parker, Jasper or Joseph — 1780
Patrick, Mary Anne — 1847
Patrick, Miles — 1795
Patrick, William — 1827-39
Pattison, Robert — 1662-83 — London
Randall, A. — 1866-74
Siedle, A. — 1855
Strelly, John — 1684
Sullivan, George — 1866-74
Sullivan, George J. — 1847
Withers, John — 1778
Wogden, Stephen — 1713-24 — London

Hadlow D5
Crundwell, Joseph — 1855-74
Gibbett, — — 1750

Halstead B4
Evens, Robert — 1720
Evens, Robert — 1795

Harbledown K4
Goodchild, Richard — 1756 — Canterbury

Hawkhurst F8
Banister, A. — 1882
Barham, George — 1837-74
Huggett, John — 1832-9
Hukins, G. N. — 1851

Headcorn G6
Tippen, James — 1866-74

Herne L3
Banks, John — 1763

Herne Bay — Herne Street L3
Carter, Jonathan — 1874

Murton, Peter	1838	
Pegden, Vincent	1826-8	
Smith, William	1845-55	

HOLLINGBOURNE G4
Farmer, R. W.	19thC

HORSMONDEN E6
Gullven, Thomas	18thC

HUNTON E5
Neale, Richard	1866-74

HYTHE L7
Abrahams, Henry	1806	
Bates, John	1847	
Chapman, D.	1805	
Gilbert, Thomas	19thC	
Godden, Thomas	1784-1824	
Harden, Charles	1793-1861	
Jenkins, George	1845	
Judge, Thomas	1874	
Mercer, John	1736-79	
Mercer, John Jnr	1738-97	
Mercer, Thomas	1751-1805	Folkestone
Mercer, William	1730-91	Folkestone
Miller, William	1826	
Pay, James	1866-74	
Price, Richard	1858-74	
Silk, John II	1795	Elmsted & Stowting

KINGSTON L5
Bean, Edward	1792-1800	
Robbins, William	1787-1832	Canterbury & Barham

LAMBERHURST E7
Ballard, Isaac	1838
Ballard, James	1826-55
Ballard, John	1885
Ballard, John Thomas	1844-8
Ballard, Joseph	1839-74
Reeves, Benjamin	1774-90

LEE A2
Bennett, Elizabeth	1839-55
Dale, W. F.	1866-74
Wheeler, William	1866-74

LEEDS G5
Norton, William	1826-39

LENHAM H5
Danell, James	18thC
Payne, John	1731-95
Payne, M.	18thC
Payne, William	1803

Tippen J.	1851-5	

LEWISHAM A2
Brock, John	1839	London
Burton, S. & Son	1851	
Cluer, Obadiah	1682-1709	
Fowkes, Gabriel	early 18thC-1759	

LYDD J9
Foster, William	1845
Hill, John	1847=55
Gardner, John	1826-8

LYNSTED H4
Punnett, Jane, Mrs	1866-74

MAIDSTONE F4
Apps, William	1839-74	
Bailey, William	1785	
Baker, Henry I.	1622	
Baker, Henry II	17th-18thC	West Malling
Barling, Joseph	1847-74	
Barnard, J.	1865	
Barnsby, James	1865-74	
Bartholomew, —	1582	
Bartlett, F. J.	1865	
Bartlett, Samuel S.	1821-55	
Bartlett, W.	1851	
Bayl[e]y, Richard Snr	1780-5	Folkestone
Bayl[e]y, William	1784	
Beeching, Stephen	1791	
Beeching, Elizabeth	1791-5	
Beeching & Edmell	19thC	
Benzie, M. A., Mrs	1866	
Bishop[p], John I	1650-1710	
Bishop[p], John II	1680-1733	
Bishop[p], Martha	1720-33	
Bristow, Richard	1874	
Brooke, Thomas	1604	
Brown, Thomas	1840	
Buckland, William Friend	1830-9	
Burch, Robert	1848-52	
Burch, William	1795-1813	
Burch, William Jnr	1815-66	
Cadwell, Charles	1800	
Cogger, William	1838	
Cornell, George	1874	
Cottam, Joseph	1694-9	
Craythorn, John	1747	
Currie, Archibald John	1826-40	
Cutbush, Charles	1734-59	
Cutbush, Edward	1652	
Cutbush, Edward Jnr	1699-1702	
Cutbush, George	1724	
Cutbush, John I	1680	

Cutbush, John II	1722-34	
Cutbush, John III	1765	
Cutbush, John IV	1772	
Cutbush, Richard	1761-79	
Cutbush, Robert I	1652	
Cutbush, Robert II	1702-72	
Cutbush, Thomas I	1678	
Cutbush, Thomas II	1702	
Cutbush, William	1772-91	
Dann, W.	1826-66	
Dennis, George	1670-84	
Dennis, Thomas	1672-85	Canterbury
Dennis, Thomas Jnr	1698	
Dobson, —	19thC	
Dold, Alexander	1847-51	
Dold, P.	1855	
Fairey, James	1840	
French, A. W.	1812-72	
French, Richard Vigor	1800-52	
French, R. V. & Son	1852	
French, Stephen	1847	
French, Susanna & William Apps	1848-74	
Fuchter, Fidel	1874	
Giles, John Scott	1845-55	
Giles, Nicholas	1807-40	
Gill, George	1737	
Gill, William	1704-70	
Goffe, Thomas	1709	
Greene, William	18thC	
Greenhill John I	1607-1632	
Greenhill, John II	1608-1636	
Greenhill, John III	1655-1712	
Hall, Thomas	1761-8	
Harris, James	1754	Town Malling
Harris, Joseph	1743-70	
Harris, Walter	1693-1704	Canterbury
Hedges, John & Son	1832	
Heitzman, Charles	1839	
Heitzman, George	1832-47	
Hickmott, Charles	1865	
Hodges, John & Son	1839	
Hodgeskynne, —	1578	
Hughes, John	1697-1709	
Hughes, John Jnr	1727	
Hughes, William	1732	
Jackson, Richard	1760	
Jackson, William	1796-1800	
Jackson, William	1796-1823	
Jackson, William Richard Radley	1823-32	
Kollsall, —	1719	
Ladbrook, Daniel	1858	
Leech, William	19thC	
Lingham, H.	1855	
McKellow, John	1823-62	
McKellow, J., Miss	1851-5	
Manwaring, Richard	1838-58	
Martin, John	1832-72	
Mercer, William	18thC	
Mesure, Lionel	1845	
Moore, John	1784	
Munn, Thomas	1858	
Pankhurst, A.	1866	
Pankhurst & Munn	1851	
Parkhurst, Alexander	1858	
Patrick, John	1838-47	
Price, Thomas R.	1845-74	
Russell, Frederick	1872	
Sawyer, Joseph Pobjoy	1839	
Scott, Nicholas	1780-95	
Scott, Giles	1800	
Simpson, George	1839	
Southgate, William	1750	
Stevenson, William	1756-61	
Stone, H.	1855	
Sutton, John Windram	1826-39	
Sutton, Thomas	1775-1823	
Swinhogg, —	17thC	
Swinyard, George	1750	
Tate, James	1840-7	
Twichell, George Hearn	1839	
Waghorne, James	1826-47	
Walmesley, William	1790-1800	
Ward, Edward	1874	
Whaley, William	1779	
Wimble, Jeremiah [Nehemiah]	1720-50	Ashford
Winterhalter, D. G.	1865-74	
Woollett, John	1782-90	London

MARDEN F6

Emberson, James	late 18th/early 19thC	
Jobson, Frederick William	1866-74	
Knight, William	1760	
Snashall, George	1847	
Waddle, —	1858	

MARGATE N2

Collins, Richard	1780-1827	
Crow, Francis	1800-23	
Dowsett, Charles	1799-1826	London
Fagg, John	1800-51	
Foster, Henry	1874	
Hurst, Edward	1715-20	
Lee, L.	1797	
Levi, Emanuel	1804	

Page, Henry	1839		
Palmer, H. W.	1813		
Pope, John	1766-9		
Prebble, Charles	1823-51		
Prebble, James	1826-74		
Solomon, Edward	1769-1800		
Solomon, Nathaniel	1783-95		
Somes, William	1838-74		
Swan, —	1754		
Tuck, John	1874		
Woodruff, Charles	1838-66		
Woodruff, William	1874		

MILTON (NEAR SITTINGBOURNE) H3

Shilling, E.	1866-74	
Shilling, E.	1800-38	Sittingbourne
Shilling, James	1823-8	
Shilling, Harriet Mrs	1847-55	
Stevens, Robert	1781-95	

MILTON (GRAVESEND) D2

Parker, Augustus	early 18thC

MINSTER N3

Dixon, William	1770

NEW ROMNEY J8

Daniel, James	1779-88
Gilbert, Jeffery	1814-74

NORTHFLEET D2

Brenerton, S.	1847-74

QUEENBOROUGH H2

Welch, D.	1847

RAMSGATE O3

Abrahams & Son	1790	
Ansell, John	1845	
Barnett, Israel	1826-55	
Bates, Henry	1874	
Bates, John	1849	
Bing, John	1781-90	
Bodley, Sidney	1874	
Collyer, Allen	1874	Dover
Fagg, Jacob	1826-8	
Fehrenbach, B.	1845	
Goatley, —	1790	London
Harland, Christopher	1858	
Harland, E.	1855	
Haycock, Silas Henry	1874-94	
Hunter, Thomas	1823-51	
Hunter, William D.	1855	
Kingston, J. T.	1855-74	
Le Plastrier, Robert Louis	1832	
Levi, Noah	1789-1802	
Lion, Isaac	1826-40	
Lyon, Isaac	1768-1840	
Mont, Antonio	1874	
Mount, William Henry	1847-66	
Murdoch, J. G.	1874	
Oclee, James	1790	London
Palmer, Robert	1874	
Raggett, James	1823-7	
Roce, J. H.	1855	
Rose, James H.	1866-74	
Standven, Thomas	1840-51	
Standven, Thomas Emery	1832-47	

RIVER M6

Langley, Charles	1851-74

ROCHESTER F3

Asher, Mordecai	18thC	
Baurle, L. & F.	1874	Chatham
Bell, Frank Godfrey	1839	
Bell, Frederick & Thomas	1839	
Blackmore, John	1716	
Booth & Co	1794	
Booth, James	1766-1822	London
Bradstreet, Robert	1737-97	
Bull, —	1753	
Dove, E. C.	1874	
Ellis, Goodman	1599	
Greenwood, John Snr	1793-5	
Greenwood, John Jnr	1815	
Greenwood, William	1823-47	
Heitzman, Anthony	1838-74	
Hick, Charle F.	1845	
Hills, W. & G.	1874	
Hills, William	1847-74	
Jenkins, J. & M.	1784	
Jenkins, Mason	1787-1802	
Jenkins, Thomas	1787-9	
Lamb, Simon	1669-1700	London
Lyon, Simon	1866-74	
Mason & Jenkins	1795	
Mayhew, Joseph	1866-74	
Nye, Thomas John	1832-9	
Phillip, Myer	1767	
Poynard, Noah	1847	
Press, J.	1874	
Reed, James	1849	
Scott, James	1823	
Southey, —	1790	London

ROMNEY J8

Bayl[e]y, John	1785-95	Ashford
Swaby, Jacob	1795	

Sandgate L7
Eastes, William	1845	
Miller, William Frederick		1838
Scott, John	1776-95	

Sandwich N4
Abrahams, Samuel	1783-1803	
Booth, Joseph	1711	
Boxer, Thomas	1832	
Brice, John	1723-55	
Brice, William	1752-1803	
Cockling, William	1725-55	
Coleman, Henry S.	1810-67	
Cumming, Joseph	1858-62	
Deacon, Arthur William Scripps	1862-1909	
Drayson, Douglas	1862-91	
Fagg, —	1753-70	
Gardner, John	1726-52	
Gardner, Joseph	1691	
Gardner, William	1733-58	
Harflete, Cornelius	1747	
Jenkinson, S.	1760	
Jenkinson, Thomas	1696-1755	
Long, William	19thC	
Martin, Celestin	1862-82	
Maspoli, Monti & Co	1838-40	
Maspoli, Peter	1845-7	
Maspoli, Vittore	1841	
Monti, Maspoli & Co	1839	
Monti, Peter	1817-95	
Pain[e], Mark	1791-1862	Ash
Pain[e], Steven	1858	
Pegden, Thomas	1802-45	
Ralph, George	1832	
Ralph, William	1826	
Romney, George	1891	
Sleath, John	1747	
Smith, Samuel	1752	
Vennall, James	1820-39	
Warren, Edward D. E.	1808-61	
Wilmshurst	1695	

Selling J4
Shilling, Thomas	1767

Sevenoaks C4
Ashdown, G.	1866	
Baker, John	mid-18thC	
Barcham, Asher	1800	
Barnett, William	1771	
Bowra, James	1685-1745	
Bowra, George	1838-51	
Bowra, Thomas	1720-73	
Bowra, William	1724-80	
Bowra, William Snr	1789-1802	
Bowra, William Jnr	1823-51	
Bowra, William (E.)	1823-66	
Brooks, Eli	1847-55	
Brown, —	1793	
Cooper, John	1845	
Fleming, Richard	1793	
Gatward, Joshua	1784	London
Harris, John	1855-66	
Hope, W. D.	1874	
Jones, Samuel	1826-51	
Payne, John	18thC	
Read, Edwin	1874	
Virgo, James	1874	
Whitehead, Edward	1823-38	
Whitehead, Edward & Son	1847-66	
Whitehead, James	1823-38	

Sheerness H2
Abraham, Isaac	1802	
Barnard, Frederick	1826-55	
Beal, Richard	1847-55	
Forster, John	1790-1823	
Forster, William Henry	1832-8	Gravesend
Kayser, Joseph	1839-47	
Kuner & Casey	1866	
Munn, William	1838-47	
Munn, William S.	1851-74	
Nicholas, Robert	1839-47	
Pursey, William	1847	
Selfe, William	1855	

Sittingbourne H3
Acres, Edward	1767
Baker, A.	1874
Baldwin, J.	1770
Barnard, James	1800-47
Blaxland, John	1866
Caryer, Jesse	1874
Crittenden, John Chapman	1796
Forster, G.	1802-13
Shilling, Edward	1839
Smith, E. S.	1851-5
Smith, John P.	1839-47
Taylor, T.W.	1874
Young, F.	1803

Smarden G6
Avery, W.	19thC
Hogben, Thomas	1740-92
Perrin, Thomas	1705-34
Perrin, Thomas	1760-80
Vennall, Thomas	19thC

Stalisfield H5
Whatman, Arthur	1691	

Staplehurst F6
Ballard, John Thomas	1844-8	
Vaughan, J.	1855	

St Mary Cray B3
Bowra, —	1839	
Whitehead, J.	1839	

Stowting K6
Silk, John II	1779	Elmsted & Hythe
Smith, E. S.	1851-5	
Smith, John P.	1839-47	
Taylor, T. W.	1874	
Young, F.	1803	

Strood F3
Corringham, Richard	1874	
Intross, A.	1800	
Patrick, George	1866-74	
Patrick, John	1851-5	
Patrick, L. Mrs	1866	
Paul, James	1847	Gravesend
Paynard, James	1808	
Pirkis, George	1838-55	

Sutton Valence G5
Daniel, James	18thC	New Romney
Harris, John	1847-51	

Sydenham & Upper Sydenham A2
Russell, R.	1866-74	
Seddon, Alfred	1851-74	Forest Hill

Tenterden G7
Allen, Thomas	1823	
Birch & Masters	1836-47	
Birch, William	1823-39	
Burch, William	1823	
Goddard, Henry	17300-67	Dover
Hopkins, William	1760	London
Hukins, George Hopper	1824-1910	
Hukins, James	1800-82	
Jackson, John	1796	
Jackson, Owen	1760-1803	Cranbrook
Kingsnorth, John	1688	London
Kingsnorth, Thomas	1714/15	
Masters, John	1818-87	
Monk [Munk], James	1800	
Ollive, Thomas	1752-1829	Cranbrook
Strickland, William	1790-1803	
Thatcher, Thomas	Ante 1772	
Woolley, W.C.	1789-1810	
Wraight & Woolley	1790-1810	

Tonbridge D6
Ashdown, Frank	1874	
Barcham, Asher	1800	
Beard, J.	1800-12	
Brookstead, John	1671-95	London
Corner, Alfred	1845-96	
Gatward, Joseph	1802	London
Gillott, William	1845-7	
Godden, George	1847	
Harris, Stephen	1692-1755	
Harris, William	1684-1725	
Hormer, J.	1802	
Hosmer, —	1790	
How, Silas Samuel	1826-66	
Muddle, Nicholas	1745-75	
Muddle, Thomas	1739-77	
Noakes, John Thomas	1874	
Ollive, Samuel	1760-94	
Smith, C.	1866	
Smyth, Charles	1838-74	
Sutton, Charlotte	1823	
Thomson, Mark Graystone	1847-74	

Tunbridge Wells D6
Bianchi, B.	1858
Braby, J.	1866
Camfield, Thomas	1823-47
Caney, William	1839
Crundwell, Samuel	1847-74
Damper, William	1838-55
Damper, J., Mrs	1866
Eldridge, A.	1866
Field, Benjamin	1823-51
Field, Benjamin Jnr	1851
Harris, Charles	1874
Hascock, John	1874
Hislop, William	1874
Huxley, Thomas	1832
Jackman, George	1855-74
Kemp, Alfred	1851-74
Kinlan, Thomas	1866
Loof, Edward Fry	1855-74
Loof, William	1823-55
Loof, William Jnr	1845-7
Manning, W. H.	1874
Minnes, James Henry	1874
Moon, George	1839
Ninnes, J. & W.	1855-66
Ninnes & Loof	1851
Page, Walter	1874
Payne, T. E.	1874
Rae, Alexander	1810
Ruffer, William	1874

Thorpe, F. 1874
Vigor, L. 1874
Wood, Thomas 1776

WESTERHAM B5
Andrews, William 1847 Edenbridge
Bodie, — 1771
Bradbourne, John 1744-64
Cornish, Richard 18thC
Fowle, Edward 1838-79
Fowle, Humphrey 1769
Harris, James 1769
Palmer, William 1826
Richards [Richardson],
   Thomas 1730

WEST MALLING, TOWN MALLING, EAST MALLING E4
Baker, Henry II late 17thC/1745
Baker, Thomas 1768/late 18thC
Dean, James 1874
Godden, Henry 1823-47
Godden, John 1784
Hoad, Henry 1858-74
Hubble, Daniel 1847-74
Ousmar, O. 1838
Pett, Robert 1729
Pett, Thomas 18thC
Usher — 1820
Usher, G. 1838
Usher, C. 1838
Usher, Oliver 1832-51
Usher, William 1845-74

WHITSTABLE K3
Aldersley, George 1838
Andrews, John Edward 1866-74
Fenn, Charles 1866-74
Pollard, Samuel 1838
Wellby, John 1845-51

WINGHAM M4
Brice, John 1708/9
Brice, Symon 1641-52
Elvis, Benjamin 1866-74
Godden, John 1750
Jefferys, James 1847-55 Canterbury

WOOLWICH B1
Alman, W. 1817-24
Barrell, George 1838-74
Beaven, John Richard 1838-51
Beedle, James Francis 1874
Bennett, G. W. 1851-5 Blackheath
Bishop, S. R. 1866
Brook, — 19thC

Bromley, T. 1851-5
Cass, John 1823
Charlton, John 1838-47
Coutts, John Francis 1874
Hart, H. 1847
How, Peter 1847-66 Eltham
Isaacs, M. 1817-24
Isaacs, John 1817-24
James, Thomas 1847
Lamb, H. T. 1855
Lee, I. J. 1851
Littlewood, B. 1866-74
Littlewood, Benjamin Jnr 1874
Memess, Robert 1826-8 London
Nye, Edwin 1874
Palmer, Thomas 1845-7
Percival, John late 18thC-1811
Percival, Mary
   and Thomas James 1817-55
Percival, James 1839
Pilkington, J. 1826-39
Poulton, E. 1874
Rumsey, John 1723
Siedle, L. 1851-66
Spurge, William 1826-55
Tindall, George Henry 1845-7
Walter, W. J. 19thC
Walter, William John 1847
Warren, G. 1866
Warren, R. J. 1866-74
Webber, John 1800-47
Webber, John II 1866-74
Wilcox, John 1838-74
Wright, Richard 1845-51

WROTHAM D4
Down, John G. 1847-55

WYE K5
Bradshaw, Henry Charles 1832
Pollard, J. 1866
Price, Daniel 1832-66
Price, T. 1855
Quested, Thomas 1780-7
Smith, John 1803
Webb, Abraham 1741 Ashford

YALDING E5
Chittenden, Isaac 1806
Crundwell, S. 1866
Engeham, John 1789 Canterbury
French, John 1832-53
Gladdish, Thomas 1720
Gladdish, William 1788-94

BIBLIOGRAPHY

Documentary Sources
Canterbury Cathedral Archives, including: Archdeaconry and Consistory Court of Canterbury, Probate Records, Wills, Probate Inventories, Bishop's Transcripts, Churchwardens' Accounts and various other parish records, Register of Marriage Licences in the Canterbury Dioceses
Canterbury City Council Records, including Apprentice and Freemen Lists
Kent County Archives Offices at Maidstone and Rochester, including Freemen Lists, Corporation Records and Council Minutes
Local Archives at Dover, Sevenoaks, Hythe, Folkstone, Lydd and Faversham
Maidstone Museum Records, including the Clement Taylor Smythe Mss
Sandwich Borough Records

Printed Sources
Antiquarian Horology
Archaeologia Cantiana
Chalklin, C. W. *Seventeenth Century Kent*. 1965
Dart, J. *History and Antiquities of the Cathedral Church of Canterbury*. 1727
Hubbard, G. E. *The Old Book of Wye*. 1951
Kentish Post and *Kentish Gazette*
Keys, S. K. *Historical and Further Historical Notes of Dartford*. 1933
Nineteenth-century local directories, including *Pigot* and *Bradshaw*
Pevsner, N. *Buildings of England: East Kent* and *West Kent* 1968 (by J. Newman)
Woodruff, C. E. *The Sacrist Rolls of Christchurch Canterbury*. 1936

Horology, General
Baillie, G.H. *Watchmakers and Clockmakers of the World*. VolI, 3rd edition 1951. (See Loomes, B. for Vol II)
Barder, R. C. R. *English Country Grandfather Clocks*. 1975
Barder, R. C. R. *The Georgian Bracket Clock 1714-1830*. 1993
Beeson, C. F. C. *English Church Clocks*. 1971
Bird, A. *English House Clocks, 1600-1850*. 1973
Britten, F. J. *Old Clocks and Watches and their Makers*. 8th & 9th editions. 1986
Bruton, E. *The Longcase Clock*. 1976
Cescinsky, H. & Webster, M. *English Domestic Clocks*. 1913, reprinted 1976
Cescinsky, H. *Old English Master Clockmakers and their Clocks*. 1938
Clutton, C. & Daniel, G. *Clocks and Watches in the Collection of The Worshipful Company of Clockmakers*. 1975
Clutton, C. & Daniel, G. *Watches*. 1986
Dawson, P., Drover, C. & Parkes, D. *Early English Clocks*. 1982
Edwardes, E. L. *Story of the Pendulum Clock*. 1977
Edwardes, E. L. *The Grandfather Clock*. 4th edition 1980
Hana, W. J. F. *English Lantern Clocks*. 1979
Lee, R. A. *The Knibb Family of Clockmakers*. 1964
Loomes, B. *British Clocks Illustrated*. 1992
Loomes, B. *Complete British Clocks*. 1978
Loomes, B. *Country Clocks and their London Origins*. 1976
Loomes, B. *Grandfather Clocks and their Cases*. 1985
Loomes, B. *The Early Clockmakers of Great Britain*. 1981
Loomes, B. *White Dial Clocks*. 1981. Revised as *Painted Dial Clocks*. 1994
Loomes, B. *Watch and Clockmakers of the World*, Volume II, 2nd edition 1989.
Pearson, M. *The Beauty of Clocks*. 1979
Roberts, D. *British Longcase Clocks*. 1990
Roberts, D. *Collector's Guide to Clocks*. 1992
Robinson, T. *The Longcase Clock*. 1990
Rose, R. E. *English Dial Clocks*. 1988
Symonds, R. *Thomas Tompion — His Life and Works*. 1951
Ullyett, K. *In Quest of Clocks*. 1970
White, G. *English Lantern Clocks*. 1989

Regional Clockmaking
Bates, I. C. *Clockmakers of Northumberland and Durham*. 1980
Beeson, C. F. C. *Clockmaking in Oxfordshire 1400-1850*. 1989
Bellchambers, J. K. *Devonshire Clockmakers*. 1962
Bird, C. & Y. *Norfolk & Norwich Clocks & Clockmakers*. 1996
Daniell, J. A. *The Making of Clocks and Watches in Leicestershire and Rutland*. 1951
Dowler, H. G. *Clockmakers and Watchmakers of Gloucestershire*. 1984.
Elliott, D. J. *Shropshire Clock and Watchmakers*. 1979
Haggar, A. L. & Miller, L. F. *Suffolk Clocks and Watchmakers*. 1974
Loomes, B. *Lancashire Clocks and Clockmakers*. 1975
Loomes, B. *Westmoreland Clocks and Clockmakers*. 1974
Loomes, B. *Yorkshire Clockmakers*. 2nd edition 1985
Mason, B. *Clock and Watchmaking in Nottinghamshire*. 1979
Miles Brown, H. *Cornish Clocks and Clockmaking*. 1970
Peate, I. C. *Clockmaking in Wales*. 1975.
Penfold, J. *Clockmakers of Cumberland*. 1977
Pickford, C. *Bedfordshire Clock and Watchmakers*. 1991
Ponsford, C. N. *Devon Clocks and Clockmakers*. 1985
Ponsford, C. N. *Exeter Clocks and Clockmakers*. 1978
Smith, J. *Old Scottish Clockmakers*. 2nd edition 1975
Tebbutt, C. *Stamford Clocks and Watches*. 1975
Tribe, T. & Whatmoor, P. *Dorset Clocks and Clockmakers* 1981

Index

This index does not include references to the clockmakers listed in Chapter 7 and Appendix II, but it does include those (eg apprentices) mentioned under the entries for other makers, and also makers mentioned in Appendix I.

Abrahams, Samuel 293
Acres, Edward 276
Act of Parliament clocks. *See* tavern clocks
Aldridge, Edward 288
Aldridge, Thomas 283
Allett, George 58
anchor escapement 39
Andrews, Thomas 88, 276, 279, 285, 289, 290, 291
Annott, Charles 134
Apps, Robert 99, 166
Apps, William 142
Arnold & Dent 89
Arnold, John 89, 119, 268, 269
Arnold, John Roger 89
Ash, St Nicholas Church 17, 26, 96, 113, 143
Ashford, St Mary's Church 18, 43, 72
Avenal, William 258

Baker family of Maidstone and Town Malling 47-53
Baker, Henry 41
Baker Henry I 47
Baker, Henry II 50-53
Baker, Thomas 53
Baldwin, John 35, 39, 247, 249, 273
ballance clocks 35
Banks, John 268
Barr, — 289
barrel engine 216, 260
Barrett family of Maidstone 17, 18
Barrett, John 17, 26
Barrett, Thomas 17, 38, 207
Barrett, William 15, 20, 21
Bartlett & Sons 24
Battle (Sussex) Church 213, 217
Bayl[e]y, Richard 39, 281, 282, 286, 287, 291, 292

Bayl[e]y, Richard II 292
Bayl[e]y, John 292
Beard, William 134
Becket, James 242
Bekelys, Nicholas 14
Belleyeterre, John 14
Berry, John 225, 262
Bethersden marble 41, 62
Bethersden, St Margaret's Church 15, 20, 73, 99, 249
Biddenden, All Saints Church 22
Bing, John 287
Bird, William 261
Birling Church 33, 95
Bishop, John 35, 40
Blacksmiths' Company 34
Bliss, Ambrose 38
Bliss, Thomas 130
Body, Obadiah 233
Booth, James 171, 194, 274
Borden, Church of St Peter and St Paul 22
Bourn, Loder 256, 259
Bowen, David 300, 302
Boys, Lawrence 256
bracket clocks 39
brass casting 63
Brice, John 33, 265, 279, 282
British Museum 118, 149, 197
Buckley, John 287, 289, 292, 293, 294, 295, 297, 299
Bull, — 289
Burbage, Thomas 60
Butterly, Stephen 23

Cackett, Thomas 259, 268
Cadwell, Charles 239
Caffinch, Lydia 77
Canterbury
　Christchurch Cathedral 12-13, 43
　Church of St Andrew the Apostle 14, 103

Church of St George The Martyr 79, 80
City Accounts 14
Museum 17, 22, 39, 80, 98, 116, 128, 135, 149, 164, 188, 209, 224, 231, 252
Poor Priests Hospital 22
silk-weaving in 41
St Dunstan' Church 14
St Mary Magdalene Church 42
workhouse 224
Carrington, Thomas 114
Chalk, William 289
Chalklin, John 125, 197, 266, 267, 270, 271
Chalklin, Mrs 274, 277, 278
Chalklin, William 46, 278, 280
chamber clocks 34
Charing, Church of St Peter and St Paul 22
Charlson, C. 261
Chen, Edward 78
Chilham, St Mary's Church 22, 41
Claris, W. 296
Clement, William 39, 261, 279
clepsydra (water-clocks) 12
clock cases 43-46
　lacquer 41, 44
　Kentish cresting 46
　mahogany 44, 46
　makers 44
　marquetry 44
　pagoda-topped 44
　walnut 44
Clockmakers' Company 35, 79
clockmaking tools 60, 63, 216
clocks
　domestic 34-35
　earliest mechanical 12
　imported mass-produced 46
clock-watches 34
Cobham Church 33

Cockey, Edward 225, 262
Collins, Richard 284, 288
convex moulding 40, 44
Coster, Salomon 35
Cottam, Joseph 166
Cramp, Richard 119, 271, 278, 280, 281
Cranbrook Church 33, 236
Craythorn, John 258, 264
Crouch, William 280
Crow, Francis 283, 286
Cutbush, Charles 54, 58
Cutbush, Edward 35, 54, 57
Cutbush family of Maidstone 54-58
Cutbush, George 54
Cutbush, John 54
Cutbush, John II 55
Cutbush, John III 55
Cutbush, Richard 55, 57, 58, 231, 246
Cutbush, Robert 55, 171
Cutbush, Thomas 54
Cutbush, William 54, 57

Danell, James 284
Dartford
   Holy Trinity Church 22, 117, 141
   Museum 117
Davis, John 43, 254
Deale, Jacobus 42, 60, 73
Deale, Thomas 35, 39, 42, 59-66, 73
   inventory 63-66, 180
   will 62-63
Denison, E. B. 33
Dennis, George 68
Dennis, Thomas 38, 68, 106
dividing plate 63
Dobson, — 261
Dodd, John 35, 60, 67
Dodd, William 67
Dolphin, — 289
Dover
   Castle 14
   Church of St Martin le Grand 15
   Church of St Mary the Virgin 15
Dowsett, Charles 205, 302
Dunstable Priory 12

East Malling, St James Church 23
Edenbridge Church 33
Edward III 13
Egerton, St James Church 24

Elgar, Richard 119
Elham, St Mary's Church 13
Elliott, John 300
Ellis, Goodman 28
Ellis, James 38, 89, 102, 143, 254
Emanuel, Levi 287
engraved clock dials 43
engraving
   rococo 43, 53
   tools 60
Evans, Henry 122
Exeter Cathedral 12
Eynesford Church 33

false plates 46
Farnborough, Church 33
Feild, James 224
Feilder, Thomas 243
Ferguson, James 199
files 63
Finch, John 246
Flint, William 20, 46
Folkestone
   Church 33
   Museum 33
Fordham, Thomas 120
Fowkes, Gabriel 23, 41
French, Susanna 89, 99, 142
Frodsham, Charles 89
Fromanteel, Ahasuerus 35
Fromanteel, John 35
Fry, Rowland 23

Gardner, John 33, 227
Gardner, William 227
Gate, Aucher [Arthur] 134
gavelkind 9, 10
Gee, John 78
Gibbons, Grinling 30, 41, 43
Gilbert, William 125
Giles, Nicholas 220
Gill, William 33, 110, 157, 166
Gillett & Johnson, Croydon 17
Glassenbury Park, Goudhurst 24
Glazier, William 183
Goddard, Henry 169, 270, 282, 288, 289
Godden, John 275
Godinton Park 17, 42
goldsmiths 39
Goodchild, Richard 264, 270
Goodlad, Richard 25, 78, 79
Gotesborgs Historiska Museum, Sweden 131
Goudhurst, St Mary's Church 24
Goulden, William 44, 88, 126,
164, 183, 188, 209
Gray, Stephen 89
Greenhill family of Ashford, Canterbury and Maidstone 17, 8, 67-82
Greenhill, John 20, 35, 38, 39, 40, 60, 67, 77, 131, 158
Greenhill, John I 68
Greenhill, John II 70
Greenhill, John III 70, 71
Greenhill, Richard 17, 20, 21, 24, 25, 35, 39, 60, 63, 67, 73, 122, 131, 143, 146, 156
   inventory 81-82
Greenhill, Richard I 71, 77
   will 73-77
Greenhill, Richard II 73, 76
Greenhill, Robert 67
Greenhill, Samuel 25, 35, 76, 77, 78, 80
Gregnion, Daniel 267
Gregsby [Grigsby], Edmund 191
Groombridge Church 33
guilds 10

Hall, Thomas 94
Hardeman, Samuel 197
Harris, James 263
Harris, Joseph 43, 271
Harris, Walter 70, 71, 78, 79, 148
Harris, Walter Jnr 71, 79
Hayman, Robert 134, 142
Headcorn, Church of St Peter and St Paul 24
Henry, Prior of Eastry 12, 13
Herne, St Martins Church 24, 79, 147
Hever, St Peter's Church 26
Higgs, Robert 257
horologium 12
Hughes, John 125, 145
Huguenot refugees 41
Huniset, Thomas 264
Hurt, Arthur 77
Huygens (Van Zulichem), Christiaan 35
Hythe
   Corporation 14
   St Leonards Church 14

immigrants, Flemish 41

Jackson, Owen 46, 145, 182, 269, 271, 274, 275, 288, 302
Jackson, Thomas 269
James Feild 224
Jeffery, George 55, 58, 108, 194

Jeffery, William 54, 55
Jenkins, Thomas 284, 294, 296
Jenkinson, Thomas 41, 109, 256, 257, 262, 263
Johnson, Charles 238
Jones, Evan 120
Jordan, James 39, 40

Kempe, John 14
Kent
   county boundary before 1889 9
   Dutch influence 41, 42
   early turret clocks 15-17
   population 10
King, Peter 265
Kings Langley 14
Kingsnorth, John 21, 35, 233
Kingsnorth, Thomas 20, 21
Kissar, Samuel 254, 255
Knibb, Joseph 39
Knight, George 66
Knight, William 267
Knole Park, Sevenoaks 31
Knowlton, St Clement's Church 26, 30, 43

Lade, Michael 254, 256, 258, 261, 262, 265, 296
Lambe, Simon 30
lantern clocks 34, 35-38
Lee, Samuel 257
Leeds
   Castle 26
   St Nicholas Church 33
Lenham Church 33
Lepine, Charles 147, 209
Levi [Levy], Benjamin 296, 301, 302
Levi, Emanuel 302
Levi, Noah 297
Loker, John 27
longcase clocks 39-46
Lydd, All Saints Church 15, 20, 26, 134
Lyminge, Church of St Mary and St Ethelburga 28
Lyncolne, John 14
Lynsted Church 33

Maidstone
   All Saints Church 33, 145
   conduit 15, 165
   Museum 23, 24, 29, 50, 57, 71, 92, 93, 94, 95, 100, 103, 116, 122, 132, 142, 145, 149, 163, 169, 221, 227, 247
   Town Hall 29,
Manley. John 283
Marsh, John 292
Maspoli, Peter 193
Masters, John 104
Mayfield (Sussex) Church 214
Mepstead, John 17
Mercer, John 35, 39, 256
Mercer, John Jnr 289
Mercer, Thomas 265
Mercer, William 266
Merton College, Oxford 12, 13
Milles, Isaac 78
Minster-in-Thanet, St Mary's Church 29
monasteries, Dissolution of 9, 15, 34
Moore of Clerkenwell 139
Moran, — 262
Mountfort, Zachary 86
Muddle, Edward 30, 108, 171
Muddle, Nicholas 200
Mudge, Thomas 119
Mummery, Thomas 205

Nantes, Edict of 41
Narborough, Sir John 30
Nash, John 272, 273
Nash, William 119, 151, 267, 270, 272, 273, 274, 289, 293, 302
Nethersole, Margaret 77
New Romney Church 33
Nicholls, Hammond 298, 300
Nicole, John. *See* Lyncolne, John

Oldridge, Thomas 286
Oliver, Thomas Reader 202
Ollive, Samuel 277, 284, 286
orologium magnum 12, 13
Osborne of Birmingham, painted dials 197

Paine, Thomas 200
painted dials 43
Papworde, Thomas 27
Parnell, Thomas 289, 296
Parquait, — 259
Payne, William 302
Peckham, John 257
Pegden, Thomas 183
pendulum
   conversion to 21, 22, 26, 33
   invention of 35
Perkins, Richard 253

Philpot, Jane 59
Pinchbeck, Christopher 270
Pluckley, St Nicholas Church 29
Pope, John 270, 273, 291
Potter, Christopher 86, 88
Potts, William 188, 295
Prestige, Sir John 17
Punnett, Thomas 20, 21
Pysing, John 264

Quare, Daniel 264, 265
Queenborough Castle 13, 14
Quellin, Arnold 42
Quested, Thomas 286, 293

Rainham Church 33, 145
Ranger, Thomas 52
Rawling, Edward 260
Rayner, Stephen 94
refugees
   Flemish 9
   religious 10
religious persecution 34, 41
Richard of Wallingford 13
Robart, — 273
Roberts, James 301
Rochester
   Corn Exchange 29, 30, 171, 194
   Guildhall 29, 30, 183

Salisbury Cathedral 12, 13
Science Museum, London 14
Sedgwick, Giles 23
Shilling, Thomas 274
Shindler, Elizabeth 137
Shindler, Thomas 35, 116, 258, 287
Shindler, Thomas II 260, 262
Shindler, Thomas III 261, 262, 268
Shovel, Rear Admiral Sir Cloudesly 30
Silke, John 269, 285, 293
Silke, Robert 35
silversmiths 39
Sims, George 259, 264, 275, 276
Sims, Henry 239, 272
Sleath, John 227
Smarden, St Michael's Church 31
Smith, Benjamin 227, 236, 271, 272, 277, 279
Smith, John 302
Snatt, John 285
Somersall, George 259
Somersall, Mandevile 259
Southgate, William 55

spring-driven table clocks 34
St Albans Abbey 13
Stacey, William 108, 274
Stads Museum, Amsterdam 224
Staple, St James Church 17, 31, 79, 99
Stevens, Daniel 178
Stevens, Robert 287
Stevenson, William 264
Strelley, Francis 230
Strickland, William 250, 300
Stubbs, Thomas 23, 178
Sturry, St Nicholas Church 31
sundials 12
Sutton, Thomas 46, 57, 58, 239
Swaby, Israel 289
Swan, — 263
Swinyard, George 247

Taillour, Johes [John Taylor] 14
tavern clocks 46, 53, 93, 120, 182, 184, 252
Thatcher, George 35, 108, 122, 178, 276, 280
Thurnham, Church of St Mary the Virgin 33
Thwaites of London 23
Tiddeman & Co 294
Tiddeman, Edward 239, 298
timepiece/alarm clocks 46
Toke, Captain Nicholas 42
Tompion, Thomas 58, 205

Tonbridge Church 33, 200
turret clocks 12-14, 79
    birdcage movement 24, 26, 31, 33
    electric winding 28
    end-to-end trains 24, 26, 31, 33
    inside dial 21
    outside dial 21, 23
    side-by-side trains 25, 26, 29

Ventisan, Thomas 175
verge and foliot escapement
    conversion to anchor 21, 25, 26 33
verge escapement 35, 39
Victoria and Albert Museum 173
Vidion, John 39, 127, 278
Virginia Museum 212

wall clocks 38, 46, 50
Walmesley, William 231
Ward, — 262
Warren, James 39, 46, 236, 283, 299
Warren, James Jnr 85
Warwell, John 293
watches 38-39
watchmakers
    early Kent 38
Watts, John 39, 102, 260, 263, 268, 299

Webb, Abraham 167, 257
West Malling Church 210
Westminster Palace 14
Weston, Abraham 108
Whaley, William 57
wheel-cutting engine 260
Wightman, James 106
Wilmshurst, Thomas 263, 269
Wilson of Birmingham, painted dials 46, 85
Wimble, George 257
Wimble, Nehimiah (Jeremiah?) 233
Wimble, John 20, 27, 28, 40, 233
Windmills, Joseph 30
Wingham, Church of St Mary the Virgin 33, 42, 113
Wittersham Church 139
Wood, Thomas 282
Woodgrave, — 289
Woollett, William 43, 155, 250
Woolley, Thomas 21
Woolley, W. C. 231, 300
Wraight & Woolley 20, 22, 46
Wraight, Thomas 20
Wren, Sir Christopher 43
Wye, Church of St Gregory and St Martin 14

Yalding Church 33
Young, Thomas 23

Picture Credits

The following have loaned photographs for use in this book; the author would like to thank all of them and also those individuals who wish to remain anonymous:
Antique Collectors' Club: 6/1-2, 6/9-19; Christopher Buck: 3/6-7; Michael Bundock: 1/3-5, 1/7, 1/10-12, 1/16-21; Canterbury Auction Gallery: 7/53-54, 7/101-102, 7/117, Colour Plate IX; Canterbury Musem: 7/10-11, 7/104, 7/123-124; Christies: 7/95; The Clock Shop, Weybridge: 7/87; Dreweatt Neate, Newbury: 4/4, 7/97; Freeman & Lloyd: 7/68, Colour Plate XI; W. J. Hana: 6/20, 7/121; Edgar Horn: 7/19; Horological Workshops: 7/74; Richard Hyde: 7/75-76; Brian Loomes: 7/55-57, 7/72-73; Richard Luck: 2/10-11, 7/108; Maidstone Museum: 1/13-14, 2/2-5, 3/1, 4/1, 6/6-7, 7/7, 7/9, 7/12-14, 7/16, 7/23-24, 7/31, 7/49, 7/93, 7/103, 7/115, 7/126-127; Gerald Marsh: 6/8; The Old Clock Shop: 3/2-5, 3/8, 3/11-12, 4/2-3, 7/8, 7/18, 7/20, 7/34, 7/35-38, 7/42, 7/66, 7/69-70, 7/84, 7/88-90, 7/97-98, 7/106; P. A. Oxley: 2/9, 7/1, 7/4, 7/21, 7/58-59, 7/80, 7/82, 7/85-86; Phillips, Exeter: 7/116; Raffety: title page; Derek Roberts: 7/50, 7/92, 7/109, 7/120, Colour Plates III, IV, XI; T. R. Robinson: 7/43-45; R. Ross: 7/60-61; Trustees of the Science Museum, London: Colour Plate II; Sotheby's: 5/1-2, 7/3, 7/26, 7/71, 7/125, Colour Plate VII; Strike One Ltd: 7/2, 7/41, 7/62-63, 7/65, 7/77, 7/79, 7/81, 7/94, 7/130-131, Colour Plate VIII; Rupert Toovey & Co: 7/25.

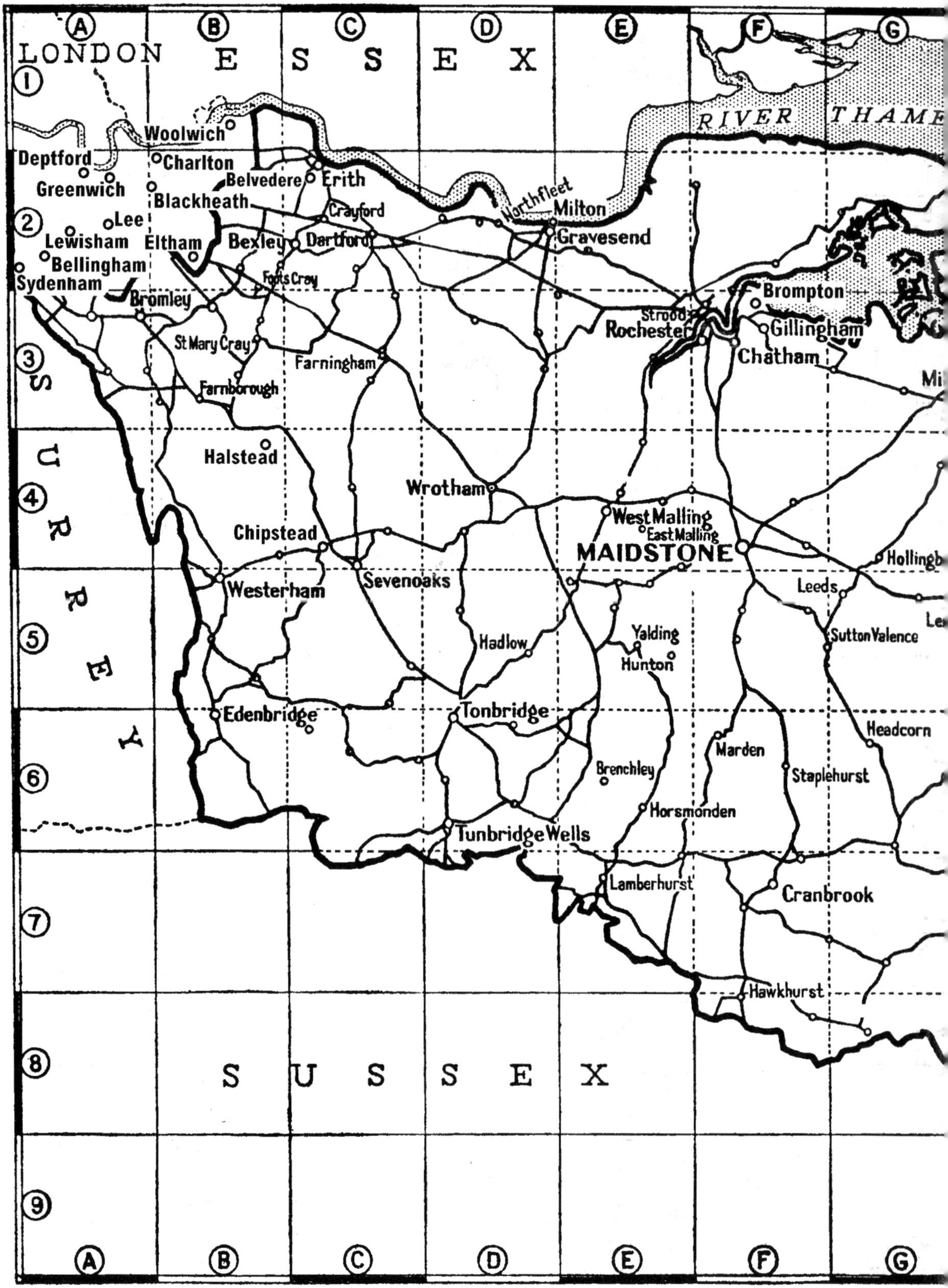
A
B
C
D
E
F
G
LONDON
ESSEX
1
RIVER THAMES
Woolwich
Deptford
Charlton
Greenwich
Belvedere
Erith
Blackheath
Northfleet
Milton
Crayford
2
Lee
Lewisham
Eltham
Bexley
Dartford
Gravesend
Bellingham
Sydenham
Foots Cray
Bromley
Brompton
Strood
Rochester
Gillingham
St Mary Cray
Chatham
3
Farningham
Farnborough
SURREY
Halstead
4
Wrotham
West Malling
East Malling
Chipstead
MAIDSTONE
Hollingb
Westerham
Sevenoaks
Leeds
5
Yalding
Sutton Valence
Hadlow
Hunton
Tonbridge
Edenbridge
Headcorn
Marden
6
Brenchley
Staplehurst
Horsmonden
Tunbridge Wells
Lamberhurst
Cranbrook
7
Hawkhurst
8
SUSSEX
9